3000 800062 36199
St. Louis Community College

D0463914

Biofuels for Transport

Biofuels for Transport

Global Potential and Implications for Sustainable Energy and Agriculture

Worldwatch Institute

London • Sterling, VA

First published by Earthscan in the UK and USA in 2007

Information in this book is accurate as at February 2006, with the exception of biofuel and petroleum production numbers, which are accurate as at March 2007.

ISBN: 978-1-84407-422-8

Typeset by JS Typesetting Ltd, Porthcawl, Mid Glamorgan
Printed and bound in the UK by Antony Rowe, Chippenham
Cover design by Susanne Harris

For a full list of publications please contact:

Earthscan
8–12 Camden High Street
London, NW1 0JH, UK
Tel: +44 (0)20 7387 8558
Fax: +44 (0)20 7387 8998
Email: earthinfo@earthscan.co.uk
Web: **www.earthscan.co.uk**

22883 Quicksilver Drive, Sterling, VA 20166-2012, USA

Earthscan publishes in association with the International Institute
for Environment and Development

A catalogue record for this book is available from the British Library

Library of Congress Cataloging-in-Publication Data

Biofuels for transport : global potential and implications for energy and agriculture / Worldwatch Institute.
p. cm.
ISBN-13: 978-1-84407-422-8 (hardback)
ISBN-10: 1-84407-422-6 (hardback)
1. Biomass energy. 2. Waste products as fuel. I. Worldwatch Institute.
TP339.B5435 2007
333.95'4--dc22

2006029028

The paper used for this book is FSC-certified and totally chlorine-free. FSC (the Forest Stewardship Council) is an international network to promote responsible management of the world's forests.

Contents

PART II NEW TECHNOLOGIES, CROPS AND PROSPECTS

PART III KEY ECONOMIC AND SOCIAL ISSUES

List of Figures, Tables and Boxes

FIGURES

Tables

Boxes

Acknowledgements

The following institutions and individuals are gratefully acknowledged.

Sponsor

German Federal Ministry of Food, Agriculture and Consumer Protection (BMELV)

Collaborating agencies

German Agency for Technical Cooperation (GTZ)
German Agency of Renewable Resources (FNR)

Editor

Lisa Mastny (Worldwatch Institute)

Project manager

Suzanne C. Hunt (Worldwatch Institute)

Lead researchers and contributors

Jim Easterly (Easterly Consulting)
Andre Faaij (Utrecht University)
Christopher Flavin (Worldwatch Institute)
Ladeene Freimuth (Freimuth Group)
Uwe Fritsche (Öko-Institut)
Mark Laser (Dartmouth College)

Lee Lynd (Dartmouth College)
Jose Moreira (Brazilian Reference Centre on Biomass, or CENBIO)
Sergio Pacca (University of São Paulo)
Janet L. Sawin (Worldwatch Institute)
Lauren Sorkin (Worldwatch Institute)
Peter Stair (Worldwatch Institute)
Alfred Szwarc (ADS Technology and Sustainable Development)
Sergio Trindade (SE2T International, Ltd)

OTHER CONTRIBUTORS

Thanks to the following people for contributing time, materials, and/or review comments: Don Abraham (Social Technologies); Clayton Adams; Carsten Agert (Fraunhofer-Institut für Solare Energiesysteme); Weber Amaral (Brazilian Biofuels Programme); Robert Anex (Iowa State University); Eliana Antoneli (Brazilian Biofuels Programme); Daniel Aronson (Petrobras); Dirk Assmann (GTZ); Sergio Barros (US Department of Agriculture, or USDA); Christoph Berg (FO Licht); Göran Berndes (Chalmers University of Technology); Davis Bookhart (Charm21); Barbara Bramble (National Wildlife Federation, or NWF); Thomas Breuer (University of Bonn); Steve Brisby (California Air Resources Board); Neil Brown, (US Senate Committee on Foreign Relations); Robert Brown (Iowa State Bioeconomy Initiative); Heloisa Burnquist (University of São Paulo); William Burnquist (Centro de Tecnologia Copersucar, or CTC); Jake Caldwell (Center for American Progress); Matt Carr (Biotechnology Industry Organization); Eduardo Carvalho (União da Agroindústria Canavieira de São Paulo, or UNICA); Frank Casey (Defenders of Wildlife); Christine Clashausen (GTZ); Suani Teixeira Coelho (Regional Government of São Paolo); Ana Unruh Cohen (Center for American Progress); Aimee Delach (Defenders of Wildlife); Mark A. Delucchi (University of California, Davis); Neeraj Doshi (Tufts Fletcher School of Law and Diplomacy); Reid Detchon (Energy Future Coalition); Brent Erickson (Biotechnology Industry Organization); Emanuel Filho, (Petrobras); Elke Foerster (GTZ); Hilary French (Worldwatch Institute); Lew Fulton (United Nations Environment Programme, or UNEP); Rubens Gama (Brazilian Embassy); Jana Gastellum (Energy Future Coalition); Simon Godwin (DaimlerChrysler); Alex Goyes (International Food and Agricultural Trade Policy Council); Nathanael Greene (US Natural Resources Defense Council, or NRDC); Ralph Groschen (Minnesota Department of Agriculture); Kathleen Hadley, (National Center for Appropriate Technology); Brian Halweil (Worldwatch Institute); Charlotte Hebebrand (International Food and Agricultural Trade Policy Council); Russel Heisner (BC International); Jan Henke (Institute for World Economics); Christian Henkes (GTZ); Chris Herman (US Environmental Protection Agency, or EPA); Bill Holmberg (American Council on Renewable Energy); Monique Hoogwijk (Ecofys); Rob Howse (University of Michigan); Roland Hwang (NRDC); Paul Joffe (NWF); Francis Johnson (Stockholm Environment Institute); John Karhnak (Raytheon UTD); Dennis Keeney (Institute for Agriculture and Trade Policy, or IATP); Birger Kerckow (FNR); Jim Kleinschmidt (IATP); Ricardo Külheim (GTZ); Regis Verde Leal (University of Campinas); Deron Lovaas (NRDC); Warren Mabee (International Energy Agency, or IEA); James Martin (Omnitech International); Joseph Mead (World Energy Alternatives); Rogerio Miranda (Winrock International); Ian Monroe (Winrock International); David Morris (Institute of Local Self-Reliance); Danielle Nierenberg (Worldwatch Institute); Peter O'Connor (Renewable Energy and International Law Project); Sillas Oliva Filho (Petrobras); Leslie Parker (Renewable Energy and International Law Project); Rodrigo G. Pinto (Worldwatch Institute); Luiz Prado (LaGuardia Foundation); Luke Pustevjosky (Fieldstone Private Capital Group); Guido Reinhardt (IFEU); Michael Renner (Worldwatch Institute); Fernando Ribeiro (UNICA); Thereza Rochelle (Brazilian Biofuels Programme); Rodrigo Rodrigues (Brazilian Ministry of Agriculture); Friederike Rother (GTZ); Katja Rottmann (Worldwatch Institute); Larry Russo (US Department of Energy); Liane Schalatek (Heinrich Böll Foundation); Hosein Shapouri (USDA); Ralph Simms (Massey University); Edward Smeets (Utrecht University); Ron Steenblik (International Institute for Sustainable Development, or IISD); Steve Suppan (IATP); Freyr Sverrisson (energy consultant); Holger Thamm (Office of Angelika Brunkhorst, Member of German Parliament); Ibrahim Togola (Mali-Folkecenter); Daniel De La Torre Ugarte (University of Tennessee); Boris Utria (World Bank); Friedrich Wacker (German Embassy); Michael Wang (Argonne National Lab); David Waskow (Friends of the Earth, or FoE); Carol Werner (Environmental and Energy Study Institute); and Jetta Wong (Environmental and Energy Study Institute).

Preface

The world is on the verge of an unprecedented increase in the production and use of biofuels. Rising oil prices, national security concerns, the desire to increase farm incomes, and a host of new and improved technologies are propelling many governments to enact powerful incentives for using these fuels, which is, in turn, sparking a new wave of investment.

Today, the question is not whether renewable biofuels will play a significant role in providing energy for transportation, but rather what the implications of their use will be – for the economy, for the environment, for global security and for the health of societies. Decisions made in the next few years will help determine whether biofuels have a largely positive impact or whether the gains from biofuel use will be coupled with equally daunting consequences.

Rapidly growing interest in biofuels is being spurred by the realization that they represent the only large near-term substitute for the petroleum fuels that provide more than 95 per cent of the world's transportation energy. And today's principal biofuels – ethanol and biodiesel – have the big advantage of being easily integrated within the vast established infrastructure for petroleum fuels. As a result, biofuels now stand as a potential solution to some of the most intractable issues facing the world, including rising oil prices, increasing national and global insecurity, rising climate instability, worsening local and global pollution levels, and deepening poverty in rural and agricultural areas.

Humanity relied on bioenergy in the form of wood fuel long before oil was ever discovered. And fuels made from renewable resources, such as plant oils and sugars, have been used to power motor vehicles for more than a century. But it is only during recent years that interest in these fuels, and in the newer 'next generation' of biofuels, has exploded. Production of fuel ethanol increased 22 per cent in 2006 alone, while biodiesel production jumped by 80 per cent (though it represents a much smaller share of the market).[1] Among those who have joined the swelling biofuels bandwagon are farmers, national security hawks, agribusiness leaders, venture capitalists and environmentalists.

The large ethanol industries first developed in Brazil and the US during the 1980s have entered a new growth phase in the past few years, setting a powerful example and exporting both policy ideas and technologies around the globe. Policy-makers in many countries are searching for biofuel strategies that can help them to achieve a range of sometimes conflicting objectives, including boosting farm incomes, creating jobs, reducing local pollution, addressing climate change and cutting oil imports.

Although biofuels now comprise only a small portion of the transportation fuel used globally, their production has already begun to affect commodity markets. In 2005, approximately 15 per cent of the US corn crop was used to provide about 2 per cent of that country's non-diesel transport fuel.[2] In Brazil, about 50 per cent of the sugar cane crop was dedicated to producing about 40 per cent of the non-diesel transport fuel, and in Europe, more than 20 per cent of the rapeseed crop was tapped to provide about 1 per cent of all transport fuel.[3] Among the other countries that have made major commitments to biofuels during recent years are China, Colombia, India, the Philippines and Thailand.

In order to assess the ramifications of the large-scale development of biofuels around the globe, the German Federal Ministry of Food, Agriculture and Consumer Protection (BMELV), through the Agency of Renewable Resources (FNR), has commissioned the German Agency for Technical Cooperation (GTZ) and the Worldwatch Institute to undertake a comprehensive survey of biofuels for transportation, guided by the principles of sustainable agriculture, energy and transport. Building on the input from detailed country studies from Brazil, China, Germany, India and Tanzania, this book aims to bring the results of focused analysis into the wider international debate.

This book is intended as a resource for decision-makers, describing existing production methods and policies for biofuels and assessing options for their future development. Parts I and II discuss current and future feedstock options, production technologies and potentials. Parts III and IV address key economic, social and environmental concerns that will be raised by the large-scale production of biofuels. Part V details the fuel, engine, vehicle and infrastructure technologies that may be deployed to facilitate greater use of biofuels in the world's transportation fuel markets. Part VI assesses the various policy frameworks being used to promote biofuels, as well as new ideas under active discussion. Part VII provides recommendations for decision-makers and Part VIII describes five country studies.

While the potential market for biofuels is enormous, this volume concludes that a wide range of issues remain to be addressed. In particular, the transition from extracting oil from beneath the ground to cultivating fuel feedstock on the surface could lead to competition for scarce resources and place additional strain on the Earth's already-stressed life-support systems. The convergence of the energy, food and fibre markets will further complicate global investment decisions and will probably increase food prices – a trend that could be beneficial to farmers, but could make it more difficult to satisfy the food needs of the world's urban poor. Additionally, if farmers are pushed to expand their cropping into new territory to meet growing biofuel demand, this could result in soil erosion, aquifer depletion and the loss of biologically rich ecosystems, including tropical forests. Government policy decisions, and the resolve to see them properly implemented, will be critical in determining the net ecological impacts of expanded biofuel use.

The book also finds that the potential benefits of biofuels will only be realized if a host of new environmentally sustainable technologies are employed, ranging from new crops and farming methods to advanced conversion technologies and highly efficient vehicles. One of the most important and anticipated innovations is the development of cellulosic ethanol derived from plant stalks, leaves and even wood. Synthetic diesel, made from an even broader range of energy crops or waste streams, also holds great promise. These technologies, which are close to being introduced commercially, will make it possible to produce biofuels from agricultural and forestry wastes, as well as from non-food crops such as switchgrass that can be grown on degraded lands. Wise and innovative policies will be needed to steer the biofuel industry in these directions.

The broader social and economic impacts of biofuels will likewise be determined largely by policy decisions. One of the great promises of biofuels – and the main political engine behind them – is to increase farm incomes and strengthen rural economies. Indeed, if farmers not only produce our food and fibre, but also a growing portion of our energy, biofuels could transform agriculture more profoundly than any development since the green revolution. Since biofuel feedstock must be gathered across wide areas, it will never be as centrally produced as petroleum products are. However, as the market grows and biofuels become a large-scale commodity, an increasing share of the income will probably go to larger farms and agribusinesses. Conscious decisions will need to be made if smaller-scale biofuel production is to be successful. The ability of small farmers to benefit from biofuels will also be determined, in part, by broader decisions about land reform and tax policies.

Another potential benefit of biofuels is the role they could play in reducing the threat of global climate change. The transportation sector is responsible for about one quarter of global energy-related greenhouse gas (GHG) emissions, and that share is rising. At least for the near term, biofuels combined with energy efficiency improvements offer the only option for dramatically reducing demand for oil and associated transport-related warming emissions. But while a dramatic increase in biofuel production and use could reduce emissions from transport significantly, there is also the possibility that such a ramping up could intensify the threat of a warming world. The overall climate impacts of biofuels will depend upon several factors, the most important being changes in land use, choice of feedstock and management practices. The greatest potential for reducing GHG emissions lies in the development of next-generation biofuel feedstocks and technologies.

International trade and the rules that govern it will play a major role in shaping biofuel development. To date, biofuels have been nurtured by national policies that favour domestically produced biofuels at the expense of imports. Domestic production can replace expensive oil imports, help unburden developing countries from staggering energy import bills, stabilize their currencies and encourage foreign investment. However, such domestically oriented policies are now limiting the development of the international biofuel market. Brazil, in particular, hopes to turn ethanol into a major export business. Such trade will undoubtedly serve as a tremendous spur to biofuel development, particularly in tropical countries blessed with inexpensive sugar cane and plant oils. A global biofuel market could serve to stabilize supply with staggered harvests and to smooth out variations in yield across climatic zones and hemispheres. But such a market could also discourage biofuel development in poorer countries and those with less favourable growing conditions.

This book concludes that biofuels have a large potential to substitute for petroleum fuels and – together with a host of other strategies, including the development of far more efficient vehicles – can help the world achieve a more diversified and sustainable transportation system in the decades ahead. However, these promises will only be achieved if policies are enacted that steer biofuels in the right direction – policies that will need to be adjusted and refined as the state of knowledge advances and as the risks and opportunities of biofuel development become clearer.

Suzanne C. Hunt
Biofuels Project Manager
Worldwatch Institute, Washington DC

Christopher Flavin
President
Worldwatch Institute, Washington DC

January 2007

Note to Readers

The views expressed in this study do not necessarily reflect those of the German Federal Ministry of Food, Agriculture and Consumer Protection (BMELV).

All units in this report are metric, unless otherwise indicated. Monetary amounts are expressed in Euros and in US dollars at the exchange rate of €1 (Euro) to US$1.21, as of 30 January 2006.

SUGGESTED CITATION

Worldwatch Institute (2006) *Biofuels for Transport: Global Potential and Implications for Energy and Agriculture*, prepared by Worldwatch Institute for the German Ministry of Food, Agriculture and Consumer Protection (BMELV) in coordination with the German Agency for Technical Cooperation (GTZ) and the German Agency of Renewable Resources (FNR), published by Earthscan, London

List of Acronyms and Abbreviations

°C degrees Celsius
3-D three dimensional
ACP Africa, the Caribbean and the Pacific countries
ADL Arthur D. Little
ADM Archer Daniels Midland
ANP Agência Nacional do Petróleo, Gás Natural e Biocombustíveis (Brazil)
ASEAN Association of Southeast Asian Nations
ASTM American Society for Testing and Materials
ASU air separation unit
ATFS American Tree Farming Systems
ATR auto-thermal reforming
B2 2 per cent biodiesel
B20 20 per cent biodiesel
BCI BC International
BMELV German Federal Ministry of Food, Agriculture and Consumer Protection
BSES Bureau of Sugar Experiment Stations (Australia)
Bt *Bacillus thuringiensis*
BTG Biomass Technology Group (The Netherlands)
BTL biomass to liquid
Btu British thermal unit
C-5 five-carbon
C-6 six-carbon
C_2H_5OH ethanol
$C_3H_8O_3$ glycerine
$C_6H_{12}O_6$ sugar
CAAE Chinese Academy of Agricultural Engineering
CAFÉ Corporate Average Fuel Economy
CAFI Consortium for Applied Fundamentals and Innovation
CAFTA Central American Free Trade Agreement
CAP Common Agricultural Policy of the European Union
CBD United Nations Convention on Biological Diversity

CBI	Caribbean Basin Initiative
CBP	consolidated bioprocessing
CCFM	Canadian Council of Forest Ministers
CDM	Clean Development Mechanism
CEN	European Committee for Standardization
CENBIO	Brazilian Reference Centre on Biomass
CETESB	São Paulo State Environment Agency
CFU	World Bank Carbon Finance Unit
CH_4	methane
CHP	combined heat and power
CIFOR	Centre for International Forestry Research
CIS	Commonwealth of Independent States
CME	coconut methyl ester
CNG	compressed natural gas
CNRS	Centre National de la Recherche Scientifique
CO	carbon monoxide
CO_2	carbon dioxide
CSA	Canadian Standards Association
CSD	United Nations Commission for Sustainable Development
CTA	Air Force Technology Centre
CTC	Centro de Tecnologia Copersucar (Copersucar Technology Centre)
CTL	coal to liquid
DDC	Detroit Diesel Company
DDG	dried distillers grain
DDGS	dried distillers grain with solubles
DEFRA	UK Department of Environment, Food and Rural Affairs
DEG	Deutschen Investitions- und Entwicklungsgesellschaft (part of KfW banking group)
DIN	Deutsches Institut für Normung (German Standards Institute)
DME	dimethyl ether
DOE	US Department of Energy
DPF	diesel particulate filter
E5	5 per cent ethanol
EBA	Everything but Arms initiative
EC	European Commission
EEA	European Environment Agency
EEB	European Environmental Bureau
EGS	environmental goods and services
EIA	US Energy Information Administration
EJ	exajoule
EMAS	Eco-Management and Audit Scheme
EPA	US Environmental Protection Agency
EPAct	US Energy Policy Act

ESALQ	University of São Paulo's Agricultural School Luiz de Queiroz
ETBE	ethyl tertiary butyl ether
EU	European Union
EU-ETS	European Union Emissions Trading System
EUGENE	European Green Electricity Network
FAME	fatty acid methyl ester
FAO	United Nations Food and Agriculture Organization
FAPESP	Research Support Foundation of the State of São Paulo
FARRE	Forum de l'Agriculture Raisonne Respectueuse de l'Environnement
FBDS	Fundação Brasileira para o Desenvolvimento Sustentável
FCV	fuel cell vehicle
FFE	fossil fuel equivalent
FFV	flexible-fuel vehicle
FIPEC	Research Financing Foundation of Banco do Brazil
FNR	Fachagentur Nachwachsende Rohstoffe (German Agency of Renewable Resources)
FoE	Friends of the Earth
FSC	Forest Stewardship Council
F-T	Fischer-Tropsch
FTPT	Tropical Foundation of Technological Research André Tosello
G8	Group of 8 industrialized nations (Canada, France, Germany, Italy, Japan, Russia, UK and US)
GAP	good agricultural practice
GATT	General Agreement on Tariffs and Trade
GBF	German Research Centre for Biotechnology
GDP	gross domestic product
GEF	Global Environment Facility
GFDL	Geophysical Fluid Dynamics Laboratory
Gha	gigahectare
GHG	greenhouse gas
GISS	Goddard Institute for Space Studies
GJ	gigajoule
GM	genetically modified
GMIO	genetically modified industrial micro-organism
GMO	genetically modified organism
GRAIN	Global Restrictions on the Availability of biomass in the Netherlands
GSP	General System of Preferences
GT	gas turbine
GTCC	gas turbine combined cycle
GTL	gas to liquid
GTZ	Gesellschaft für Technische Zusammenarbeit (German Agency for Technical Cooperation)

GVEP	Global Village Energy Partnership
GW	gigawatt
GWh	gigawatt hour
GWth	gigawatt-thermal
ha	hectare
HC	hydrocarbon
HHV	higher heating value
HIPC	highly indebted poor country
HTU	hydrothermal upgrading
IAA	Institute of Sugar and Alcohol (Brazil)
IAC	Agronomic Institute of Campinas
IATP	Institute for Agriculture and Trade Policy
ICEV	internal-combustion engine vehicle
ICRISAT	International Crop Research Institute for Semi-Arid Tropics
IEA	International Energy Agency
IEEP	Institute for Energy and Environmental Protection
IFC	International Finance Corporation
IFEU	Institut für Energie- und Umweltforschung
IFI	international financial institution
IFOAM	International Federation of Organic Agriculture Movements
IFPRI	International Food Policy Research Institute
IIASA	International Institute for Applied Systems Analysis
IISD	International Institute for Sustainable Development
ILO	International Labour Organization
IMAGE	Integrated Model to Assess the Global Environment
IMF	International Monetary Fund
IPCC	Intergovernmental Panel on Climate Change
IRR	internal rate of return
ISEC	Institute for Social and Economic Change
ISO	International Organization for Standardization
ITTO	International Timber Trade Organization
J	joule
JI	Joint Implementation provisions
kg	kilogram
km	kilometre
KWh	kilowatt hour
KWST	Kraul & Wilkening and Stelling
KWth	kilowatt-thermal
LCA	life-cycle analysis
LNG	liquefied natural gas
LPG	liquefied petroleum gas
MDF	medium-density fibreboard
MDG	United Nations Millennium Development Goal

MERCOSUR	Mercado Commun del Sur (Latin American trade bloc)
MFN	most favoured nation
mm	millimetre
MoE	Tanzanian Ministry of Energy and Minerals
MOU	memorandum of understanding
MSW	municipal solid waste
MTBE	methyl tertiary-butyl ether
MW	megawatt
MWh	megawatt hour
MWth	megawatt-thermal
N	nitrogen
N_2O	nitrous oxide
NAFTA	North American Free Trade Agreement
NATT	Centre for Absorption and Transfer of Technology (Brazil)
NGO	non-governmental organization
NILE	New Improvement for Lignocellulosic Ethanol
NO_2	nitrogen dioxide
NOVEM	Dutch Energy Agency
NO_x	nitrogen oxides
NRDC	US Natural Resources Defense Council
NREL	US National Renewable Energy Laboratory
NWF	National Wildlife Federation
O_3	ozone
OECD	Organisation for Economic Co-operation and Development
OEM	original equipment manufacturer
OPEC	Organization of Petroleum Exporting Countries
PAH	polycyclic aromatic hydrocarbon
PAN	peroxyacetyl nitrate
PCB	polychlorinated biphenyl
PCF	Prototype Carbon Fund
PEFC	Pan-European Forest Certification
PM	particulate matter
PPM	process and production method
PPO	pure plant oil
psi	pounds per square inch
PV	photovoltaic
PVC	polyvinyl chloride
R&D	research and development
REEEP	Renewable Energy and Energy Efficiency Partnership
REN 21	Renewable Energy Global Policy Network
RFS	Renewable Fuels Standard
RIVM	Dutch National Institute of Public Health and the Environment

RME	rapeseed methyl ester
RSPO	Roundtable on Sustainable Palm Oil
RTFO	Renewable Transport Fuel Obligation (UK)
RVP	Reid vapour pressure
SAN	Sustainable Agriculture Network
SASTA	South African Sugar Technologists' Association
SBT	Southern Online Bio Technologies
SCM	Subsidies and Countervailing Measures agreement
SEFI	Sustainable Energy Finance Initiative
SFI	Sustainable Forestry Initiative
SHF	separate hydrolysis and fermentation
SL	sustainable livelihoods
SME	soybean methyl ester
SMR	steam-methane reforming
SO_2	sulphur dioxides
SO_x	sulphur oxide
SR	short rotation
SRC	short-rotation coppice
SRES	*Special Report on Emissions Scenarios* (IPCC)
SRF	short-rotation forestry
SRWC	short-rotation woody crops
SSA	Standard for Sustainable Agriculture
SSCF	simultaneous saccharification and co-fermentation
SSF	simultaneous saccharification and fermentation
SUV	sport utility vehicle
SVO	straight vegetable oil
TaTEDO	Tanzanian Traditional Energy Development and Environment Organization
TB	treated biogas
TBT	Technical Barriers to Trade Agreement
TEL	tetra-ethyl lead
TERI	The Energy Resources Institute
UFOP	Union zur Foerderung von Oel- und Proteinpflanzen e.V. (German Union for the Promotion of Oil and Protein Plants)
UK	United Kingdom
UN	United Nations
UNCTAD	United Nations Conference on Trade and Development
UNDP	United Nations Development Programme
UNEP	United Nations Environment Programme
UNFCCC	United Nations Framework Convention on Climate Change
UNICA	União da Agroindústria Canavieira de São Paulo (São Paulo Association of Sugar and Alcohol Manufacturers)
US	United States

USAID	US Agency for International Development
USDA	US Department of Agriculture
VEETC	Volumetric Ethanol Excise Tax Credit
VOC	volatile organic compound
VWP	Vereinigte Werkstätten für Pflanzenöltechnologie
WEC	World Energy Council
WTO	World Trade Organization
WWF	World Wide Fund for Nature (*formerly* the World Wildlife Fund)
WWT	wastewater treatment

Part I

Status and Global Trends

1

Current Status of the Biofuel Industry and Markets

A GLOBAL OVERVIEW

The liquid biofuels most widely used for transport today are ethanol and biodiesel. Ethanol is currently produced from sugar or starch crops, while biodiesel is produced from vegetable oils or animal fats. The growth in the use of biofuels has been facilitated by their ability to be used as blends with conventional fuels in existing vehicles, where ethanol is blended with gasoline and biodiesel is blended with conventional diesel fuel.

Ethanol currently accounts for 86 per cent of total biofuel production.[1] About one quarter of world ethanol production goes into alcoholic beverages or is used for industrial purposes (as a solvent, disinfectant or chemical feedstock); the rest becomes fuel for motor vehicles.[2] Most of the world's biodiesel, meanwhile, is used for transportation fuel, though some is used for home heating.

Global fuel ethanol production more than doubled between 2001 and 2006, while production of biodiesel, starting from a much smaller base, expanded nearly sixfold (see Figures 1.1 and 1.2).[3] In contrast, the oil market increased by only 10 per cent over this period (in absolute terms, however, world petroleum production increased by some 80 million litres a year from 2001 to 2006, compared to some 5 million litres annually for biofuels).[4] In 2006, biofuels comprised about 0.9 per cent of the world's liquid fuel supply by volume, and about 0.6 per cent by transport distance travelled. Yet, as a percentage of the increase in supply of liquid fuels worldwide from 2005 to 2006, the surge in production of the two biofuels accounted for 17 per cent by volume and 13 per cent by transport distance travelled.[5]

HISTORY OF BIOFUEL PRODUCTION PROGRAMMES

Biofuels have been used in automobiles since the early days of motorized transport. American inventor Samuel Morey used ethanol and turpentine in the first internal combustion engines as early as the 1820s. Later that century, Nicholas Otto ran

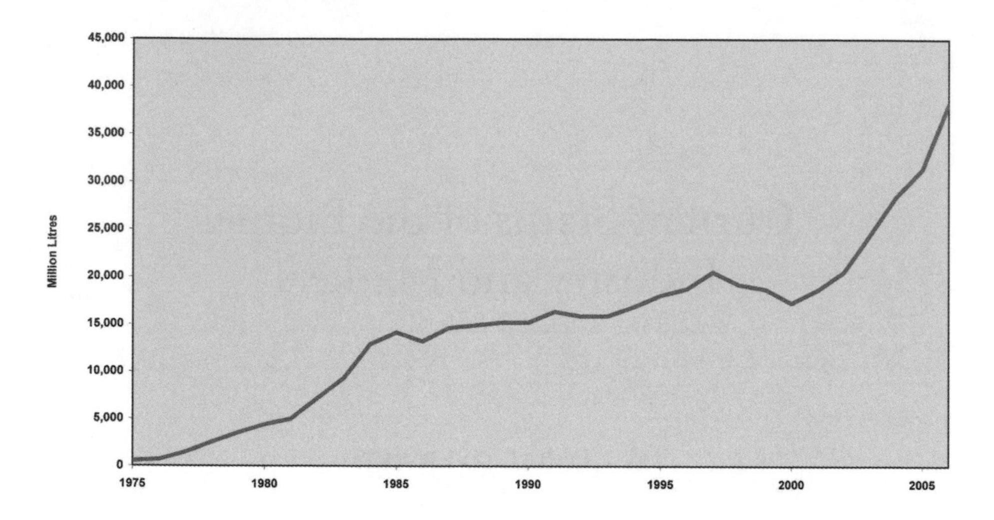

Figure 1.1 *World fuel ethanol production, 1975–2006*

Source: F. O. Licht

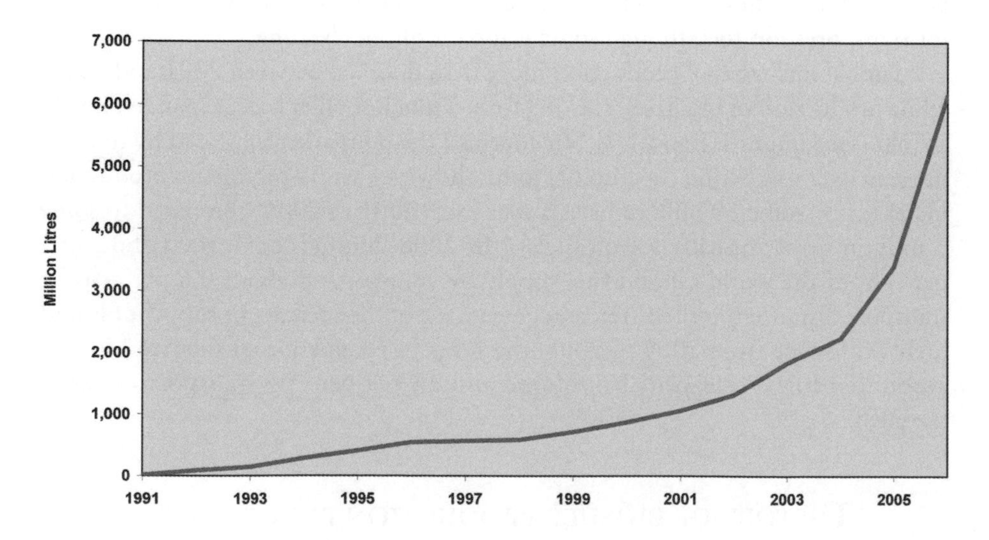

Figure 1.2 *World biodiesel production, 1991–2006*

Source: F. O. Licht

his first spark-ignition engines on ethanol, and Rudolph Diesel used peanut oil in his prototype compression-ignition engines. Henry Ford's Model T could even be calibrated to run on a range of ethanol–gasoline blends. However, just as automobiles were becoming popular at the beginning of the 1900s, the fuel market was flooded with cheap petroleum fuels.[6]

Biofuels represented only a small proportion of total fuel during the early 20th century. They were supported by policies in several European countries, especially France and Germany, where at times they neared 5 per cent of the fuel supply. In tropical areas with irregular supplies of petroleum and in enclosed settings such as mines, biofuels were often the favoured fuel; during World Wars I and II, ethanol was also used to supplement petroleum in Europe, the US and Brazil. However, military demobilization in the post-war period and the development of new oil fields in the 1940s brought a glut of cheap oil that virtually eliminated biofuels from the world fuel market.[7]

The oil crises of the 1970s prompted countries to again seek alternatives to imported oil. Brazil, which had maintained a small fuel ethanol industry since the 1930s, expedited a national ethanol programme called *Proálcool* with an eye to alleviating its great national debt and expanding its agricultural industry. Especially after the second oil crisis of 1979, when oil prices reach their historic zenith, the Brazilian government prioritized ethanol production, supporting expanded sugar cane acreage, new ethanol distilleries and ethanol-only cars. By the mid 1980s, ethanol was displacing almost 60 per cent of the country's gasoline.[8]

Also motivated by the high and volatile oil prices of the 1970s, the US launched its own fuel ethanol programme at the end of the decade, using corn to produce a proportionally small but increasing amount of ethanol. The Brazilian and US ethanol industries still produce the vast majority of the world's fuel ethanol – almost 90 per cent in 2005.[9]

The oil crises prompted other countries to promote biofuels as well, although these efforts were less successful. In China, the government encouraged peasants to cultivate oil plants that would provide insurance against disruptions in the supply of diesel fuels; but it abandoned these efforts after the price of oil fell in the mid 1980s.[10] In 1978, the Kenyan government initiated a programme to distil ethanol from sugar cane, mixing it in a 10 per cent blend with gasoline; but this programme faltered due to drought, poor infrastructure and inconsistent policies.[11] Zimbabwe and Malawi initiated larger programmes in 1980 and 1982, respectively; but only Malawi has consistently produced fuel ethanol since then.[12]

In Europe, a trade dispute triggered a rise in biodiesel production, starting in 1992. The European Union (EU) agreed to prevent gluts in the international oilseeds market by confining production to just under 5 million hectares. European governments helped to create a new market for farmers on the remaining 'set-aside' land, primarily by reducing the taxes on biodiesel – a policy that has led to a rapid increase in European biodiesel production, particularly in Germany.[13]

More recently, environmental standards have become important drivers for biofuel markets. In the US, the Environmental Protection Agency (EPA) began requiring cities with high ozone levels to blend gasoline with fuel oxygenates, including ethanol. When state governments learned in the late 1990s and early 2000s that the most common oxygenate, methyl tertiary-butyl ether (MTBE), was a possible carcinogen that was seeping into groundwater, 20 states passed laws to phase it out, creating a surge in demand for US ethanol in the early 2000s.[14]

In Brazil, the auto industry's 2003 introduction of so-called flexible-fuel vehicles (FFVs), which can run on any combination of gasoline or ethanol, has given drivers the freedom to choose whichever of the fuels is cheaper. Consumer demand for these vehicles has surged, and by early 2006, more than 75 per cent of new cars sold in the country were FFVs.[15] Combined with high petroleum prices, these cars have led to a dramatic increase in Brazilian ethanol production.[16]

CURRENT BIOFUEL PRODUCTION

The US and Brazil dominate world ethanol production, which reached a record 38.2 billion litres in 2006 (see Table 1.1).[17] Close to half the world's fuel ethanol was produced in the US in 2006, nearly all of it from corn crops grown in the northern Midwest, representing 2 to 3 per cent of the country's non-diesel fuel.[18] More than two fifths of the global fuel ethanol supply was produced in Brazil in 2006, where sugar cane grown mostly in its centre-south region provides roughly 40 per cent of the country's non-diesel fuel.[19]

The remainder of ethanol production comes primarily from the EU, where Spain, Sweden, France and Germany are the big producers, using mainly cereals and sugar beets. China uses corn, wheat and sugar cane as feedstock to produce

Table 1.1 *World fuel ethanol production, 2006*

Country or region	*Production (million litres)*	*Share of total (percentage)*
United States	18,300	47.9
Brazil	15,700	41.1
European Union	1550	4.1
China	1300	3.4
Canada	550	1.4
Colombia	250	0.7
India	200	0.5
Thailand	150	0.4
Australia	100	0.3
Central America	100	0.3
World Total	38,200	100.0

Source: see endnote 17 for this chapter

a large amount of ethanol destined mostly for industrial use. In India, sugar cane and cassava have been used intermittently to produce fuel ethanol.[20]

In 2005, many new ethanol production facilities began operating, were under construction or were in the planning stage. For example, US ethanol production capacity increased by nearly 3 billion litres during 2005, with an additional 5.7 billion litres of new capacity under construction going into 2006.[21]

Biodiesel has seen similar growth, almost entirely in Europe (see Table 1.2).[22] Biodiesel comprises nearly three-quarters of Europe's total biofuel production, and in 2006 the region accounted for 73 per cent of all biodiesel production worldwide, mainly from rapeseed and sunflower seeds.[23] Germany accounted for 40 per cent of this production, with the US, France and Italy generating most of the rest.

The rapidly changing character of worldwide biofuel production capabilities is illustrated by recent trends in the US. US biodiesel production, mainly from soybeans, was 1.9 million litres (500,000 gallons) in 1995; by 2005, it had jumped to 284 million litres (75 million gallons); and in 2006 it tripled, to 852 million litres (224 million gallons).[24] At mid-2006, US biodiesel production capacity stood close to 1.2 billion litres per year from 42 facilities, and more than 400 million litres per year of additional production capacity were under construction at 21 new plants.[25, 26] Meanwhile, the EU was home to approximately 40 biodiesel plants, and this capacity was also growing rapidly, both in Germany, which has been the

Table 1.2 *World biodiesel production, 2006*

Country or region	*Production (million litres)*	*Share of total (percentage)*
Germany	2499	40.6
United States	852	13.8
France	625	10.2
Italy	568	9.2
Czech Republic	153	2.5
Spain	142	2.3
Malaysia	136	2.2
Poland	114	1.9
United Kingdom	114	1.9
Australia	91	1.5
Austria	85	1.4
Denmark	80	1.3
Philippines	68	1.1
Brazil	68	1.1
China	68	1.1
Others	490	8.0
Europe Total	4504	73.2
Americas Total	1113	18.1
World Total	6153	100.0

Source: see endnote 22 for this chapter

clear leader in world biodiesel production, and also in Austria, the Czech Republic, France, Germany, Italy, Spain and Sweden.

WORLD PETROLEUM USE AND IMPLICATIONS FOR BIOFUELS

The doubling of petroleum prices, from about US$30 per barrel in early 2004 to about US$60 per barrel at the end of 2005, and subsequent price increases in 2006 have substantially heightened worldwide interest and investment in biofuels.[27] The expected growth in global demand for liquid fuels, together with increasing geological limitations in the supply of oil, have led many to assume that oil prices will remain high in years to come. In 2006, the US Energy Information Administration, for instance, upgraded its forecasted price for a barrel of oil in 2025 to US$54 from the previous year's projection of US$33.[28]

Substantial growth in energy use in many developing countries has also begun, most notably in China and India, whose populations are by far the world's largest. From 2002 to 2004, world oil demand increased by 5.3 per cent, while China's demand alone increased by a staggering 26.4 per cent; demand in other Asian countries increased by 5.8 per cent combined. This growth has come as oil consumption in many industrialized countries continues to rise. From 2002 to 2004, the demand for oil increased by 4.9 per cent in the US, 10.2 per cent in Canada and 6.3 per cent in the UK (demand in Germany and Japan, meanwhile, dropped by 1 per cent and 2.6 per cent, respectively).[29]

While there are dramatic differences in per capita gasoline and diesel fuel consumption between industrialized and developing countries, economic growth and lifestyle changes in populous countries such as China and India will probably put tremendous pressure on world petroleum supplies in the coming decades. Currently, US per capita gasoline consumption is a staggering 180-fold higher than in India and 45-fold higher than in China, and US per capita diesel consumption is 17-fold higher than in India and about 11-fold higher than in China (even German per capita consumption of these fuels is dramatically higher than in developing countries – for gasoline, about 45-fold higher than in India and 11-fold higher than in China, and for diesel, about 19-fold higher than in India and 12-fold higher than in China.) However, if all the residents of China were to consume oil at the same per capita rate as people in the US, they would require an amount greater than the current total production of oil worldwide (for national-level information on per capita gasoline and diesel fuel consumption, see Appendix 1).[30]

Although per capita energy consumption in the developing world is currently low, as these countries experience economic growth and increased demand for oil, they will require new energy supplies to meet their transportation needs. For countries that depend upon imported petroleum fuels, there is a clear need for alternative transport fuel supplies, either domestically produced or imported. There is an opportunity for biofuels to play an important role, particularly as petroleum costs continue to rise and as the unpredictability of future oil availability triggers national security concerns (for national-level information on biofuel production in relation to petroleum consumption, production and imports, see Appendix 2; for a comparison of biofuel production as a percentage of petroleum use, see Appendix 3).

RECENT DEVELOPMENTS IN THE BIOFUEL INDUSTRY

The recent proliferation of biofuel programmes around the world can be attributed to a combination of factors. Countries that seek to bolster their agricultural industries (long the main driver of biofuel programmes) have been joined by an increasing number of nations that are concerned about such factors as high oil prices, political instability in oil-exporting countries, climate-altering greenhouse gas emissions and urban air pollution. Continuing developments in biorefining technology have also brought greater attention to biofuels as a potentially large-scale and environmentally sustainable fuel.

A diverse range of countries around the world has recently sought new ways of promoting the use of biofuels. For example:

- In Japan, the government has permitted low-level ethanol blends in preparation for a possible blending mandate, with the long-term intention of replacing 20 per cent of the nation's oil demand with biofuels or gas-to-liquid (GTL) fuels by 2030.
- In Canada, the government wants 45 per cent of the country's gasoline consumption to contain 10 per cent ethanol by 2010. Ontario will be the centre of the ethanol programme, where the government expects all fuel to be a 5 per cent blend of ethanol by 2007.[31]
- An EU directive, prompted by the desire for greater energy security, as well as the requirements of the Kyoto Protocol, has set the goal of obtaining 5.75 per cent of transportation fuel needs from biofuels by 2010 in all member states. In February 2006, the EU adopted an ambitious Strategy for Biofuels with a range of potential market-based, legislative and research measures to increase the production and use of biofuels. Germany and France, in particular, have announced plans to rapidly expand both ethanol and biodiesel production, with the aim of reaching the EU targets before the deadline.[32]
- In the US, high oil prices and agricultural lobbying prompted the recently enacted Renewable Fuels Standard (RFS), which will require the use of 28.4 billion litres (7.5 billion gallons) of biofuels for transportation in the country by 2012. Many US government fleet vehicles that run on diesel fuel are now required to use B20 blends under new guidelines implementing the 1992 Energy Policy Act. Many in the industry believe that these targets represent a floor, rather than a limit, to biofuel production.[33]

- In Brazil, the government hopes to build on the success of the *Proálcool* ethanol programme by expanding the production of biodiesel. All diesel fuel must contain 2 per cent biodiesel by 2008, increasing to 5 per cent by 2013, and the government hopes to ensure that poor farmers in the north and northeast receive much of the economic benefits of biodiesel production.
- Elsewhere in Latin America, as of 2006, Colombia will be mandating the use of 10 per cent ethanol in all gasoline sold in cities with populations over 500,000. In Venezuela, the state oil company is supporting the construction of 15 sugar cane distilleries over the next five years as the government phases in a national E10 blending mandate. In Bolivia, 15 distilleries are being constructed, and the government is considering authorizing blends of E25. Costa Rica and Guatemala are also in the trial stages for expanding production of sugar cane fuel ethanol.[34] Argentina, Mexico, Paraguay and Peru are all considering new biofuel programmes as well.[35] As the world's leader in fuel ethanol, Brazil has helped many of these countries to learn from its example (see Box 1.1).[36]
- In Southeast Asia, Thailand, eager to reduce the cost of oil imports while supporting domestic sugar and cassava growers, has mandated an ambitious 10 per cent ethanol mix in gasoline starting in 2007.[37] For similar reasons, the Philippines will soon mandate 2 per cent biodiesel, to support coconut growers, and 5 per cent ethanol, probably beginning in 2007.[38] In Malaysia and Indonesia, the palm oil industries plan to supply an increasing proportion of the countries' diesel.
- Chinese and Indian planners have also sought to expand the national supply of ethanol and biodiesel. In India, a rejuvenated sugar ethanol programme calls for E5 blends throughout most of the country, a level that the government plans eventually to raise to E10 and then E20. In China, the government is making E10 blends mandatory in five provinces that account for 16 per cent of the nation's passenger cars.[39]
- In Africa, efforts to expand biofuels production and use are being initiated or are under way in numerous countries, including Kenya, Malawi, Zimbabwe, Ghana, Ethiopia, Benin, Mozambique, Senegal, Guinea Bissau, Ethiopia, Nigeria and South Africa.[40]

For more detailed information on international policies and initiatives under way to foster biofuel development, see Chapter 17.

Along with the rapid increase in government-supported biofuel programmes, recent advances in technology have brought new interest in biofuels. In addition to producing ethanol from so-called 'cellulosic' feedstock, technologies are currently being developed that will be able to convert abundant cellulosic biomass supplies to a variety of potential diesel fuel or gasoline substitutes (see Chapters 4 and 5).

For example, a conversion system that uses high temperatures and low-oxygen conditions to convert solid biomass into combustible gases can be coupled to a gas-to-liquid (GTL) conversion process to produce liquid fuel. The 'Fischer-Tropsch'

Box 1.1 Brazil's ethanol experience

In response to the oil crises of the 1970s, the Brazilian government turned to one of the country's oldest industries: sugar cane. By making it a national priority to build distilleries that ferment sugar into ethanol, and requiring that this fuel be mixed into all gasoline, Brazil became a global leader in the transition away from oil.

In the 1990s, rising sugar prices in Brazil coincided with lower petroleum prices, causing a drop in ethanol production and subsequent shortages. This forced the country to import ethanol and imperilled the national ethanol programme, *Proálcool*. Several key government initiatives and market changes worked in tandem to turn this situation around. Brazil phased out sugar and ethanol quotas, as well as a constrained government subsidy programme that had limited new capacity investments. It worked with farmers to help reduce sugar cane production costs and improve yields, mandated the use of ethanol in government vehicle fleets, and fostered sales and use of flexible-fuel vehicles – in addition to requiring 20 to 25 per cent ethanol blends in all regular gasoline sales.

Along with these changes, Brazil's industry has reduced ethanol production costs in a variety of ways, particularly through the increased use of sugar cane processing residues (bagasse) as fuel to produce the steam and electricity needed to process cane. The industry is also recycling organic-rich liquid effluent from cane processing (vinasse) and using it as a fertilizer and irrigation supply for cane production, thereby increasing cane yields and reducing feedstock costs. In addition, rising petroleum prices have increased the market value of ethanol. The end result is that Brazil has become the world's largest exporter of ethanol fuel while also meeting a growing share of its domestic fuel needs.

In June 2005, Brazilians could purchase ethanol for half the price of gasoline per litre (or about 75 per cent as much as gasoline costs per unit of energy, after adjusting for the lower energy content of ethanol per litre compared to gasoline). When they do, they are not sending money to oil producers overseas, but to Brazilians. Since the 1970s, Brazil has saved almost US$50 billion in imported oil – nearly ten times the national investment through subsidies – while creating as many as 1 million rural jobs.

Source: see endnote 37 for this chapter

(F-T) process, a technology originally developed by German researchers during the 1920s, uses chemical reactions with catalysts to convert the combustible gases from a biomass gasifier into a liquid fuel that can substitute for diesel fuel. Researchers from DaimlerChrysler, Volkswagen and Shell have recently collaborated to develop a marketable version of this technology.[41]

In Canada and the US, the governments have supported groundbreaking research into enzymes that could refine abundant low-value plant fibres into ethanol. The enzyme company Novozymes, with funding from the US National Renewable Energy Laboratory, announced in 2005 that it could reduce the cost of some of these enzymes by 10 to 30 times and promised further reductions in the near future. Abengoa, a multinational ethanol company, has already begun building a facility in Spain that will utilize these enzymes.[42] In Canada, Iogen Corporation operates a pilot plant to convert straw into ethanol using enzymatic technology, and has now teamed up with Shell, Volkswagen and DaimlerChrysler to build a pre-commercial straw-to-ethanol plant in Europe.

2

Liquid Biofuels: A Primer

Introduction

In order to better understand biofuel markets and the prospects for expanded biofuel use, it is helpful to examine some of the basic characteristics of these various fuel alternatives. This chapter explores the main biofuel options on the market today, including ethanol (produced from sugars and starches), ethyl tertiary butyl ether (ETBE) and lipid-derived biofuels (straight vegetable oil and biodiesel). It discusses the various fuel blends and compares the production costs of different biofuel options.

Carbohydrate-derived biofuels

Ethanol produced from sugars

A variety of common sugar crops can be used as the feedstock for producing ethanol fuel, including sugar cane stalks, sugar beet tubers, and sweet sorghum stalks – all of which contain a large proportion of simple sugars. Once these sugars have been extracted they can be fermented easily into ethanol. Starch crops such as corn, wheat and cassava can also be hydrolysed into sugar, which can then be fermented into ethanol.

Left alone in low-oxygen conditions, the sugar in plants naturally ferments into acids and alcohols (particularly ethanol) over time; however, people have used yeast for thousands of years to expedite this process. Ethanol production starts by grinding up the feedstock so it is more easily and quickly processed. Once ground up, the sugar is either dissolved out of the material or the starch is converted into sugar. The sugar is then fed to yeast in a closed anaerobic chamber. The yeast secretes enzymes that digest the sugar ($C_6H_{12}O_6$), yielding several products, including lactic acid, hydrogen, carbon dioxide (CO_2) and ethanol (C_2H_5OH).[1]

Brazilian facilities are the most significant producers of sugar-based fuel ethanol in the world. The fermentation units are usually integrated within existing sugar mills, where the co-products of refining sugar cane include various grades of sugar, molasses, CO_2 and the fibrous residue of crushed sugar cane stalks, called bagasse.

In Brazil, the bagasse residue is typically used as a boiler fuel to produce steam, which is used to provide process heat and often to generate electricity for use in the ethanol production process (in many cases, the excess electricity is sold to the electric grid).[2]

Even after thousands of years of development, the process of fermentation has become much more efficient in recent decades. In particular, the discovery during the 1960s of 'continuous fermentation', which permits the recycling of yeast, was a revolutionary step, substantially increasing the speed of the process and reducing the costs of heating and cooling required in single-batch processing.[3] Fermentation facilities in the US and Europe have tended to use this continuous fermentation process, while many in Brazil are only now beginning to adopt it.[4]

Ethanol produced from starches

Producing ethanol from feedstocks containing large amounts of starch (such as corn, wheat and cassava) adds an extra step to the process. Starches – polymers that are often thousands of sugar molecules long – must first be catalysed into simple sugars. This stage, called saccharification, requires additional energy and adds to the cost of ethanol production.

The two common methods for refining starches into sugars differ primarily in the pre-treatment of the feedstock. The 'wet-milling' process soaks grains in water, usually with a sulphurous acid, to separate the starch-rich endosperm from the high-protein germ and high-fibre husks. These wet mills tend to be larger and produce a number of co-products in addition to ethanol. By comparison, the 'dry-milling' process involves simply grinding the unprocessed heterogeneous seed into granules. These mills require less investment, but produce fewer co-products.[5]

The co-products of milling processes have been crucial to their viability. Wet mills co-produce corn oil, gluten feed, germ meal, starches, dextrin and sweeteners, such as high fructose corn syrup. Sold mostly as processed foods and feeds, these products together comprise more than one quarter of a wet mill's economic output.[6] The primary co-product of dry mills is dried distillers grain (DDG), a fibrous, high-protein residue (28 per cent protein) that livestock producers buy as food for animals that can digest high proportions of fibre, primarily cattle. DDG can provide about 20 per cent of a dry mill's income. Both wet and dry mills sometimes also sell the CO_2 released during fermentation, often to the carbonated beverage industry.[7]

Virtually all of the new starch ethanol facilities being built or expanded in the US are dry-milling operations since these are less costly and complex to develop than wet mills. But wet mills offer more flexibility and diversity. Wet milling facilities have the ability to switch between the production of ethanol and the production of corn syrup and/or fructose, similar in a general way to sugar cane factories, which can switch between the production of ethanol and refined sugar, depending upon the markets for ethanol versus sweeteners. Because they produce a variety of products, wet mills more closely represent the 'biorefineries' that may play an important role in the future (for more on biorefineries, see Chapter 5).[8]

Distilling fuel ethanol

Unlike ethanol produced for beverages, fuel ethanol must be distilled to have only a very small amount of water content. After removing the yeast and by-products, ethanol distillers dehydrate the 5 to 12 per cent solution of ethanol into a concentrated product of 95 to 99.8 per cent ethanol.

The allowable water remaining in ethanol fuel depends upon the specifications for particular end uses. Ethanol that is blended with gasoline needs to be dehydrated to have only trace amounts of water (less than about 1 per cent) because water can cause problems with the fuel. Unblended 'neat' ethanol, used in warm climates, can contain small amounts of water since winter freezing conditions do not occur. For instance, neat ethanol fuel sold in Brazil contains about 4 per cent water.

Ethanol as a fuel

The chemical equation for a molecule of ethanol is C_2H_5OH. It can blend with gasoline, which contains a variety of larger molecules ranging from C_5H_{12} to $C_{12}H_{26}$.

A litre of ethanol contains about two-thirds as much energy as a litre of gasoline. However, pure ethanol has a high octane value, which improves the performance of gasoline by reducing the likelihood that engine knock problems will occur (engine knock occurs when the fuel combusts too soon in an engine cylinder when a vehicle is working hard to accelerate, go up a hill or pull a heavy trailer; if the fuel ignites too soon, the combustion is not efficient in moving the vehicle forward). Adding ethanol to gasoline increases its octane level.

Since ethanol molecules contain oxygen (unlike gasoline molecules), ethanol fuel is referred to as an 'oxygenate'. The oxygen in ethanol can improve the fuel combustion process, helping to reduce the emission of pollutants such as carbon monoxide, ozone-forming unburned hydrocarbons and carcinogenic particulates. However, for related reasons, ethanol combustion also reacts with more atmospheric nitrogen, which can marginally increase emissions of ozone-forming nitrogen oxide (NO_x) gases. Since ethanol also contains a negligible amount of sulphur compared to petroleum, blending ethanol in gasoline helps to reduce the sulphur content of the fuel, resulting in lower sulphur oxide (SO_x) emissions, which can contribute to both acid rain and cancer concerns (for more on ethanol's emissions characteristics, see Chapter 13).

Since the biomass used to produce ethanol is created by photosynthesis, the carbon dioxide created by the combustion of ethanol is generally just recycling carbon back to the air. The net reduction in greenhouse gases related to ethanol's displacement of petroleum fuel can vary substantially depending upon the amount of fossil fuel used in the ethanol fuel production process (for further discussion of biofuel climate impacts, see Chapter 11).

Ethanol has solvent characteristics that, particularly in high-concentration blends, can cause corrosion of certain types of metal or deterioration of some rubber or plastics used in hoses and gaskets. In general, vehicle manufacturers have been able to readily use engine and fuel handling components that avoid concerns related to these solvent characteristics for ethanol blends at low levels. Another concern with ethanol is its potential to separate from gasoline under certain conditions. In cooler climates in particular, water contamination can trigger a 'phase separation' of ethanol and gasoline. And blending even small amounts of ethanol with gasoline raises the fuel's vapour pressure, which can affect engine performance and contribute to ozone emissions (for further discussion, see Chapters 13 and 15).

Ethyl tertiary butyl ether (ETBE)

Due to potential concerns in blending ethanol with gasoline, one approach that oil refiners can take is to combine ethanol with isobutylene to produce an additive called ethyl tertiary butyl ether (ETBE).[9] A typical 15 per cent blend of ETBE with gasoline has a biofuel content of about 6.3 per cent by volume. The isobutylene component of ETBE comes from fossil fuels and thus reduces ethanol's displacement of gasoline.

ETBE has virtually the same chemical properties as methyl tertiary-butyl ether (MTBE), an oxygenate derived entirely from fossil fuels. Both are toxic and possibly carcinogenic. But ETBE differs in that it is less water soluble than MTBE. Thus, while spills have led to surprisingly high concentrations of MTBE in groundwater, ETBE may disperse less rapidly and be easier to clean up. However, ETBE, like ethanol, has historically been more expensive than MTBE.[10]

ETBE has an advantage over ethanol in that it does not raise the fuel vapour pressure of gasoline blends. It also blends more completely with gasoline and will not separate if exposed to water in pipelines, ships or trucks. Partly as a result, petroleum companies in many European countries have more readily embraced ETBE use, since it is easier to manage fuel vapour pressure specifications and it is less complicated to transport blends of ETBE and gasoline.

Ethanol in fuel blends

In most countries, ethanol has been introduced to markets as a straight blend with gasoline. Rather than manufacturing ETBE, refiners often modify the 'base' gasoline to have a lower vapour pressure to accommodate the increase in vapour pressure caused by adding ethanol. In order to minimize the potential for phase separation during storage or transport, ethanol is typically 'splash blended' (without stirring) in the tanker trucks that deliver gasoline to vehicle refuelling stations.

Most major automobile manufacturers warranty their cars to run on ethanol blends of up to 10 per cent; however, current European standards allow only for blends of up to 5 per cent ethanol, although countries may prefer blends of ETBE. Sweden is the primary country in Europe that uses 5 per cent blends of ethanol in gasoline; but Spain, too, has allowed some marketing of E5 blends, while France and Spain primarily use ETBE blends. About 30 per cent of all gasoline sold in the US is E10 (10 per cent ethanol). All gasoline sold in Brazil contains 20 to 25 per cent ethanol, a level achievable because automakers use components that are resistant to the solvent characteristics of ethanol.

Cars with specially designed engines are able to run on even higher proportions of ethanol fuel. In Brazil, ethanol-only vehicles run on 'neat' hydrous ethanol, which is available at more than 90 per cent of gas stations. In Brazil, the US, and Europe, flexible-fuel vehicles (FFVs) that can run on low- and high-level ethanol blends are an increasingly popular option. In colder climates, where the low vapour pressure of pure ethanol can cause cold-start problems, neat ethanol is not considered viable. Instead, a blend of E85 is available for FFVs in the US (at less than 1 per cent of gas stations) and Sweden. In the winter months, the proportion of ethanol in the US is actually adjusted to E70, again to avoid cold-start problems.

A few companies have developed additives that allow for blending 5 to 15 per cent ethanol in diesel fuel. The preferred path to make the blend has been the use of an additive package that may contain a surfactant (to make the emulsion possible and stable), a lubricant (to compensate for the lubricity loss) and a cetane enhancer (to compensate for reductions in cetane, a measure of fuel ignitability under compression). These ethanol–diesel blends (sometimes referred to as E-diesel) are likely to be limited to niche applications for fleet vehicles due to technical and safety constraints associated with the relatively high volatility of ethanol–diesel blends.

LIPID-DERIVED BIOFUELS

Straight vegetable oil (SVO)

Straight vegetable oil (SVO) can be extracted from nearly any oilseed crop, such as rapeseed, sunflower, soybean and palm, for potential use as a fuel in diesel engines. Used cooking oil from restaurants and animal fat from meat slaughterhouses are also potential feedstocks for use in diesel fuel applications. After filtering out particles and removing water, this purified biomass-derived oil can burn directly in some internal combustion engines as well (see Chapter 15).

Oils are usually extracted from plant seeds by first cutting them into flakes and then immersing them in a chemical solvent (hydraulic crushing processes have typically proven too energy intensive). The non-oil components of the seed are often sold as a high-protein meal for animal feed or used as a fertilizer.[11]

Due to differences in the properties of SVO compared to diesel fuel (primarily its high viscosity, especially at cooler temperatures), 'neat' SVO cannot be used in normal diesel engines. In order to run on SVO, these engines must either be refitted (often by attaching a mechanism for pre-heating the oil), or they must be dedicated engines, such as the Elsbett engine. Vehicle manufacturers generally will not warranty their engines for operation with SVO. Moreover, development of modern engines has led in the direction of increased electronic engine and combustion control systems, which are generally not compatible with operation using SVO (see Chapter 15).[12]

Largely because SVO tends to coagulate at colder temperatures, it has been difficult to blend it with conventional diesel fuel.[13] However, different types of plant oil have different properties that affect engine performance. Some tropical oils with more saturated, shorter-chained fatty acids, such as coconut oil, can be blended directly with diesel fuel, offering the potential for the use of SVO–diesel blends in unmodified engines in tropical locations (see Chapter 3).

Where saturated tropical oils are more readily available, and where warm temperatures prevent the oil from thickening, SVO may be a viable fuel. In temperate countries, technical barriers generally limit the use of SVO to niche markets. However, fuel quality standards have been defined for pure rapeseed oil in Europe, and there has been some experience with the use and handling of the fuel in daily operation. For example, efforts are currently under way in Ireland to evaluate the ability to use low-level blends of certain pure vegetable oil types in existing vehicles.[14] And the Vereinigte Werkstätten für Pflanzenöltechnologie (VWP) company has provided an enhanced service offer for SVO use focused on filter exchange and maintenance.[15]

Biodiesel

Compared to SVO, biodiesel is a more blendable form of lipid-based biofuel. Biodiesel is made by chemically combining the oil with an alcohol (such as methanol or ethanol) in a process known as transesterification. The resulting biodiesel is an alkyl ester of fatty acid, which contains an alcohol group attached to a single hydrocarbon chain comparable in length to that of diesel ($C_{10}H_{22}$ to $C_{15}H_{32}$).[16]

This reaction can happen by heating a mixture of 80 to 90 per cent oil, 10 to 20 per cent methanol and a catalyst.[17] The catalyst is usually an acid or a base; but bases such as NaOH and KOH are the most common, in part because, with them, transesterification can happen at a lower temperature. Typical processes produce a volume of biodiesel equivalent to the volume of the original plant oil.[18]

Methanol has been the most commonly used alcohol in the commercial production of biodiesel, in part because it has typically been less expensive than ethanol. But there are also technical concerns in using ethanol, such as greater difficulties in separating the glycerine by-product from the biodiesel, as well as a propensity for higher process energy costs.

Glycerine molecules ($C_3H_8O_3$) are the primary co-product of biodiesel production. Even though glycerine offsets only about 5 per cent of the cost of producing biodiesel, it is valuable to the cosmetics, ink, lubrication and preservative industries, and is particularly noted as an ingredient in soap.[19] The 'meal' left in the seed after oil has been removed is currently sold as an animal feed.[20]

Due to the wide variety of oils and fats that can be used to produce biodiesel, there is a greater range in the characteristics of biodiesel fuels than for ethanol fuel (ethanol is actually one very specific molecule, whereas biodiesel is a mix of molecules that varies somewhat, depending upon the initial oil or fat source used to produce the fuel). Some oils are shorter or more saturated – characteristics that affect the viscosity and combustibility of the biodiesel.

Rapeseed oil is the dominant feedstock used to make biodiesel in Europe, with some sunflower oil also used. In the US, biodiesel has generally been made from soybean oil because more of this is produced domestically than all other sources of fats and oils combined. In tropical and subtropical countries, numerous plant oils are candidates for biodiesel production, including palm oil, coconut oil and jatropha oil.

Biodiesel contains 88 to 95 per cent as much energy as diesel fuel. But biodiesel can also improve diesel lubricity and raise the cetane value; thus, the fuel economy of biodiesel approaches that of diesel. Moreover, the alcohol component of biodiesel contains oxygen, which helps to complete the combustion of the fuel, reducing air pollutants such as particulates, carbon monoxide and hydrocarbons. Like ethanol, biodiesel contains practically no sulphur, so it can help to reduce emissions of sulphur oxides (see Chapter 13).

Biodiesel blends are sensitive to cold weather and may require special anti-freezing precautions, similar to those taken with standard number 2 diesel. Long-term storage of biodiesel can be a concern because it may oxidize, although additives can ensure stability. Biodiesel acts like a detergent additive, loosening and dissolving sediments in storage tanks and also causing rubber and other components to fail; these concerns are typically minimal at low-level blends of biodiesel, and at higher blend levels problems can be avoided with some attention to the materials used in engine fuel injectors and the overall fuel handling system (see Chapter 15).

Biodiesel as a blend

Although conventional diesel engines operate readily with up to 100 per cent biodiesel fuel, using blends above 20 per cent may require modest costs to replace some rubber hoses that are sensitive to the solvent character of biodiesel. In the US, blends of 20 per cent biodiesel (B20) have been used extensively in vehicle fleets. European fuel standards permit 5 per cent biodiesel blended with diesel, and blends of 2 per cent biodiesel (B2) have been used with diesel fuel in numerous countries around the world.

PRODUCTION COSTS FOR BIOFUELS

Over the last century, biofuels have almost always been more expensive than petroleum fuels. Government incentive programmes have generally been necessary to allow biofuels to play a role in the marketplace.

Adjusting for the difference in energy content, the price of ethanol produced from sugar cane in Brazil is competitive with gasoline when crude oil prices rise above the €30 (US$35) per barrel range, while ethanol produced from corn in the US is competitive when crude oil prices exceed roughly €45 (US$55) per barrel. European Union (EU)-produced biodiesel is competitive with oil prices at about €75 (US$90) per barrel, while EU-produced ethanol becomes competitive between about €60 and €80 (US$75 to $100) per barrel.[21] Outside of Europe, disparities in fuel prices between biodiesel and diesel have been greater than for ethanol and gasoline. This is primarily because plant oils grown in temperate regions are more expensive to produce than sugar or starch crops since they are less productive per hectare of cropland. Another contributing factor is that, in most countries, diesel fuel is less expensive per litre than gasoline.

Despite being generally more expensive than gasoline, biofuels have often appeared cheaper at the pump. This is, in part, because they contain less energy than petroleum fuels; but it is primarily because of government tax credits. In Germany, biodiesel has been €0.15 to €0.20 (US$0.18 to $0.24) cheaper than conventional diesel, thanks to an exemption from the €0.47 (US$0.59) tax on diesel.[22] In the US, E85 ethanol has often been cheaper than gasoline, thanks to an €0.11 (US$0.13) per litre excise tax credit for ethanol. These tax credits apply only to the biofuel portion of the fuel. Table 2.1 illustrates the difference in the price of gasoline and diesel with and without taxes, compared to the estimated production costs for ethanol and biodiesel in the US, the EU and Brazil.[23]

Feedstock costs account for the majority of a biofuel's eventual price, while processing costs and a small proportion for transport represent most of the rest. For ethanol, feedstock comprises 50 to 70 per cent of the production cost, while for biodiesel, which requires less extensive processing, feedstock can be 70 to 80 per cent of the production cost.[24]

Biofuels produced in more fertile tropical countries are typically less expensive than biofuels produced in more temperate regions. Yields are higher, while land and labour costs are also typically lower in tropical countries. Figures 2.1 and 2.2 illustrate the current cost ranges for ethanol and biodiesel production (at the factory gate).[25] The cost of producing ethanol from sugar cane in Brazil, for example, is roughly half that of producing ethanol from grain or sugar beets in Europe (and

Table 2.1 *Production costs of ethanol and biodiesel and prices of petroleum-based fuel in major biofuel-producing countries, 2004*

	Ethanol	*Gasoline*	*Biodiesel*	*Diesel*
		(Euros per energy-equivalent litre[a])		
US	€0.36 (corn)	€0.45 (with tax) €0.32 (without tax)	€0.50 (soy)	€0.47 (with tax) €0.31 (without tax)
European Union	€0.70 (wheat)	€1.09 (with tax) €0.34 (without tax	€0.56 (rapeseed)	€1.06 (with tax) €0.33 (without tax)
Brazil	€0.27 (sugar cane)	€0.69 (with tax) €0.33 (without tax)	€0.52 (soy)	€0.40 (with tax) €0.32 (without tax)

Notes: [a] Biofuel prices are accommodated for differences in energy content. Ethanol is assumed to contain 0.67 the energy of 1 litre of gasoline, and biodiesel is assumed to contain 0.9 the energy of 1 litre of diesel.

Source: see endnote 23 for this chapter

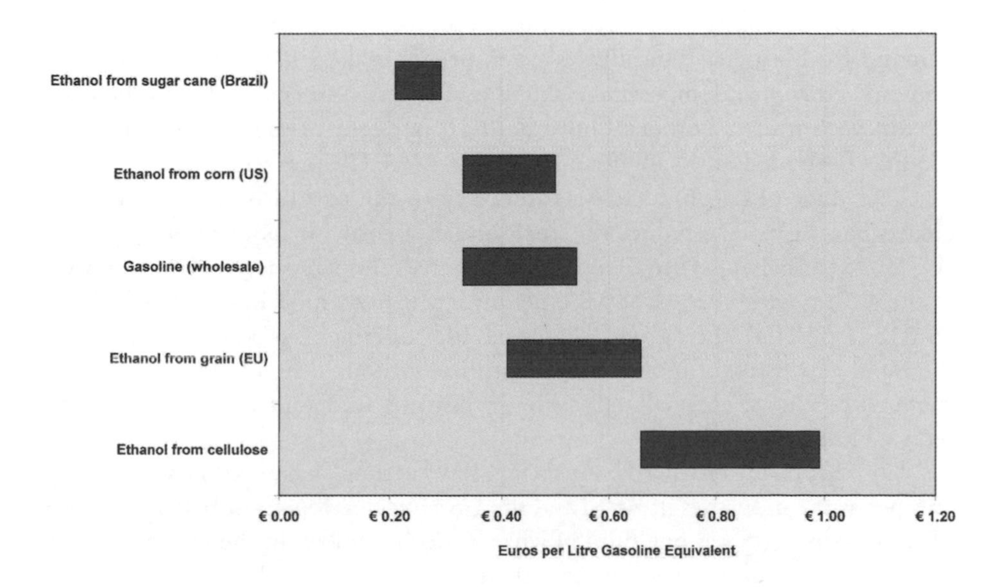

Figure 2.1 *Cost ranges for ethanol and gasoline production, 2006*

Source: Fulton et al (2004); Gardner (2006); US DOE and EIA (2006b)

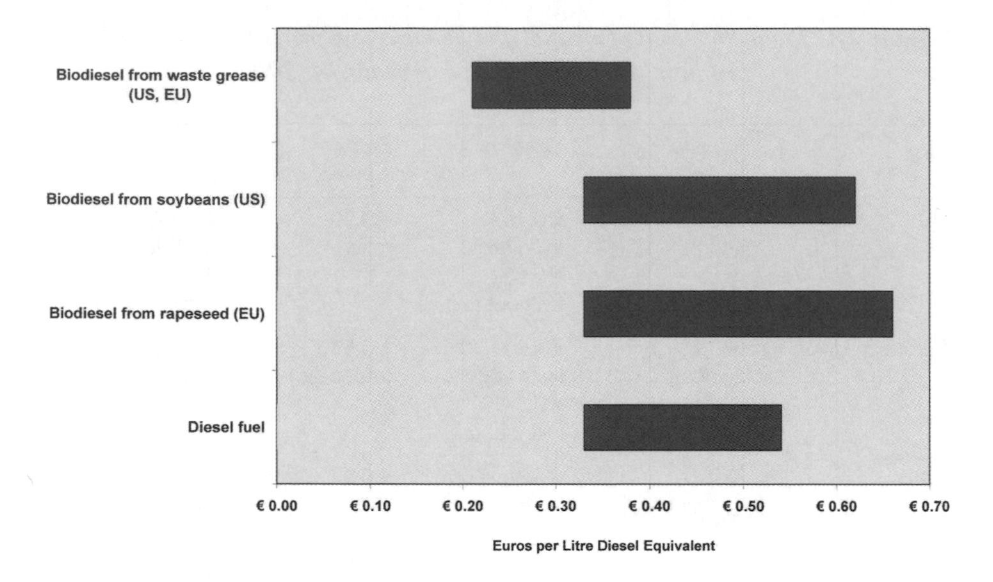

Figure 2.2 *Cost ranges for biodiesel and diesel production, 2006*

Source: Fulton et al (2004); US DOE and EIA (2006b)

cheaper than the retail price of gasoline on an energy equivalent basis throughout 2005). This highlights an opportunity for tropical countries to help meet the global demand for biofuels. Typically, however, policy-makers have worked to foster domestic ethanol and imposed tariffs and trade barriers to ensure that the benefits remain within their borders. Only about 10 per cent of the ethanol currently produced is traded across international borders (see Chapter 9).

The value of co-products is another key to the cost of biofuels. Glycerine credits have helped to reduce the final biodiesel costs by about €0.04 to €0.08 (US$0.05 to $0.10) per litre. The co-product credit for large dry-mill operations is in the vicinity of €0.025 (US$0.03) per litre, while wet mills in the US sell about €0.035 (US$0.04) of co-products for each litre of ethanol produced.[26] However, the rapid expansion of biofuel production has saturated the market for some of these co-products, especially glycerine in Europe, undercutting their ability to reduce biofuel prices.

Conventionally produced biofuels, especially ethanol, have become significantly cheaper as the industries in Brazil and the US have developed. In Brazil, the price of ethanol in 2005 was one third of what it was in 1980.[27] In the US, the cost of processing ethanol declined by more than half between the 1970s and the early 2000s.[28] But still, feedstock remains the primary determinant of the cost and the primary limitation of the potential fuel supply. The variety of current feedstocks is discussed in more detail in Chapter 3.

3

First-Generation Feedstocks

Introduction

The various biomass feedstocks used for producing biofuels can be grouped into two basic categories: the currently available 'first-generation' feedstocks, which are harvested for their sugar, starch and oil content and can be converted into liquid fuels using conventional technology, and the 'next-generation' feedstocks, which are harvested for their total biomass and whose fibres can only be converted into liquid biofuels by advanced technical processes.

The focus of this chapter is on the first-generation feedstocks, including sugar crops such as sugar cane, sugar beets and sweet sorghum; starch crops such as corn, wheat, barley, cassava and sorghum grain; and oilseed crops such as rapeseed, soybeans, palm oil and jatropha. The chapter also briefly addresses other oil sources for biodiesel, including sunflower, mustard seed, waste vegetable oil, micro-algae and animal oils. It concludes with a discussion of the varying production potentials for these current feedstocks and their overall suitability for expanded biofuel production.

'Next-generation' feedstocks, which have much greater potential for expanding the supply of biofuels for transportation energy, are discussed in detail in the next chapter.

Relative feedstock yields

In countries that have fostered the development of biofuels, the primary impetus has typically been to subsidize or otherwise support the agricultural sector. This remains a central priority to biofuel initiatives around the world, which have generally promoted agricultural crops that are already produced at a large scale for the human food and animal feed markets.

To date, only a relatively few types of crops have provided the vast majority of the world's fuel ethanol and biodiesel. Nearly all of Brazil's ethanol production is derived from sugar cane, currently the highest volume ethanol feedstock worldwide. In the US, more than 90 per cent of the ethanol comes from corn, the world's second largest fuel ethanol crop and one of the most important agricultural crops globally. In Europe, about 70 per cent of the biodiesel is made from rapeseed, the world's second largest source of plant oils, with most of the remainder coming from sunflower seeds. And nearly all of the biodiesel produced in the US comes from soybeans, the world's largest source of plant oil (for both food and fuel uses). Interestingly, the two crops with the largest planted area worldwide – wheat (214 million hectares) and rice (148 million hectares) are not significant in biofuel production: only a modest amount of wheat is used for ethanol fuel, and no rice is used (due to higher priority demands for food markets).

A key variable in the choice of an appropriate feedstock is the amount of biofuel that can be produced per hectare. In general, starches such as corn and wheat, which are grown predominantly in temperate regions, have lower yields than sugars, such as sugar cane, which is grown in more tropical areas. In 2002, total sugar cane yields in Brazil reached 6500 litres per hectare, more than double the corn production yield in the US (see Table 3.1).[1] Likewise, oilseed crops grown in temperate areas, such as soybeans and rapeseed, have lower yields than more tropical oilseed plants, such as palm. Higher yields per hectare confer an advantage to tropical areas in the production of conventional biofuels (see Figure 3.1).[2]

Table 3.1 *Typical biofuel production per hectare of farmland yield by crop and by region, 2002*

Crop	*Typical yield (litres per hectare of cropland)*				
	US	*European Union*	*Brazil*	*India*	*Malaysia*
Ethanol source:					
Sugar cane			6500	5300	
Sugar beet		5500			
Corn	3100				
Wheat		2500			
Barley		1100			
Biodiesel source:					
Palm oil			5000		6000
Rapeseed		1200			
Sunflower seed		1000			
Soybean	500	700	400		
Jatropha				700	

Source: see endnote 1 for this chapter

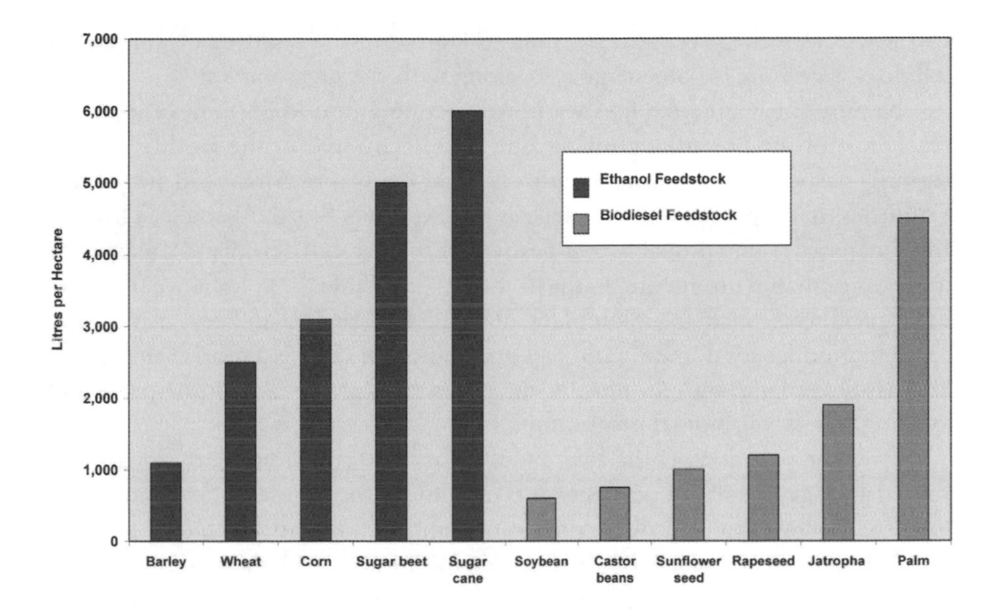

Figure 3.1 *Biofuel yields of selected ethanol and biodiesel feedstocks*

Source: Fulton et al (2005)

SUGAR CROPS

Sugar cane

Sugar cane is the most significant crop for producing biofuels today, supplying more than 40 per cent of all fuel ethanol. The majority of the world's sugar cane comes from the centre-south region of Brazil, where a long growing season, natural rainfall and appropriate soils provide fertile conditions for cane production. On a smaller scale, ethanol is also produced from sugar cane grown in Australia, China, India, Indonesia, Pakistan, South Africa and Thailand.

In recent years, nearly half of Brazil's annual sugar cane harvest – or 2.75 million out of 5.5 million planted hectares – has gone to producing ethanol. This represents about 0.5 per cent of the country's total agricultural land area.[3] The sugar cane is produced in two distinct regions: the centre-south and the north–northeast, which have very different climates, production systems and harvesting periods. Average yields per hectare reach about 85 tonnes in the centre-south, which is well suited for sugar cane, and about 70 tonnes in the north–northeast. The average sugar cane yield nationwide is 82.4 tonnes per hectare.[4]

Sugar cane stalks contain so much sugar that the plant is currently the lowest cost source of biofuel. Cane plants produce a large amount of fibre in their stalks and leaves as well, making it possible to 'co-harvest' a significant amount of cellulosic feedstock for bioenergy uses, along with the sugar harvest.[5]

Because sugar cane requires warm weather and around 850mm of rainfall a year, most of the potential growing land is concentrated in the world's tropical regions, particularly in Latin America (23 per cent) and Africa (18 per cent).[6] Countries that export raw sugar – such as top exporters Brazil, Australia, Thailand and Guatemala – are probably best positioned to have extra cropland capacity for sugar cane ethanol production in the near term (see Table 3.2).[7] Somewhat smaller sugar producers, such as Colombia, Cuba, the Philippines and Swaziland, may begin to produce for domestic and regional markets if there is a near-term increase in ethanol demand; indeed, most of these countries are already in the process of forming significant biofuel programmes.

Over the past decade, the area under sugar cane cultivation has grown at an annual average rate of 1.4 per cent.[8] It is expanding the fastest in Thailand, at 2.7 per cent, followed by Brazil (2.5 per cent), and China and India (1.9 per cent and 1.8 per cent, respectively); it is interesting to note that these four countries are also the top four sugar cane producers in the world.[9] In Cuba, formerly the largest sugar supplier to the US, the cultivated area is decreasing at an annual rate of 5.8 per cent due to shortages in production equipment and fuel, as well as poor primary resources and deficient technical operations for production and harvesting.[10] Brazil's centre-south region contains the vast *cerrado* prairies, perhaps the largest land area in the world available for increasing agricultural acreage, and a region capable of growing highly productive sugar cane varieties. It is also a highly diverse and sensitive ecosystem (see Chapter 12 for a discussion of environmental concerns related to sugar cane expansion).

Sugar beets

Sugar beets are an important source of sugar in Europe and a valuable feedstock for biofuels in France. Europe is the principal producer, where average yields vary from 53 tonnes per hectare in Germany to 58 tonnes per hectare in The Netherlands.[11] Ukraine and Russia have the largest cultivated area; but the largest producers by volume are France and Germany. Worldwide, the total area cultivated in sugar beets is decreasing by 3.5 per cent a year; in Ukraine, it is shrinking by 7.4 per cent annually, and in Russia, by 3.3 per cent annually. France's sugar beet yield per hectare is the highest of the ten countries, with the largest cultivated area.

Sugar beets generate good yields in many temperate settings; but compared to tropical sugar cane, they are a more chemical- and energy-intensive crop.[12] Due to concerns about the potential survival of pests in the soil, the beets cannot be cultivated more than once every three years on the same field, and yields depend strongly upon climatic conditions. Since the plant root must be processed to obtain the sugar, producing ethanol from sugar beets is more energy intensive and costly

Table 3.2 *Top 20 sugar cane producers worldwide, 2004*

Country	Production	Raw exports	Refined exports	Total exports
		(thousand tonnes)		
Brazil	26,400	10,820	4420	15,240
India	15,150	0	250	250
China	10,096	10	57	67
Thailand	7010	2281	2579	4860
Mexico	5330	7	7	14
Australia	5178	4017	140	4157
Pakistan	4023	0	214	214
US	3590	0	261	261
Colombia	2680	620	580	1200
South Africa	2560	765	305	1070
Cuba	2450	1900	0	1900
Philippines	2340	202	0	202
Argentina	1925	45	156	201
Guatemala	1850	1125	210	1335
Indonesia	1730	0	0	0
Vietnam	1250	0	100	100
Egypt	960	0	0	0
Peru	959	61	0	61
Sudan	830	55	160	215
Swaziland	628	283	2	285
Subtotal	96,939			
World	107,890			

Source: see endnote 7 for this chapter

than producing it from cane. In general, harvesting and processing sugar beets is a heavily mechanized operation.

Since beets are a more expensive feedstock than sugar cane, their economic sustainability has often depended upon government protection through subsidies and tariffs on imported sugar (in particular, sugar made from cane). Recent shifts in European subsidy schemes have prompted beet growers to seek other markets for their crops, including ethanol, though some have considered switching to another crop entirely, such as wheat or rapeseed (see Chapters 9 and 19 for more on changing European Union agriculture sector supports.).[13]

Sweet sorghum

Although currently not a significant ethanol feedstock, sweet sorghum deserves particular attention as a multi-use crop. Farmers can harvest the seeds at the top of the plant for food and the sugars in the stalk for fuels. In settings where land is particularly scarce, this co-harvesting of sorghum may be particularly efficient.

As with sugar cane, the sugar in sweet sorghum is found in the plant's main stalk. The crop grows particularly well in drier, warmer climates, although it can also be grown in temperate areas. With its drought tolerance and ability to produce sugar, sweet sorghum could receive increasing attention as a feedstock for ethanol production.

STARCH CROPS

Corn

Corn (also known as maize) is the second largest source of biofuel feedstock today, primarily because of its dominance in the US for ethanol production. Corn ethanol production is centred in several states in the US corn belt, including Illinois, Iowa, Minnesota, South Dakota and Nebraska. A much smaller amount of ethanol is produced from corn grown in north-eastern China and South Africa. In the US, 98 per cent of the corn crop is treated with synthetic nitrogen fertilizers, and 97 per cent of cornfields are treated with herbicides.[14]

Producing ethanol from grain starches is more land intensive than producing it from sugar cane, because corn crops have lower fuel yields per hectare. As a result, while the US and Brazil produce comparable amounts of ethanol, the US must use almost twice as much land to fuel production (about 5 million hectares versus 2.7 million to 3 million hectares). In comparison to sugar cane, corn starch must also undergo additional processing to convert it into sugars before it can be fermented to ethanol fuel. However, one advantage of corn as a feedstock is its longer 'shelf life' before processing; while corn can be stored for long periods after harvesting, sugar cane must be processed very quickly (usually within 24 to 48 hours).

Wheat

The global volume of ethanol fuel produced from wheat is considerably lower than the quantity produced from sugar cane and corn. In the US, less than 3 per cent (approximately 445 million litres) of the installed ethanol fuel capacity comes from wheat.[15] Spain and Germany also produce shares of domestic ethanol from wheat, both producing between 130 million to 270 million litres annually.[16] Canada and France are smaller producers, generating 70 million litres and 58 million litres of ethanol from wheat, respectively, in 2002.[17] As with corn, only the kernel portion of the cereal (which contains the starch) is used to produce ethanol. Overall, the ethanol yield per hectare of wheat is lower than for corn and sugar crops.

Because wheat is an important food source, most of the wheat produced in the world today is consumed as human food. The total cultivated land area is increasing only slightly, showing just 0.03 per cent annual growth over the last ten years.[18] Wheat yields per hectare vary considerably depending upon weather and climate factors, averaging around 5.7 tonnes in the European Union (EU) (15 countries), 3.8 tonnes in China, 2.7 tonnes in India, and 2.4 tonnes each in the

US and Russia.[19] The average wheat yield worldwide has increased by 1.7 per cent annually over the last decade.

Two relatives of wheat – rye and barley – are also used for ethanol production. As crops, they are resistant to drier and cooler conditions and can grow in more acidic soils. As feedstocks for ethanol, they are significant primarily in Northern Europe. Demand for rye as both a food and a feed has declined in recent years, though new ethanol plants have stimulated some additional planting.[20]

Cassava

Cassava, or tapioca, is the most cultivated crop in sub-Saharan Africa. It is the second most grown crop in Africa overall, fourth in Southeast Asia, fifth in Latin America and the Caribbean, and seventh in Asia. While more than 60 per cent of the world's cassava is grown in Africa, the highest yields are achieved in Asia due to less disease, fewer pests and relatively intensive crop management, including irrigation and fertilizers. Because of its high tolerance, cassava is typically cultivated in marginal areas with poor soils and high risk of drought.[21]

Cassava has long been considered a candidate for ethanol production, and it is beginning to be a more significant feedstock. Brazil considered ethanol production from cassava during the 1980s, but yields were lower than sugar cane, the crop was more labour intensive, and the processing was considerably more complex – and commercial production of ethanol from cassava never took off. However, cassava is a highly productive crop. Thailand has begun using it on a larger scale for its ethanol programme, while Nigeria has placed cassava at the centre of its planned ethanol programme.[22]

Sorghum grain

Sorghum grain is used only to a limited extent for the production of fuel ethanol. It is a distant second to corn for ethanol production in the US. Over the past decade, the area cultivated in sorghum worldwide has been decreasing by 0.1 per cent a year; in the US, India and Sudan, it decreased by 3 per cent, 2 per cent and 1 per cent, respectively.[23] However, the annual cultivated area is growing in Nigeria and Niger by 2 per cent and 1 per cent, respectively.

Only 4 per cent of the world's sorghum production is processed into ethanol, most of which is used for non-fuel purposes. Of the remainder, 44 per cent is used in animal diets, 43 per cent is consumed as human food, and 7 per cent becomes waste crop. The amount of crop waste generated in sorghum cultivation is 3.8 million tonnes and could be a potential source for 'next-generation' biofuel production.

OILSEED CROPS

Oilseed crops provide the primary feedstocks for producing biodiesel. Of the major oilseeds cultivated today, soybean production is by far the world's largest, followed by rapeseed and cottonseed (see Table 3.3).[24] The dominant feedstock used in biodiesel, however, is rapeseed (primarily in Europe).

Table 3.3 *World production of major oilseed crops, 2004–2005*

Crop	*Production (million tonnes)*
Soybean	215.3
Rapeseed	46.1
Cottonseed	45.2
Peanut	33.1
Sunflower seed	25.7
Palm kernel	9.5
Copra	5.4
Total	380.3

Source: see endnote 24 for this chapter

In temperate regions, oilseed crops typically generate lower yields per hectare than starchy cereal feedstocks such as corn and wheat. But because oil seeds require less processing, they generally have more favourable energy balances overall. Oilseed crops grown in tropical areas can be especially productive.

Oilseed species vary considerably in their oil saturation and fatty acid content, characteristics that significantly affect the properties of the biodiesel produced. Highly saturated oils produce a fuel with superior oxidative stability and a higher cetane number (a measure of a diesel fuel's quickness to ignite), but which performs poorly in low temperatures (see Chapters 2 and 15). For this reason, vegetable oil with a high degree of saturation is more suited for use in warmer climates. Figure 3.2 compares the degree of vegetable oil saturation for a variety of oils that could be produced in Brazil (higher iodine numbers correspond to a greater percentage of polyunsaturated oil content[25]).[26]

Rapeseed

As noted earlier, rapeseed is the primary feedstock for biodiesel production in Europe. Commonly grown in rotation with cereal crops, it is a relatively productive oilseed and accounts for the highest output of biodiesel per hectare in the EU when compared to soybeans and sunflower seed (see Table 3.1). Like most oilseeds grown

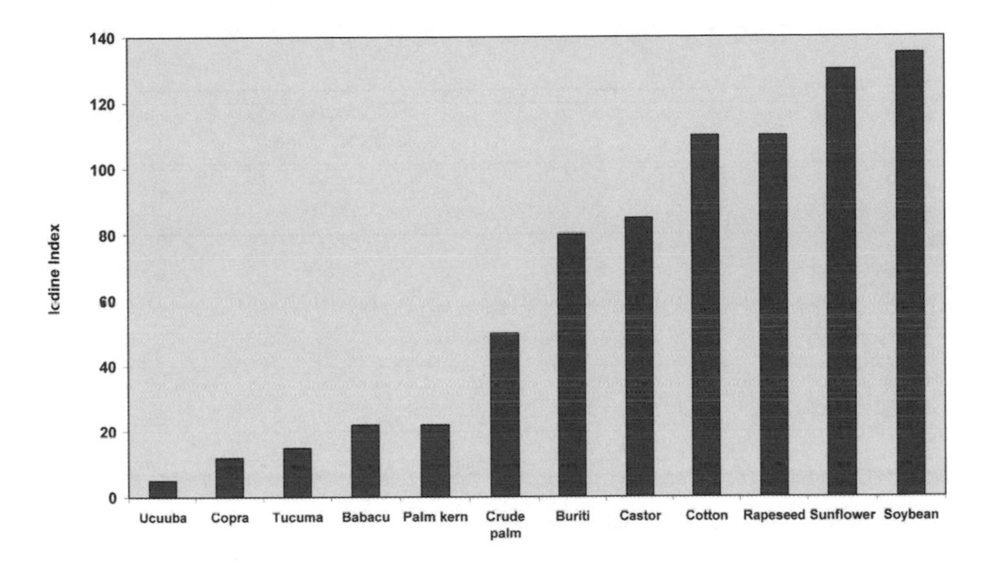

Figure 3.2 *Degree of saturation of selected vegetable oils in Brazil*

Source: Kaltner et al (2005, p42)

in temperate climates, however, rapeseed yields a lower quantity of fuel per hectare than starchy crops such as wheat and sugar beet.

To avoid the spread of plant disease, at least two years should be left between the cultivation of rapeseed and other cruciferous crops, such as broccoli, cauliflower, cabbage and Brussels sprouts. This restriction, along with soil quality considerations, tends to limit the expansion opportunities for rapeseed cultivation.[27] Nonetheless, the global cultivated area is growing by 2 per cent annually. In China, the world's largest rapeseed producer, the area planted is expanding rapidly, while in India, the third largest producer, growth is minimal.[28] Australia, the world's seventh largest rapeseed producer, has seen recent increases in cultivated area of 12 per cent per year. In 2006, the rapeseed harvest in Germany is expected to use 1.4 million hectares of land, up from 1.32 million hectares in 2005; roughly 48 per cent of total rapeseed oil production in 2005 in Germany was used for biodiesel production.[29]

In Europe, 1.4 million hectares of rapeseed was planted specifically for biodiesel use in 2005.[30] The continent's biodiesel producers typically have special arrangements with their governments to produce a certain amount of feedstock for biofuel production, usually on set-aside land. About half of this production was in Germany; but France, the Czech Republic and Poland were also significant growers (see Table 3.4).[31]

Table 3.4 *Top ten rapeseed producers worldwide, 2004*

Country	*Production (thousand tonnes)*
China[a]	11,900
Canada	7001
India[a]	6800
Germany[b]	5250
France	3961
UK[a]	1612
Australia	1549
Poland	1292
Czech Republic[a]	910
US	572

Notes: [a] unofficial figure; [b] United Nations Food and Agriculture Organization (FAO) estimate.

Source: see endnote 31 for this chapter

Soybeans

Soybeans are the dominant oilseed crop cultivated worldwide, far surpassing the output of other oil crops. World production of soybeans totalled 215 million tonnes in 2004 to 2005, accounting for 57 per cent of major oilseed production. Brazil, the US and Argentina dominate world soybean production, accounting for an estimated 30 per cent of the global supply for export.[32] Primarily because of its prevalence, rather than it specific desirability as a biofuel feedstock, soy oil is increasingly being used for biodiesel production in these countries.

Although soybeans generate a relatively low yield of biodiesel per hectare when compared to other oilseed crops, they can grow in both temperate and tropical conditions. As a nitrogen-fixing crop, they also replenish soil nitrogen and require less fertilizer input, giving them a relatively favourable fossil energy balance (see Chapter 10). Soybeans are grown in rotation with corn in the US, and with sugar cane in Brazil.

Soybean harvesting tends to be highly mechanized and is controlled almost exclusively by large multinational agricultural processors. American companies Cargill and ADM are heavily invested in Brazil's soy industry. Of total soybean production worldwide, 86 per cent is used in food manufacturing and 8 per cent is consumed directly as human food or animal feed. Only a small fraction of the soybean supply is currently transformed into fuels.

Palm

Palm is an attractive candidate for biodiesel production because it yields a very high level of oil per hectare. The two largest producers are Malaysia and Indonesia, where palm oil production has grown rapidly over the last decade (by 4 per cent and 11 per cent, respectively). Nigeria has the second largest planted area; but its annual growth rate is only 1 per cent due to the low yield of palm fruit per hectare.[33] Brazil currently produces only a small share of the world's palm oil, but it has the potential to significantly expand production in the north; the African palm has been identified as the most promising species for Brazilian biodiesel use.[34]

While most palm oil is used for food purposes, the demand for palm biodiesel is expected to increase rapidly, particularly in Europe. The Netherlands is the EU's largest importer of palm oil, followed by the UK. UK imports alone doubled between 1995 and 2004 to 914,000 tonnes, which represented 23 per cent of the EU total.[35]

Jatropha

Jatropha curcas is an oilseed crop that grows well on marginal and semi-arid lands. The bushes can be harvested twice annually, are rarely browsed by livestock and remain productive for decades. Jatropha has been identified as one of the most promising feedstocks for large-scale biodiesel production in India, where nearly 64 million hectares of land is classified as wasteland or uncultivated land.[36] It is also particularly well suited for fuel use at the small-scale or village level (see Chapter 8).

The economic viability of biodiesel from jatropha depends largely upon the seed yields. To date, there has been a substantial amount of variability in yield data for the plant, which can be attributed to differences in germplasm quality, plantation practices and climatic conditions. In addition, due to absence of data from block plantations, several yield estimates are based on extrapolation of yields obtained from individual plants or small demonstration plots. D1 Oils, a British company aiming to cultivate biodiesel in the developing world, has chosen jatropha as its primary feedstock due to the plant's high oil content, its ability to tolerate a wide range of climates, and its productive lifespan of as much as 30 years.[37]

Several agencies promoting jatropha are projecting significantly improved yields as the crop is developed. In India, researchers estimate that by 2012, as much as 15 billion litres of biodiesel could be produced by cultivating the crop on 11 million hectares of wasteland.[38] Further development and demonstration work is needed, however, to determine whether these levels of productivity are feasible.

OTHER POTENTIAL OIL SOURCES FOR BIODIESEL

Oilseed crops and tree-based oilseeds

Many other plant varieties could be promising feedstocks for biodiesel production in the future. Some are already grown on a wide scale, while others are only now being evaluated for their specific characteristics, including high yields and the labour intensity of production.

Potential plant oil feedstocks currently grown or available widely include:

- *Sunflower.* The world's fifth largest oilseed crop, it accounts for most of the remaining biodiesel feedstock in Europe after rapeseed. Sunflower seed generates a higher yield of biodiesel per area when compared to soybeans, and a yield similar to rapeseed. Though slightly less productive than rapeseed, it is heartier and requires less water and fertilizer.[39]
- *Cottonseed.* The world's third largest oilseed crop, it is produced predominantly in India, the US and Pakistan, which are together responsible for 45 per cent of world production and 50 per cent of the total cultivated area.[40]
- *Peanut.* The world's fourth largest oilseed crop, it accounts for 8.7 per cent of major oilseed production. The major producers are China, India and the US, which together account for 70 per cent of world production. China and India represent 56 per cent of the world's cultivated area.
- *Mustard seed.* A relative of rapeseed and canola, it provides a potentially valuable non-food feedstock. The plant's roots, stems and leaves contain glucosinolates that break down in the soil into a variety of active but biodegradable chemicals, which provide a pesticide effect. Removal of the plant oil leaves a co-product meal – residual press cake – with a strong potential market (and environmental) value as an organic pesticide. To create a viable biodiesel feedstock, however, genetic engineering would probably need to be applied to boost the oil content of the seeds and to increase the effectiveness of the residue for pesticide use.[41]
- *Coconut.* The favoured feedstock in the burgeoning biodiesel industry in the Philippines, it is another high-yielding feedstock that produces a highly saturated oil. Studies have shown that vehicles running on coco-biodiesel reduce emission levels by as much as 60 per cent and increase mileage by 1–2km per litre due to increased oxygenation, even with a 1 per cent minimum blend.[42]
- *Castor oil.* Identified as the second most promising species for Brazil after palm oil, the castor oil (or momona) plant is a particularly labour-intensive crop that could provide jobs in the poorer north-eastern regions of the country. India is the largest producer and exporter of castor oil worldwide, followed by China and Brazil. World demand for castor oil is projected to continue growing by 3 to 5 per cent per year in the near term.[43]
- *Waste vegetable oil.* Soybean, rape, palm and coconut are the waste oils most frequently used in biodiesel production. Their use requires additional processing to filter out residues and to handle the acids produced by high temperatures. At present, China produces most of its biodiesel using waste oil from cooking – using between 40,000 to 60,000 tonnes of cooking oil a year. It is estimated that Chinese biodiesel may be available on a large scale for the transportation sector by 2007 or 2008.[44]

Beyond these common plant oils, more than 100 native Brazilian species have been identified as having potential for biodiesel production, most of them palm tree species. India, too, is home to more than 300 different tree species that produce oil-bearing seeds. Given the large demand for vegetable oil for human food use worldwide, there is particular interest in identifying non-edible species that can be grown in arid to semi-arid regions poorly suited for food crops. The most promising non-edible sources are jatropha, pongamia, melia (neem) and shorea (sal).[45]

Poorer populations have typically collected and sold tree-based oilseeds for use as a lighting fuel. The oils are also used in soaps, varnishes, lubricants, candles and cosmetics. Non-edible oilseeds are not currently utilized on a large scale; but such oils can be an important component of local economies (see Chapter 8).[46]

Micro-algae

Micro-algae are microscopic single-cell aquatic plants with the potential to produce large quantities of lipids (plant oils) that are well suited for use in biodiesel production. Micro-algae can be grown in arid and semi arid regions with poor soil quality, with a per hectare yield estimated to be many times greater than that of even tropical oilseeds. Algae can also grow in saline water, such as water from polluted aquifers or the ocean, which has few competing uses in agriculture, forestry, industry or municipalities.[47]

Algae feedstocks received early attention from the US National Renewable Energy Laboratory in the 1980s, and interest in them has recently resurged based on their potential for cultivation near power plants. The primary nutrients for growing micro-algae are carbon dioxide (CO_2) and nitrogen oxides (NO_x), creating an opportunity for developing integrated systems that produce oil-rich micro-algae that feed on the emissions of coal, petroleum and natural gas power plants.

Recent efforts at the Massachusetts Institute of Technology in the US have demonstrated promising technology for using micro-algae to clean up power plant emissions. A private start-up company, GreenFuel, is working to commercialize this technology. Such algae colonies could reduce NO_x levels by some 80 per cent and CO_2 by 30 to 40 per cent, while also producing a large quantity of raw algae biomass.[48] By selecting the right micro-algae, the single-cell plants can produce 40 to 50 per cent oil by weight.[49] However, the cost of cultivating algae in other scenarios is likely to make it uneconomical in the near term.[50]

Animal fats

Animal fats may be gathered from cattle, pig, fish or chicken processing and are increasingly being considered for the production of biodiesel, especially to replace fuel in vehicle fleets for companies producing these raw materials.[51] The low retail price of suet (animal fat) has made this raw material attractive; however, the material is not always readily available and is often restricted because it is a sub-product, and therefore has not been produced primarily for a biodiesel programme. Chicken oil has similar potential to cattle oil, but has only recently become available on a large scale, and its use will be contingent upon inventory availability and greater investment in research.[52]

POTENTIAL AND LIMITATIONS OF CURRENT FEEDSTOCKS

The starch, sugar and plant oil feedstock that current technologies are able to convert into ethanol and biodiesel will dominate the biofuel industry in the near term. But these inputs will ultimately be limited in comparison to cellulosic feedstocks, the potential of which is covered in Part II of this book. Nevertheless, some current feedstocks have much greater biofuel potential than others. In general, crops grown in tropical settings can produce large quantities of fuel more easily than those cultivated in temperate climates.

As Figure 3.3 illustrates, the amount of land necessary to cultivate biofuel feedstock varies greatly, depending upon the type of feedstock and the conditions in different regions.[53] This figure primarily illustrates the difference in yields between sugar cane grown in tropical regions versus yields for cereal and oilseed crops grown in temperate regions. It should be noted that the land area estimates in this figure do not take into account potential contributions from agricultural residues, forest residues or perennial energy crops (which are anticipated to have much higher yields in temperate regions than cereal or oilseed crops).

Fuel production potentials of current feedstocks

Figure 3.3 shows the proportion of cropland that would be required for countries to displace 10 per cent of the transportation fuel supply with crops currently used to produce biofuels. The limitations of these crops in temperate countries are clear, while the potential opportunities for countries such as Brazil are also clear. This potential exists not only because Brazilians drive less than Americans and Europeans on average, but because tropical crops are more productive than temperate crops.

Ethanol: The limitations of starch and the risks of sugar cane expansion

Corn is the highest-yielding cereal feedstock used in ethanol production today. In 2005, 15 per cent of the US corn crop (converted into ethanol) displaced some 2 to 3 per cent of the country's gasoline.[54] However, corn typically requires large amounts of fertilizer to achieve high yields, and processing it requires more energy than extracting sugar from cane.

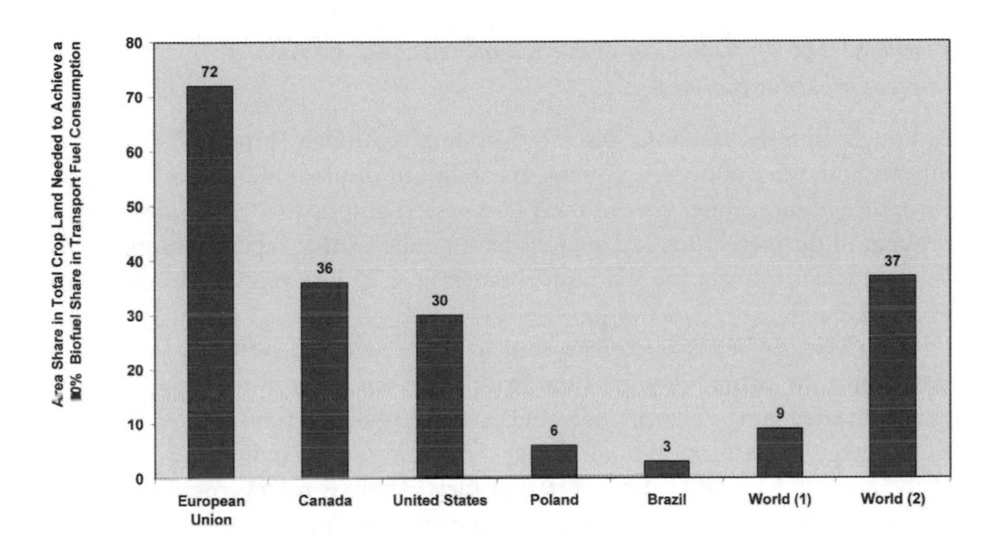

Figure 3.3 *Percentage of agricultural land required for 10 per cent biofuel shares in major biofuel-producing regions*

Notes: World area shares are calculated relative to land used for cereals, oilseeds and sugar globally (world 1) and within the five major biofuel-producing regions (world 2). All area requirements are calculated on the basis of average crop area and yield data for 2000 to 2004 and transport fuel consumption in 2004. For these calculations, the 2004 shares in the feedstock mix are assumed to remain unchanged.

Source: OECD (2006)

As such, sugar cane has much greater potential to supply a significant portion of the world's transport fuels. According to the United Nations Food and Agriculture Organization (FAO), the total area potentially available for sugar cane, excluding forests, amounts to 153 million hectares (to put this into perspective, Brazil currently devotes about 5.5 million hectares to sugar cane production).[55] Assuming the near-best productivity now achievable, as well as efficient use of sugar cane residues for energy, one particularly optimistic study concluded that 143 million hectares of this land could produce 163.9 exajoules (EJ) of primary energy per year and 90EJ of final energy per year in the form of liquid fuel (alcohol) and electricity.[56] This is an amount comparable to about 40 per cent of the world's current primary energy consumption and 60 per cent more than current transportation energy consumption. Another study concluded that sugar cane grown in tropical areas could relatively easily produce enough fuel to displace 10 per cent of global gasoline demand by 2020.[57]

The Brazilian *cerrado*, a biologically diverse area that is largely uncultivated, is by far the largest area remaining worldwide for expanding sugar cane plantations. Expansion of such cultivation would probably come at great ecological expense to the region, however (see Chapter 12).

Biodiesel: The limitations of temperate oilseeds and the risks of tropical oilseed expansion

Although rapeseed is among the most productive oilseed crops able to grow in Europe and other temperate regions, its ability to displace diesel fuel is limited. Currently, about 20 per cent of the EU rapeseed crop is used to produce just 1 per cent of the diesel fuel, and analysts have predicted that rapeseed supplies will be insufficient to meet the EU biofuel target of 5.75 per cent of transport fuels by 2010.[58]

Likewise, soybeans are constrained by their comparatively low yields. For instance, if the entire oil crop of the world's top ten soybean-growing countries were converted into biodiesel, this would generate 34 billion litres of biodiesel (1.12 EJ of energy in the form of liquid fuels) – enough to meet only about 1 per cent of the current demand for transportation fuels. There is far less room to expand soybean cultivation in the US compared to Brazil and Argentina.[59]

Higher-yielding tropical oilseeds have greater promise to use land more efficiently. The palm industries in Malaysia and Indonesia expect to ramp up their production of palm oil significantly, and could satisfy a large proportion of Europe's future demand for biodiesel. However, as with the anticipated expansion of Brazil's sugar cane industry onto further stretches of the *cerrado* prairies, large palm oil plantations are displacing wilder ecosystems and have been a primary motivation for the burning of rainforests in Southeast Asia in recent years (see Chapter 12).

Prospects for improving crop yields

Modern agriculture has changed considerably since the early 20th century, mainly due to mechanization, increased use of fertilizer and other inputs, and dramatically higher yields of major grain and fibre crops. These heightened yields were a direct result of agricultural research, such as corn and wheat hybridization, as well as governmental price support policies, which fostered continued investments in new equipment for cultivating and harvesting corn and wheat.

The areas with the greatest potential for improving crop yields are in the developing world, where traditional farming methods are less productive and farmers often operate on a small scale, without mechanized tools or sufficient inputs. But over the years, agriculture in industrialized countries has also become more productive, increasing the maximum achievable yields for crops as well.

In the past, agricultural yields in the 15 EU member states (EU-15) improved at an average rate of 1 per cent per year, in equal parts due to genetic breeding and technical farming improvements; in Germany, a 2 per cent annual increase in yields was achieved from the 1950s to the present.[60] And in the US, corn grain yields have risen steadily over the past 25 years (1975 to 2000) at an average annual rate of 1.2 per cent, even while fertilizer inputs have declined. Similar improvements have occurred in Brazil with sugar cane and soybeans (see Table 3.5).[61,62]

Table 3.5 *Yield improvements in selected feedstock crops*

Crop	*Location*	*Time period*	*Average yield or improvement*	*Percentage increase*
Sugar cane	Brazil	1990/1991–2002/2003	63 tonnes per hectare in 1990 to 66 tonnes per hectare in 2002	5
Sugar cane	Brazil (centre-south)	2005	~69 tonnes per hectare	–
Corn	US	1975–2003	86 bushels per acre to 142 bushels per acre	65
Soybean	Brazil	1940/1941–2000/2001	651kg per hectare per year to 2720kg per hectare per year	318
Soybean	US	1990–2001	31.6 bushels per acre to 37.12 bushels per acre	17
Oil palm	–	1960s–2000s	–	~200

Source: see endnote 62 for this chapter

Agriculture will continue to adapt to new technologies and circumstances, including biotechnology, which is transforming production by making available genetically altered varieties of corn, soybeans, sugar cane and other crops. The breeding of hybrids and crops that can grow in close proximity has helped achieve higher corn yields, and genetically modified (GM) varieties offer promise to increase these yields further.[63] Biotech corn hybrids accounted for 40 per cent of the total planted acreage in the US in 2004.[64] In Brazil, yields of both soybeans and sugar cane have increased through breeding and genetic modification, as well as the increased use of pesticides and fertilizers (see Box 3.1).[65]

Plant breeding has also played a major role in boosting the yields of oil palm, and new clonal hybrids promise even higher yields. More dramatic genetic modification of crops may bring still higher yields, while the next generation of lignocellulosic crops could be bred to maximize not food yields, but raw energy content (see Chapter 4 for more on breeding next-generation feedstocks). The development of very high-yield crops could be limited, however, by a lack of public acceptance for GM crops and intensified energy crop cultivation (see Chapter 12 for more on genetically modified organisms, or GMOs).[66]

Conclusion

Over the next decade, existing starch, sugar and oilseed crop varieties will continue to provide the bulk of the biomass supplies used for biofuel production. Biofuels grown in tropical areas are cheaper and can displace a larger share of petroleum

Box 3.1 Biotechnology and enhanced sugar cane production in Brazil

When Brazil launched its *Proálcool* programme in the 1970s, average ethanol production in the country was around 2000 litres per planted hectare, and the total planted area was about 1.5 million hectares. By 1999, however, ethanol yields had more than doubled to 5000 litres per hectare, and by 2004 they averaged 5900 litres per hectare. Today, yields are as high as 7000 litres per hectare under good conditions, triple the output of 30 years ago, representing an average annual increase of 3.8 per cent. Sugar cane is currently cultivated on about 5.5 million hectares, spread over the 27 states of the Brazilian federation.

The use of biotechnology has been crucial to enhanced sugar cane production in Brazil. During the 1970s, only ten different varieties of the crop were available in the country. Today, more than 550 varieties are cultivated, and the production period has increased from 150 days to 220 days. In the last decade alone, 51 new sugar cane varieties were released, the predominant 20 of which now occupy 70 per cent of the total planted area. Those varieties were produced mainly through genetic improvement. The germplasm bank of Brazil's *Copersucar* programme contains more than 3000 genotypes, including a large collection of native (wild) species that are the precursors for modern sugar cane varieties and the source of the great genetic variability found in the plant.

In 2003, a private Brazilian investment group, Votorantim, created the enterprise Canavialis, with a large laboratory for genetic selection and development of sugar cane. Each year, the laboratory releases more than 1.5 million seedlings to be planted and tested on three farms (called experimental stations) located in different parts of the country. The Votorantim group also formed a biotechnology company, Alellys, dedicated to modifying the genetic composition of sugar cane varieties produced by Canavialis. New varieties that are more productive and resistant to diseases are under development.

Source: see endnote 65 for this chapter

than biofuels produced with more temperate feedstocks. European countries will probably find it preferable to import biofuels rather than attempt to grow all of their own. The US may be able to produce more indigenous biofuel, but will ultimately face similar limitations.

Sugar cane, in particular, stands out as a feedstock that could provide a large amount of transportation fuel, while palm oil, jatropha, sorghum and cassava could also help to displace large quantities of petroleum. However, as elaborated upon in Chapter 12, the expansion of such tropical feedstocks, especially sugar and palm, threatens sensitive ecosystems and may not be desirable.

Although the yields of both temperate and tropical biofuel crops will probably continue to increase, they will still probably remain limited in comparison to the next generation of feedstocks. Algae, which is included as a first-generation feedstock because it can be processed with the same conversion technologies, stands alone as a feedstock with huge theoretical, though not yet economical, potential. There is much greater and more sustainable long-term potential for biofuels produced not from the sugars, starches and oils of plants, but from their more abundant fibres, as discussed in Chapter 4.

Continuing to increase the yields of conventional crops may thus become important mainly as a way of freeing up agricultural land for the production of dedicated energy crops. This potential is discussed in greater detail in Chapter 6.

Part II

New Technologies, Crops and Prospects

4

Next-Generation Feedstocks

INTRODUCTION

Cellulosic biomass such as wood, tall grasses, and forestry and crop residues are expected to significantly expand the quantities and types of biomass feedstock available for biofuel production in the future as new conversion technologies are developed that enable the production of biofuels from these feedstocks. Over the next 10 to 15 years, it is expected that lower-cost residue and waste sources of cellulosic biomass will provide the first influx of 'next-generation' feedstocks, with cellulosic energy crops expected to begin supplying feedstocks for biofuel production towards the end of this time-frame, then expanding substantially in the years beyond.

There are a variety of reasons why cellulosic biomass is considered an attractive option. The use of waste biomass offers a way of creating value for society, displacing fossil fuel with material that typically would otherwise decompose, with no additional land use required for its production. Cellulosic biomass from fast-growing perennial energy crops, such as short-rotation woody crops and tall grass crops, can be grown on a much wider range of soil types, where the extensive root systems that remain in place with these crops help to prevent erosion and increase carbon storage in soil. Energy crops can often be grown on poorer soil, particularly on sloped land where production of conventional annual food crops is not desirable due to erosion concerns. However, high biomass yields will only be achieved on good soils with sufficient water supply.

Cellulosic biomass is more difficult to break down and convert to liquid fuels; but this tenacity of the material also makes it more robust in handling (with fewer costs for maintaining feedstock quality compared to many food crops). In addition, cellulosic biomass can be easier to store for long periods of time, with less deterioration than sugar-based feedstocks, in particular. Compared to conventional starch and oilseed crops, where only a fraction of the plant material can be used for biofuel production, perennial energy crops can supply much more biomass per hectare of land since essentially the entire biomass growth can be used as feedstock.

This chapter explores the range of cellulosic biomass supplies that could be used in biofuel production in the coming years. It describes the basic characteristics of cellulosic feedstocks and the various supply options; it then addresses opportunities to increase cellulose production as an integral aspect of conventional food crop farming. The chapter concludes by addressing energy crop alternatives.

BASIC CHARACTERISTICS OF CELLULOSIC BIOMASS

Understanding the basic physical characteristics of cellulosic biomass is helpful in differentiating among the various types of biomass and their compatibility for producing different biofuels. Cellulosic biomass has three primary components: cellulose, hemicellulose and lignin. Cellulose has a strong molecular structure made from long chains of glucose molecules with six atoms of carbon per molecule (referred to as six-carbon sugar). Hemicellulose is a relatively amorphous component that is easier to break down with chemicals and/or heat than cellulose; it contains a mix of six-carbon (C-6) and five-carbon (C-5) sugars. Lignin is essentially the glue that provides the overall rigidity to the structure of plants and trees (trees typically have more lignin, which makes them able to grow taller than grasses).

For different types of plants and trees, these three main components of biomass are present in varying proportions. A typical range is 40 to 55 per cent cellulose, 20 to 40 per cent hemicellulose, and 10 to 25 per cent lignin (see Table 4.1).[1] To acknowledge its mix of components, cellulosic biomass is often referred to as 'lignocellulosic' biomass.

Different technologies for producing biofuels from cellulosic feedstocks use different components of the biomass. So-called enzymatic conversion technology focuses on processing the core sugar components of cellulose and hemicellulose into ethanol; the lignin is considered a good boiler fuel or feedstock for the production of various chemicals, fuel additives or bio-products (such as adhesives). So-called gasification systems, meanwhile, use a gasifier to convert all three main components

Table 4.1 *Physical composition of selected biomass feedstocks*

Feedstock	*Cellulose*	*Hemicellulose (percentage)*	*Lignin*
Poplar (hybrid)	42–56	18–25	21–23
Switchgrass	44–51	42–50	13–20
Bamboo	41–49	24–28	24–26
Sugar cane bagasse	32–48	19–24	23–32
Hardwood	45	30	20
Miscanthus	44	24	17
Softwood	42	21	26
Corn stover	35	28	16–21
Sweet sorghum	27	25	11

Source: see endnote 1 for this chapter

of biomass to a 'syngas', which can then be used to produce liquid fuel such as synthetic diesel, and/or other fuels and chemicals (see Chapter 5 for a more detailed discussion of biomass-to-liquid (BTL) fuels and other conversion technologies).

Table 4.2 provides information about the basic chemical content of varying types of biomass feedstock.[2] This content is important in determining a feedstock's suitability for different conversion processes. Agricultural residues, such as sugar cane leaves, tend to be bulkier (lighter weight) and typically have greater amounts of ash than do woody crops such as poplar. As a result, this feedstock tends to be more difficult to gasify. Thus, there has been more of a focus on using crop residues or tall grass energy crops for enzymatic conversion to ethanol, particularly since they also tend to have a higher intrinsic sugar content and smaller amounts of lignin. In contrast, woody crops, because of their higher lignin content, are considered somewhat more attractive feedstocks for gasification and conversion to synthetic diesel fuel. However, a given facility will also utilize the cheapest and most available feedstock in its region.

Biomass feedstocks that have higher potassium or ash content tend to be more of a problem for gasification technology since these components can create (or contain) compounds that melt at the high gasification temperatures, leading

Table 4.2 *Chemical characteristics of selected biomass feedstocks*

Feedstock	*Heating value (gross)*	*Ash*	*Sulphur*	*Potassium*	*Ash melting temperature*
	(gigajoule/ tonne)		*(percentage)*		*(degrees Celsius)*
Bamboo	18.5–19.4	0.8–2.5	0.03–0.05	0.15–0.50	
Miscanthus	17.1–19.4	1.5–4.5	0.1	0.37–1.12	1090 [600]*
Hardwood	19.0–21.0	0.5-2.5	0.01–0.04	0.04	[900]*
Softwood	19.6–22.4	0.3–1.2	0.01		
Poplar (hybrid)	19.5–19.7	0.5–1.5	0.03	0.3	1350
Switchgrass	18.3–19.0	2.5–7.5	0.07–0.12		1016
Sugar cane bagasse	18.1–19.0	2.8–5.5	0.02–0.15	0.73–0.97	
Sugar cane leaves	17.4	7.7			
Wheat straw	17.4	8.9–10.2	0.16		
Rice straw	18.9	13.4–18.7	0.18		
Corn stover	17.9–18.5	10.0-13.5	0.06–0.12		
Sweet sorghum	15.4	5.5			

Notes: Characteristics vary somewhat for specific samples and field conditions – ranges in characteristics are provided where source data was available for these ranges.

* Bracketed values for ash-melting temperatures indicate that some initial ash sintering (a sticky pre-melting condition) is observed above the temperatures indicated.

Source: see endnote 2 for this chapter

to potential problems such as slagging or fouling of heat-transfer surfaces. While these are not insurmountable barriers, they do constrain use of these feedstocks in a gasifier-based system (whereas enzymatic systems will typically be less affected by potassium and ash content). Ash that melts at a lower temperature can also be a concern for gasification systems since it tends to be easier to clean up solid particles than sticky half-melted or liquefied material.

BIOMASS RESIDUES AND ORGANIC WASTES

Biomass residues with potential energy uses are diverse. A distinction can be made between primary, secondary and tertiary residues and wastes (which are available as a by-product of other activities) and biomass that is specifically cultivated for energy purposes.[3]

Primary residues are produced during production of food crops and forest products (e.g. straw, corn stalks and leaves, or wood thinnings from commercial forestry). Such biomass streams are typically available 'in the field' and must be collected to be available for further use.

Secondary residues are generated during the processing of biomass for production of food products or biomass materials. They include nut shells, sugar cane bagasse and sawdust, and are typically available at food and beverage industries, saw and paper mills, etc.

Tertiary residues become available after a biomass-derived commodity has been used. A diversity of waste streams is part of this category, from the organic fraction of municipal solid waste (MSW) to waste and demolition wood, sludges, etc.

In general, biomass residues and wastes are intertwined with a complexity of markets. Many residues have useful applications as fodder, fertilizer and soil conditioner, or as the raw material for a variety of products, such as particleboard, medium-density fibreboard (MDF) and recycled paper. Net availability, as well as market prices, of biomass residues and wastes generally depend upon a number of factors, including market demand, local and international markets for various raw materials, and the type of waste treatment technology deployed for the remaining material. The latter is particularly relevant when fees are charged to dispose of the waste, giving some organic waste streams a (theoretical) negative value.

Typically, the net availability of organic wastes and residues can fluctuate and is influenced not only by market developments, but also by variability in weather conditions (causing high and low production years in agriculture) and other factors.

The physical and chemical characteristics of this diverse spectrum of biomass resources also vary widely. Certain streams such as sewage sludge, manure from dairy and swine farms and residues from food processing are very wet, with moisture contents over 60 to 70 per cent (these wet waste streams are typically more suited for producing biogas, rather than ethanol or biodiesel). Other streams may be more or less contaminated with heavy metals (such as waste wood from construction and demolition) or may have higher chlorine, sulphur or nitrogen content, depending upon the origin or part of the original crop. Clearly, the different properties of biomass resources lead to varying suitability for different conversion technologies.[4]

Wood residues

Forest residues

Over the last century, human efforts to control or limit forest fires have generally been quite successful. However, this has often led to an excess amount of undergrowth in forests (e.g. in the form of small diameter trees) that creates imbalances in the health of a forest. Some amount of understorey management can help to improve the health and productivity of forests, typically entailing thinning and removing the excess build-up of small diameter woody growth. However, the cost of removing this growth has often been too high to justify the investment.

Creating a market for this woody undergrowth for use in biomass-to-liquid (BTL) fuel applications may complement efforts to create healthier forests. Some amount of treetops and limbs that result from traditional logging industry activities may also be suitable as a supply of wood for biofuel production; however, the amount of woody material that should be kept in the forest for habitat and carbon storage needs must be evaluated before the wood is removed. Wood from pest or storm-damaged forests could also be a potential source of biomass for biofuel applications.

Industrial and urban woody residues

Much of the wood residues produced by the lumber industry are used to provide the energy needed for the lumber production process (such as lumber drying and cogeneration of heat and power), though some of this wood may be available for biofuel uses.

The pulp and paper industry tends to use much of its wood waste as boiler fuel. However, the 'black liquor' residue that results from the pulping process requires expensive boilers for disposal and could be a potential source of biofuel feedstock (the cellulose from wood is used to make paper, but the hemicellulose is contained in the black liquor). There has been an increasing interest in using black liquor residues for producing ethanol in the near term since there are substantial amounts of C-5 and C-6 sugars in this residue, resulting from the hemicellulose portion of the woody feedstock used for pulping. For example, the US Department of Energy has been co-sponsoring development efforts with the paper industry in an effort to foster the production of ethanol fuel from black liquor residues.[5]

Wood from urban tree trimming from backyards and rights of way has various competing uses, such as the production of mulch, or for electric power production or thermal energy needs; however, some of this wood may be available for use in biofuel production (see Chapter 6 for more discussion on competing uses for biomass resources).

Municipal solid waste

A mix of cellulosic waste material is typically present in municipal solid waste, including wood, paper, cardboard and waste fabrics. Since fees are typically charged to dispose of this waste, it could provide a supply of low or 'negative cost' biomass for some early pioneer cellulose-to-biofuels facilities in urban areas. Such facilities will have to overcome legal barriers and public reluctance to accept waste processing near populated areas.

Crop residues

Crop residues in the form of stems and leaves from conventional food crop harvests represent a substantial quantity of cellulosic biomass produced each year. In many instances, much of this residue needs to be left in the field to provide protection from erosion and to provide benefits such as micronutrient supplies, soil organic matter and enhanced soil 'tilth' (the texture, structure and pore spacing in soil). However, in cases where land is relatively flat and/or where conservation tillage methods are employed, a portion of the crop residues may be sustainably harvested. Table 4.3 provides estimates of crop residues produced for various conventional crops.[6] It indicates the total amount of residues produced before taking into account site-specific limitations on the amount of residues that can be sustainably removed. Residues for rapeseed and soybean residues are not sufficient to warrant collection, in part because current varieties of these plants deteriorate too quickly.

Driven by a need to reduce erosion, maintain soil structure and nutrients, and build soil carbon levels, agriculture has increasingly adopted more sound environmental and conservation practices. One increasingly popular option is no-till cultivation, where the soil is left undisturbed from harvest to planting (with only a narrow slot or drill hole used to plant seeds) and weeds are typically controlled with herbicides. No-till cultivation is now practised on more than 25 million hectares in the US, representing nearly 23 per cent of the country's planted cropland area in 2004.[7] On highly productive land, these practices increase the amount of crop residues that can potentially be collected for certain types of crops. For example, no-till cultivation may allow harvesting of as much as 75 per cent of corn stover (the stalks and leaves remaining on corn plants) where land slope and erosion problems are minimal.

Table 4.3 *Agricultural residues from conventional crops*

Crop	*Residue amount (dry tonnes per hectare)*	*Range in straw residues per tonne of grain harvested (tonnes)*
Corn	~10.1	0.55–1.50
Sorghum	~8.4	0.85–2.0
Cotton	~6.7	0.95–2.0
Rice	~6.7	0.75–2.5
Wheat	~5	1.10–2.5
Barley	~4.3	0.82–2.50
Rapeseed	–	1.25–2.0
Soybeans	–	0.8–2.6

Notes: Ranges reflect factors such as different crop varieties, levels of soil fertility or fertilization, rainfall and water availability, etc.

Source: see endnote 6 for this chapter

There are long-term economic and environmental concerns associated with the removal of large quantities of residues from cropland. Removing any residue on some soils could reduce soil quality, promote erosion and lead to a loss of soil carbon, which, in turn, lowers crop productivity and profitability. On other soils, some level of removal can be sustainable and even beneficial. A substantial amount of research has been conducted to evaluate sustainable levels of crop residue removal, and additional research is needed to help further establish criteria under many circumstances. Establishment and communication of research-based guidelines are necessary to ensure that removal of residue biomass is done in a sustainable manner (see Chapter 12 for more on soil quality concerns).[8]

Corn stover and wheat straw

As indicated in Table 4.3, corn crops typically produce the largest amounts of crop residues per hectare of all of the main conventional crop types (these residues, known as stover, include the stalks, leaves and cobs of the plant after the grain is harvested). Since large amounts of corn are often grown in specific geographic areas, such as the US Midwest, the aggregate supply of corn stover may be plentiful in these areas, probably representing one of the best near-term options for abundant cellulosic feedstock for ethanol production. In other places, particularly Canada and Europe, wheat straw is a more abundant potential cellulosic feedstock.

With either stover or straw, more could be harvested if no-till cultivation methods are adopted. Because they release less soil carbon, such techniques allow a greater portion of the crop residue to be harvested for biofuel use since less stover or straw would be needed to protect the soil from erosion and carbon losses.

Sugar cane residues

In Brazil, more than 80 per cent of the sugar cane harvest is cut manually.[9] Before this cutting occurs, the tops and leaves of the cane are typically burned off to make harvesting safer and more productive for workers. However, plans are advancing to mechanize the cane harvest to avoid burning of fields, a practice that causes considerable air pollution (see Chapter 12). The state of São Paulo has set deadlines for eliminating the use of fire in crop management, and 25 per cent of the state's cultivated area is now harvested mechanically.[10] Technological evolution is gradual, however, requiring development of realistic policies for retraining and redeploying labour, and for monitoring the environmental impacts of erosion and the spread of pests that follows mechanization.

As technologies for converting cellulose to biofuels are commercialized, this should create markets for the cellulosic field residues (tops and leaves) from sugar cane harvesting. This could significantly expand the supply of feedstock available for biofuel production in the tropical areas where sugar cane is produced, while facilitating significant reductions in air pollution caused by the burning of cane fields.

In addition, the commercialization of cellulose-to-ethanol technology could expand the supply of cellulose available for ethanol production by allowing the use of bagasse residues (the cellulosic plant stalks/residues left after extracting sugar from the cane stalks, which are currently burned in boilers for process heat) to be used for ethanol production instead. The lignin residues that remain after processing the bagasse could then be used as boiler fuel (depending upon the demands for process heat and alternative markets for electricity production from bagasse boilers).[11]

A fairly detailed evaluation of the amount of agricultural residues that could be sustainably collected each year for use in bioenergy applications has been done for the US, as summarized in Figure 4.1.[12] These estimates are based on current tillage practices using existing harvesting technology. The 'other residues' category includes municipal solid waste and animal fats. The combined total supply of sustainable residues available in the US is about 175 million dry tonnes per year.

Improved crop-residue collection technology

Most residue recovery operations today pick up residue left on the ground after primary crops have been harvested. Collecting these residues involves multiple passes of equipment over fields and results in no more than 40 per cent removal of stover or straw, on average.[13] This low recovery amount is due to a combination of collection equipment limitations, contour ridge farming, economics and conservation requirements. It is possible under some conditions to remove as much as 60 to 70 per cent of corn stover or straw with currently available equipment. However, this level of residue collection is economically or environmentally viable only where land is under no-till cultivation and crop yields are very high. This

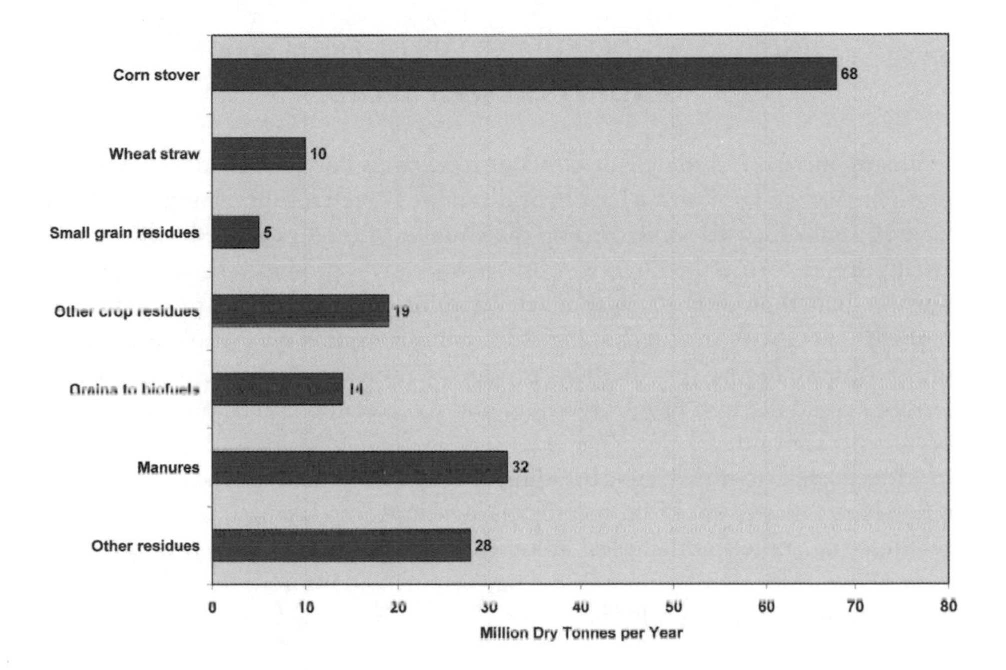

Figure 4.1 *Estimated availability of sustainable agricultural residues in the US*

Source: Perlack et al (2005)

analysis assumes that the efficiency of harvest technology and the percentage of cropland under no-till management are increased simultaneously.

Certain components of crop residues are more valuable than others; the components that should be left on the field to address sustainability and erosion concerns vary depending upon climate and growing conditions and upon the crop. Selective harvest technologies could have the ability to leave much of the desired residue components on the field for soil enhancement and erosion protection, and harvest only the portion of residues that sustainable crop management will allow.

Future residue collection technology with the potential of collecting up to 75 per cent of the residue is envisioned. These systems are likely to be single-pass systems that would reduce costs by collecting the grain and residue together. Single-pass systems will also address concerns about soil compaction from multiple pieces of residue collection equipment unless the single-pass system is heavier than the current grain harvesters. Furthermore, one-pass systems for corn and grain will need to have selective harvesting capability so that some portions of the residue stream can be reapplied to the field to meet site-specific conservation requirements.[14]

INCREASING CELLULOSE YIELDS FROM GRAIN AND/OR OILSEED CROPS

Dramatic increases in the yields of major food crops have been a direct result of research, such as corn and wheat hybridization. These crop development efforts have naturally focused on increasing the amount of food produced, while often attempting to reduce the quantity of stems and leaves produced since these were not the desired product. As markets develop for cellulose for the production of biofuels (along with new technologies for converting this type of plant material into biofuels), food crops that also produce significant quantities of cellulosic residues could increase farmer revenues and increase the overall output of each hectare of farmland.

It is possible to do selective breading of starch, sugar and oilseed crops, where new plant varieties could be specifically developed to increase the amount of cellulose (i.e. more stem and leaf volume) that is produced along with the food crop, allowing for significant cellulose harvesting in addition to the primary food crop harvesting (see Box 4.1).[15]

BOX 4.1 CROP OPTIONS TO INCREASE CELLULOSE RESIDUES

A change in the residue-to-grain ratio is a possible technology change that could occur for any crop. Indeed, previous breeding efforts have reduced the residue-to-grain ratio. Consider the case of soybeans. Most, if not all, soybean residue currently needs to be left on the ground to meet conservation practice requirements. A genetic improvement research programme on soybeans is being conducted by the US Department of Agriculture, which focuses on developing varieties that have a higher ratio of straw to beans, grow taller, have improved lodging resistance, have a better over-winter residue persistence, and are able to attain these traits without genetic transformation.

Originally, the soybean programme was geared to develop larger biomass soybeans for forage production and resulted in three varieties. A recently released variety for the southeast, Tara, has the characteristics of a 1.75 residue-to-grain ratio without sacrificing expected levels of grain yield. It is evident from data on the forage soybean varieties that the potential exists to produce 100 per cent more crop residue and thus provide more soil conservation benefits than the conventional varieties. It cannot be predicted whether farmers will adopt these new varieties; but clearly the technology will be available. Potentially, with such varieties, soybean acreage could contribute to the availability of residues for the production of biofuels – increasing biofuel production without using additional land.

Source: see endnote 15 for this chapter

It may also be possible to modify crop cultivation and management approaches with the intent of increasing the amount of cellulosic crop residue supplies that can be harvested through means such as increasing plant-spacing densities to maximize the combined yield of food and fibre. Food yields may go down somewhat with increased plant densities; but the total amount of crop residues produced per hectare would increase.

Double-cropping approaches

As markets develop for cellulose use in the production of biofuels, new approaches for growing more than one crop per year on farmland could significantly increase the amount of biomass produced per hectare of land each year. In traditional approaches for producing food crops, vigorous plant growth occurs in the spring as the stems and leaves quickly grow to create the plant structure that will then work to produce the final food crop (grain, oilseed, etc.). If the goal is to obtain maximum yields of cellulose rather than the end 'fruit' of the plant cycle, however, an entirely new strategy can be used for crop planting and harvesting. One such strategy, 'double-cropping,' has been particularly successful in Germany (see Chapter 24).

Winter wheat crops planted in the late autumn, for example, could be harvested much earlier in the following year if is not necessary to wait for the grain to form and mature. This cellulose in the form of tall, grassy wheat stems and leaves could be harvested and sold for use in the production of biofuels. A second crop could then be planted (such as legumes that fix nitrogen in the soil) early enough in the year to mature for autumn harvesting. Research and evaluation could determine the best combination of crops to grow for the first and second cycle of planting and harvesting each year to maximize the output and revenue that farmers obtain from each hectare of land; the best combination might be to produce just cellulose, or a combination of cellulose and food crops, depending upon the local climate, soil type and market considerations.

Energy crops

Large amounts of cellulosic biomass could be produced via dedicated plantations of energy crops based on the use of perennial herbaceous plant species (such as various tall grass species), or with the use of 'short-rotation' (i.e. fast-growing) woody crops, also known as SRWC.

There has been some use of SRWC for industrial purposes – for example, eucalyptus trees have been grown for pulp markets and to supply charcoal for the steel industry in Brazil, and in Europe and the US, poplar trees have been grown to provide fibre for the pulp and paper industry. In general, however, efforts to evaluate and develop energy crops are still in a relatively early stage of development when compared to conventional crops where plant breeding has been under way for many years. The relatively early phase of energy crop development reflects a situation where tremendous opportunities exist to use advanced plant science and agronomy to dramatically increase biomass yields.

There are a number of reasons why energy crop production could be quite attractive, beyond offering the potential to substantially expand the supply of biomass feedstock. The conversion of land from intensive annual crop production to perennial herbaceous species, or to SRWC, progressively increases the soil's organic matter content – whereas the conversion of land from natural cover to intensive annual crop production on farms typically decreases the organic matter content of the soil over time. The roots of the perennial crops provide protection from erosion, and the crops generally require less intensive use of fertilizers and pesticides, as well as less overall energy consumption for crop management (especially since it is not necessary to plough the land each year to do new crop planting).

Willow is a good example of a tree species suitable for use as a SRWC in temperate climates. Willow trees can achieve high biomass yields using a short-rotation coppice (SRC) approach. Short-rotation coppicing entails harvesting the aboveground growth of young trees, where the vigorous new growth of shoots and branches from the remaining tree trunks are then harvested every few years. In the case of SRC-willow energy crops, the new growth is harvested every two to five years over a period of some 20 to 25 years.

Most experience with SRC-willow systems in Europe has been in Sweden, where this crop is produced on some 14,000 hectares. A substantial amount of development work has also been conducted on SRC-willow crops in New York in the US. Willow crop yields have increased significantly due to research on genetics and breeding, with yields for some varieties doubling (or more) as a result. Hybrid poplar trees are also well suited to SRC energy crop applications.

As noted above, eucalyptus plantations have been grown in tropical regions for a variety of industrial uses. During the 1970s and early 1980s, an SRC approach was used for the initial eucalyptus plantations grown in Brazil. The tree stands were harvested every five to seven years, for up to three rotations before replanting. However, over this time-frame they found that problems occurred with diseases and pests, and the planted tree species were ultimately not as robust as newer clones and hybrids that were developed during the coppice periods. To take advantage of the fast pace of eucalyptus species improvements, the current practice in Brazil is generally to plant new improved hybrid eucalyptus varieties after the first harvest. It should also be noted that eucalyptus can absorb large quantities of water from water tables.[16]

Tall grass species such as miscanthus, switchgrass and reed canary grass are also examples of perennial crops that can be harvested every year. They have been the focus of considerable interest in Europe and North America. Some experts believe breeding could result in at least a doubling of the productivity of energy grasses. With varieties such as Bermuda grass and Pensacola Baha grass, yields have been increased by twofold and sevenfold.[17] Research suggests that future gains in switchgrass productivity, through an aggressive breeding programme, could increase average yields per hectare to more than 17 dry tonnes by 2025 and nearly 28 dry tonnes by 2050, even without using genetically modified plants.[18] Although they assume adequate soils and sufficient water, such advances will not be as complicated as breeding food crops since it is easier to breed for size rather than for a particular quality, such as taste in fruits or vegetables. While commercial production of perennials as energy crops is currently negligible, their future potential is expected to be quite large, particularly as conversion technologies that can economically produce biofuels from this cellulosic feedstock are commercialized.[19]

In general, dedicated biomass production is more expensive per unit of energy produced than the use of available residues and wastes. Typical cost ranges for perennial woody crops under North-Western European conditions are €3 to €6 (US$3.6 to $7.3) per gigajoule (compared to some €1 to €2, or US$1.2 to $2.4, per gigajoule for imported coal). Biomass production costs of dedicated production systems are especially dependent upon the costs of land and labour and the (average) yield per hectare. Typically, land costs (e.g. through land rent) can contribute about one third of the total biomass production costs under North-Western European conditions.

Both land and labour are relatively expensive production factors in Europe, a tendency that is maintained indirectly by structural agricultural subsidies under the European Union's (EU's) Common Agricultural Policy (CAP). In addition, agricultural surpluses in the EU are partially counteracted by measures to take agricultural land out of production (i.e. classified then as set-aside land). This land category could, in theory, be available for energy crop production; but the total set-aside land surface varies over the years (from 10 per cent to less than 3 per cent of the arable land) and is generally taken up in typical rotation systems of farmers, making introduction of perennial crops difficult. This partially accounts for the relative popularity of annual crops for energy purposes (such as rapeseed and interest in hemp).[20]

Table 4.4 summarizes the performance characteristics and developmental status of four potential energy crops considered for short- and long-term uses in temperate climates: short-rotation (SR) willow, short-rotation hybrid poplar, miscanthus and switchgrass.[21] Commercial experience with SR-willow has been gained in Sweden and the US, and, to a lesser extent, in the UK and other countries. In Eastern Europe there has been major interest in producing willow trees as an energy crop, where conditions are well suited and where low costs can be achieved on a somewhat longer term. Short-rotation hybrid poplar, meanwhile, is well suited to deliver both biomaterial and energy fractions as a typical multi-product system. The economics of producing the tree depends upon the production region, as well as market prices for the material produced. So far, there has been only limited

Table 4.4 *Performance characteristics and developmental status of four potential perennial energy crops*

Energy crop	*Description*	*Typical rotation*	*Typical annual yield (dry tonnes per hectare)*	*Price (Euros per gigajoule)*	*Suitable climate*
Short-rotation willow	Perennial tree crop	3–4 years	10 tonnes (next ten years); 15 tonnes (in 10 to 15 years)	€3–€6 (next ten years); €2 or less (in 10 to 15 years)	Colder and wetter
Short-rotation hybrid poplar	Tree planted for pulpwood in various countries	8–10 years	9 tonnes (next ten years); 13 tonnes (in 10 to 15 years)	€3–€4 (next ten years); €2 or less (in 10 to 15 years)	Temperate climates
Miscanthus	Perennial tall grass crop	Harvested annually	10 tonnes (next ten years); 20 tonnes (in 10 to 15 years)	€3–€6 (next ten years); €2 (in 10 to 15 years)	Both temperate and warm (yields highest in warm climates)
Switchgrass	Perennial tall grass crop	Harvested annually	12 tonnes (next ten years); 16 tonnes (in 10 to 15 years)	€3–€4 (next ten years); €2 or less (in 10 to 15 years)	Temperate climates

Source: see endnote 21 for this chapter

commercial experience with miscanthus (in Europe), and the breeding potential of the species has hardly been explored.

Habitat and mono-crop issues

Compared to conventional annual farm crops, energy crops can provide a friendlier habitat for some forms of wildlife in that all of the vegetation on energy plantations would not necessarily be removed each year. In addition, the density and height of energy crop vegetation is likely to be greater than for conventional annual row crops. One approach that could help to reduce the impacts of harvesting on wildlife would be to harvest during non-nesting periods, and to harvest in alternating strips of crops so that animals can move to an adjacent strip of crop area that will not be harvested that season (habitat and mono-crop issues are discussed further in Chapter 12).

Research is needed to determine whether some mixing of varied grass or tree species could be allowed on energy plantations in an effort to add some variety in the vegetative landscape. This would probably entail special considerations regarding harvesting and conversion technology that could accommodate the variations in feedstock. This added variety in energy crop species may well tend to decrease the maximum crop yields that are achieved; however, cost–benefit analyses could help to determine the impacts on the feedstock production costs in comparison to the wildlife habitat benefits provided by added plant diversity.

CONCLUSION

As new technology for converting cellulose to biofuels is commercialized, vast supplies of waste biomass could be available for the production of liquid transportation fuels. Potential constraints on the ability to use these resources will include competing uses for fibre, and environmental constraints regarding soil and habitat sustainability. The lowest-cost supplies, with potentially no cost or even negative costs (such as with municipal solid waste, where tipping fees are paid to dispose of the waste), will be an early source of cellulosic biomass used to produce biofuels.

Crop residues are anticipated to supply an increasing amount of biomass for use in biofuel production. Research and development will be needed to better understand soil and crop dynamics in relation to residue removal practices, and improved harvesting and handling equipment will need to be developed that can collect crop and forest residues. For crop residues, harvesters that result in minimal soil compaction and minimal interruption of primary food crop harvests are desirable (these harvesting operations typically must be completed within a tight 'window' of time, before weather damage occurs to crops and before crops begin to deteriorate).

Further research and development is needed on perennial energy crops, including willow, hybrid poplar, eucalyptus, miscanthus and switchgrass, in an effort to improve yields, refine crop management techniques, and refine harvesting and handling equipment. Given the amount of time that will be necessary to develop, demonstrate and commercially deploy energy crops, these crops need to be developed under various conditions worldwide as soon as possible in order to ensure that feedstock supplies using these crops will be available in the medium term (i.e. in the next 8 to 15 years). In order to improve the economics of energy crop production, research is also needed for related systems, such as densification of crop residues and tall grasses, optimization of feedstock storage methods, and transport systems for biomass.

The widespread establishment of perennial energy crops has the potential to significantly reduce net carbon emissions from transportation fuel use, to reduce soil loss from erosion, to reduce agro-chemical run-off, to create improved habitat, to create jobs, and to increase domestic energy supplies and national security. Trade-off analyses are needed to determine whether some amount of diversity can be included in the variety of energy crops grown on plantations in an effort to enhance habitat resulting from energy plantation development.

5

New Technologies for Converting Biomass into Liquid Fuels

Introduction

The use of 'next-generation' cellulosic biomass feedstocks has the potential to dramatically expand the resource base for producing biofuels in the future. Technology development efforts, to date, have demonstrated that it is possible to produce a variety of liquid fuels from cellulosic biomass for use in existing vehicles. So far, however, the costs of producing liquid fuels from cellulosic biomass are not competitive with petroleum-derived fuels, even with the recent rise in petroleum costs. Various government and industry-sponsored efforts are under way to lower the costs of making liquid fuel from cellulosic biomass by improving the conversion technologies.

This chapter describes the basic technology options available for converting cellulosic biomass into biofuels and evaluates the prospects for their implementation.

Basic conversion technology options

A key characteristic of cellulosic biomass resources is that they are naturally resistant to being broken down into their constituent parts, particularly in comparison to first-generation biomass feedstocks such as starch, sugar or vegetable oil. Thus, while the cost for cellulosic feedstocks themselves is expected to be lower than for current feedstock, the difficulty of converting these next-generation feedstocks to liquid fuel means that the conversion technologies are prone to being more expensive than current conversion technologies.

To overcome the recalcitrant nature of cellulosic biomass, multiple steps are generally required to convert it into liquid fuel. The two primary pathways for producing liquid fuels from biomass are thermo-chemical conversion and biochemical conversion.

Thermo-chemical conversion

There are two main thermo-chemical pathways for converting biomass into liquid fuel. The first is to gasify cellulosic biomass in a high-temperature vessel where oxygen levels are kept low enough to prevent the resulting combustible gases from burning. The so-called 'syngas' produced in this intermediate step is then converted to liquid transportation fuel using advanced catalyst conversion (such as Fischer-Tropsch technology, described later in this chapter). The production of synthetic diesel fuel has been a primary focus with this approach.

The second thermo-chemical pathway for converting biomass into liquid fuel is pyrolysis, which (similar to the gasification pathway) uses high temperatures in the absence of oxygen to convert the biomass into liquid 'bio-oil', solid charcoal and light gases similar to syngas. Pyrolysis is done at about 475 degrees Celsius (°C), whereas gasification is done at temperatures of 600°C to 1100°C.[1] Bio-oil is moderately acidic with a water content typically in the range of 20 to 25 per cent; it does not mix well with petroleum products and is generally best suited for use as a fuel for stationary electric power or thermal energy applications, rather than as a transportation fuel.

Biochemical conversion

The other basic pathway for converting cellulosic biomass into liquid fuel is biochemical. It involves breaking down the biomass into its component sugar molecules, followed by the use of fermentation organisms (specialized bacteria or yeasts) to biologically convert the sugar into ethanol fuel. This pathway requires an initial pre-treatment phase (e.g. using steam and/or acid) to break down the biomass into a liquid slurry of its component parts – cellulose, hemicellulose and lignin.

The pre-treatment conditions are typically sufficient for breaking down the hemicellulose component into its basic molecules of sugar – a mix of five-carbon (C-5) and six-carbon (C-6) sugars, as described in Chapter 4. However, breaking down the cellulose component into its sugars (glucose) is more difficult, requiring an additional step in the conversion process. This has been accomplished by using either concentrated acid and low temperatures, or dilute acid at higher temperatures. At present, however, the use of customized enzymes (known as cellulase enzymes) is generally viewed as the most promising way to break down the cellulose into glucose.[2]

Fermentation organisms are readily available for converting glucose to ethanol. However, the sugar molecules produced from hemicellulose (particularly the C-5 sugars) have required the development of customized fermentation organisms to enable their conversion to ethanol. With modern advances in biotechnology, it is possible to substantially customize and enhance the performance of enzymes for specialized conversion applications. Much of the research funding for improving the biochemical conversion pathway has focused on either improving fermentation

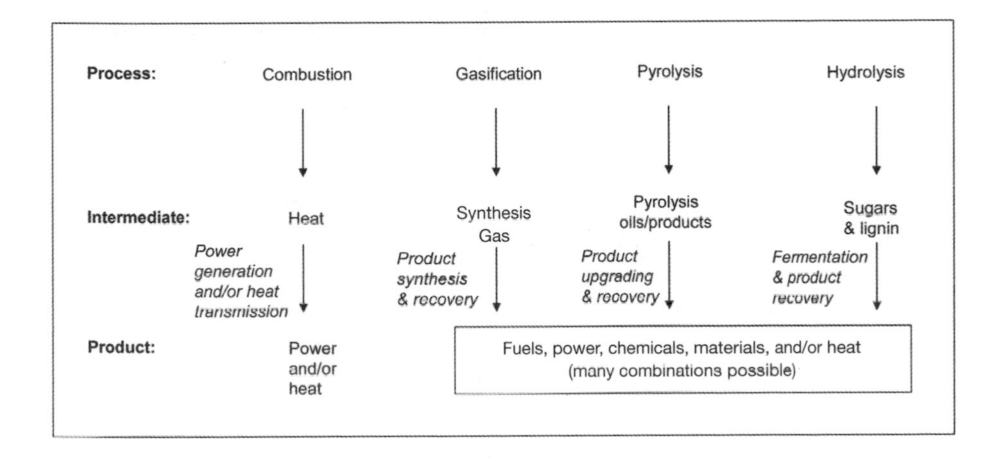

Figure 5.1 *Lignocellulose-processing pathways*

Source: see endnote 4 for this chapter

organisms or improving the performance of the cellulase enzymes used to convert cellulose to glucose and reducing their costs (note that the lignin component of the biomass is separated from the slurry and can be used in a variety of ways – for example, as boiler fuel to provide steam and electricity for the conversion process, or further processed to make products such as adhesives, fuel additives, etc.).[3]

Researchers are also developing variations and combinations of thermo-chemical and biochemical pathways for converting biomass resources into useful energy products. Four primary pathways for bioenergy production are highlighted in Figure 5.1.[4] Combustion and pyrolysis are two thermo-chemical pathways that can be used to produce electricity and/or provide thermal energy for applications such as industrial processes or space heating. Whereas gasification (another thermo-chemical pathway) and hydrolysis (a biochemical pathway) can provide a variety of products, the production of liquid fuels for transportation uses is a primary focus of this book. Thus, the following chapters focus mostly on these two pathways.

The various conversion pathways are described in greater detail in the following sections.

Converting lignocellulosic fibres and wastes into liquid fuels

Gasification and Fischer-Tropsch (F-T) synthesis

The gasification of biomass has been possible for much of the last century. Gasification followed by synthesis to liquid fuels was first applied to coal in

Germany during the 1920s, using technology developed by German researchers Franz Fischer and Hans Tropsch (hence, the technology is known as the Fischer-Tropsch, or F-T, process). During World War II, Germany as well as Japan produced diesel fuel from coal due to shortages of petroleum supplies. Later, South Africa used coal-to-liquid (CTL) technology in response to international embargoes on petroleum. For fuel synthesis, other processes are also under consideration – for example, the methanol-to-synfuel route.

Using gasification technology, biomass is converted to a mixture of carbon monoxide, carbon dioxide, hydrogen and methane – commonly referred to as synthesis gas, or 'syngas' – by heating it in the presence of limited oxygen. Syngas can be converted to a variety of fuels such as hydrogen, methanol or dimethyl ether (DME), as well as synthetic diesel and gasoline (via the F-T process). In principle, numerous process configurations exist for gasification-based conversion of biomass to fuels, depending, for example, upon the gasifier type, the gas cleaning process, the product fuel, and whether electricity is cogenerated using part of the syngas output from the gasifier.[5]

When applied to biomass, a key advantage of the gasification pathway is that it can convert all of the organic matter in biomass into gases and then liquids. In particular, the lignin component of biomass, which enzymes can hardly crack, is readily gasified and made available as a fuel feedstock. Gasification and F-T synthesis thus has the potential to produce more fuel per tonne of biomass.[6]

However, the gasification/F-T pathway has so far remained too expensive to be economical. During the 1980s, a number of biomass gasification projects sprouted in France, Sweden and Finland, which mostly produced methanol from wood and wood wastes; but lower petroleum prices and cheaper methanol eventually undercut these operations.[7] In particular, the capital costs of construction have proven prohibitive.[8] But keeping the gases and equipment clean has represented another key obstacle. The gas produced from gasification also contains tars, fine particles, alkali compounds and halogens that can clog filters, poison catalysts used in downstream fuel synthesis, or corrode the gas turbine used in electricity co-generation. Thus, techniques for cleaning the syngas are an important requirement for successful operation of these systems.[9]

As of early 2006, there were no commercial plants producing lignocellulosic biomass-derived liquid fuels via gasification; but demonstration efforts are well under way. Pilot and demonstration units producing electricity via biomass gasification were under development in Scandinavia, the US, Brazil and the European Union (EU).[10]

Pyrolysis

Variations of pyrolysis, which puts biomass under heat at very low levels of oxygen, have existed as well, but are not employed on a large scale. Depending upon the operating conditions (temperature, heating rate, particle size and solid residence time), pyrolysis can be divided into three subclasses: conventional, fast or flash. To maximize bio-oil production, fast/flash pyrolysis is used, heating the biomass at about 500°C for less than ten seconds.[11]

Several challenges must be overcome before biomass pyrolysis can become a commercially viable means to produce energy on a large scale. Pyrolysis oils have several undesirable characteristics that necessitate downstream processing. For instance, they contain suspended char and alkali metals that can damage engines. They are acidic, temperature sensitive and highly viscous, which can also cause storage and engine problems.[12] They also typically contain 20 to 25 per cent water (contributed by the water in the initial biomass and from the conversion process). With proper treatment, however, pyrolysis oils can be used in many applications, such as combustion in boilers, stationary diesel engines, industrial combustion turbines and Stirling engines (which are external combustion engines used to produce heat and power). The comparative option of upgrading these oils for vehicle engine use does not appear promising.[13]

As of early 2006, there were no large-scale biomass pyrolysis facilities, though several smaller facilities were in operation. Ensyn Technologies in North America has built several small commercial plants geared towards the production of speciality products (e.g. natural resins, co-polymers and other chemicals). DynaMotive Energy Systems in Vancouver, Canada, recently opened a 100 tonne per day facility to produce fuel for combined heat and power production at a wood products plant in Ontario.[14] Biomass Technology Group (BTG), located in The Netherlands, has demonstrated its rotary cone reactor technology at the 5 tonne per day scale, with plans to build a 50 tonne per day plant.[15] Notably, pyrolysis oils are not currently used for transportation.

It should be noted that biomass can also be converted into a liquid oil by direct hydrothermal liquefaction, which is a form of intermediate pyrolysis that occurs in the presence of water.[16] Hydrothermal treatment is based on early work performed by the Albany Laboratory of the US Bureau of Mines during the 1970s. Developers include Changing World Technologies (West Hampstead, New York), EnerTech Environmental, Inc (Atlanta, Georgia) and Biofuel BV (Heemskerk, The Netherlands).[17] In particular, the oil company Royal Dutch Shell has experimented with a process called hydrothermal upgrading (HTU), which also yields a synthetic diesel fuel from wet biomass.[18]

Acid hydrolysis

During the early part of the 20th century, the Germans developed an industrial process for converting biomass into ethanol that used dilute acid to hydrolyse cellulose to sugars. This acid hydrolysis process was soon adopted in the US, resulting in the operation of two commercial plants during World War I.[19]

Commercial processes using concentrated sulphuric acid were in operation during the 1940s, particularly in the former Soviet Union and Japan. These plants, however, were only successful during times of national crisis, when the economic competitiveness of ethanol production could be ignored.

Today, only a few acid hydrolysis facilities exist. Arkenol operates a 1 tonne per day pilot facility that uses concentrated acid hydrolysis.[20] A few acid hydrolysis facilities have also been operating in Russia, using inefficient technology to produce ethanol for industrial uses. Smaller demonstration facilities also exist at universities and government laboratories such as Lund University in Sweden, the Danish Technical University and the US National Renewable Energy Laboratory.[21]

Using acid to hydrolyse lignocellulosic fibres has not been as economical as hydrolysis of starches or the simple fermentation of sugars. Processes that hydrolyse fibres with dilute acid and pressure have tended to degrade too much of the hemicellulose sugars before they can be fermented into ethanol, causing low yields.[22] The other option, using concentrated acid at lower pressures, has required purchasing and then recycling expensive quantities of acid.[23]

Enzymatic hydrolysis and microbial digestion

The prospect of using enzymes instead of acid to hydrolyse fibres into sugars has generated much greater enthusiasm. Enzymatic hydrolysis of conventional starchy feedstocks has already replaced acid hydrolysis in ethanol facilities in the US.[24] The next step would be to apply similar enzymes to cellulose and hemicellulose fibres. Since cellulose is more difficult to digest than hemicellulose, the effectiveness of enzymes at hydrolysing cellulose has been identified as a limiting factor.

Another limiting factor has been the inability of yeasts to digest some of the sugars contained in hemicellulose. Typically, the sugar that results from hydrolysing starch and cellulose contains six carbons (glucose and mannose), which are familiar to conventional fermentative yeasts. Hydrolysing hemicellulose, however, yields a combination of both six-carbon sugars and five-carbon sugars, and normal yeasts cannot ferment the five-carbon ones. Since one third or more of the total carbohydrate content of typical biomass is comprised of five-carbon (or 'pentose') sugars, it is highly desirable that these be fermented.[25]

Research and development on conversion of pentose sugars to ethanol has been a major focus over the last 15 years, and has resulted in several promising strains. These include strains of enteric bacteria (*Esherichia coli* and *Klebsiella oxytoca*), thermophilic bacteria, the mesophilic bacterium *Zymomonas mobilis* and yeast.[26]

Iogen Corporation, a consortium of PetroCanada and Royal Dutch Shell, currently uses related micro-organisms at a test facility in Ottawa, Canada; but they can only produce ethanol for €0.44 to €0.66 per litre (US$0.53 to $0.79), in comparison to about €0.24 per litre (US$0.29) for conventional ethanol.[27] The rapid development of this technology is covered in the following section.

Status of next-generation conversion technology

During the early 2000s, the price of petroleum has risen again, and many expect it to remain higher than €35 (US$42) per barrel. Sustained high oil prices are making next-generation conversion technologies more viable, prompting further development of older sulphuric acid processes, as well as continued refinements in the newer biological and thermo-chemical pathways.

Processes that hydrolyse lignocellulosic fibres with acids are gaining some renewed interest. Arkenol, a leader in advanced sulphuric acid hydrolysis, is planning the construction of several new facilities in the US. Similarly, the primary supplier of equipment for Brazil's sugar and ethanol mills, Dedini, is developing its own sulphuric acid process for converting sugar cane bagasse fibres into ethanol.[28]

In the US, the 2005 Energy Policy Act set a priority on the production of cellulosic biofuels, setting a national target of 250,000 gallons (946,000 litres) of cellulosic ethanol production by 2012. In early 2006, the government offered €132 million (US$160 million) in incentives to build up to three industry–government funded cellulosic ethanol refineries, and the Department of Energy has a larger budget for researching biomass alternatives to petroleum.[29] In the EU, converting lignocellulosic matter into liquid fuels has been identified as a priority in the EU Biomass Action Plan.[30]

Various efforts are under way to estimate the anticipated costs for biofuels in the future as progress is made in making advanced 'next-generation' biofuels less expensive. Figures 5.2 and 5.3 summarize the results of an International Energy Agency (IEA) study that estimated the costs of biofuels after the year 2010, comparing both first-generation and next-generation technologies for producing gasoline and diesel substitutes.[31] The lowest-cost biofuels are expected to continue to be ethanol produced from sugar cane, and biodiesel produced from recycled cooking oil and waste grease. Beyond these two least-cost options, the costs for producing next-generation biofuels are expected to be in a range that should make them generally competitive with first-generation technologies. As noted earlier, the ability of next-generation technologies to use abundant cellulosic feedstocks that do not rely on food crops offers the promise of dramatically expanding the amount of biofuels that could be produced for transportation needs in the future.

EMERGING DEVELOPMENTS IN CONVERSION TECHNOLOGY

While higher petroleum prices have improved the relative competitiveness of cellulosic refining procedures, advances in conversion technology are making cellulose more economical as well. The US and the EU have been the primary laboratories for these advances, with the US focusing more on the biochemical pathway and EU countries focusing more on the thermo-chemical pathway, particularly integrated gasification and F-T synthesis processes.

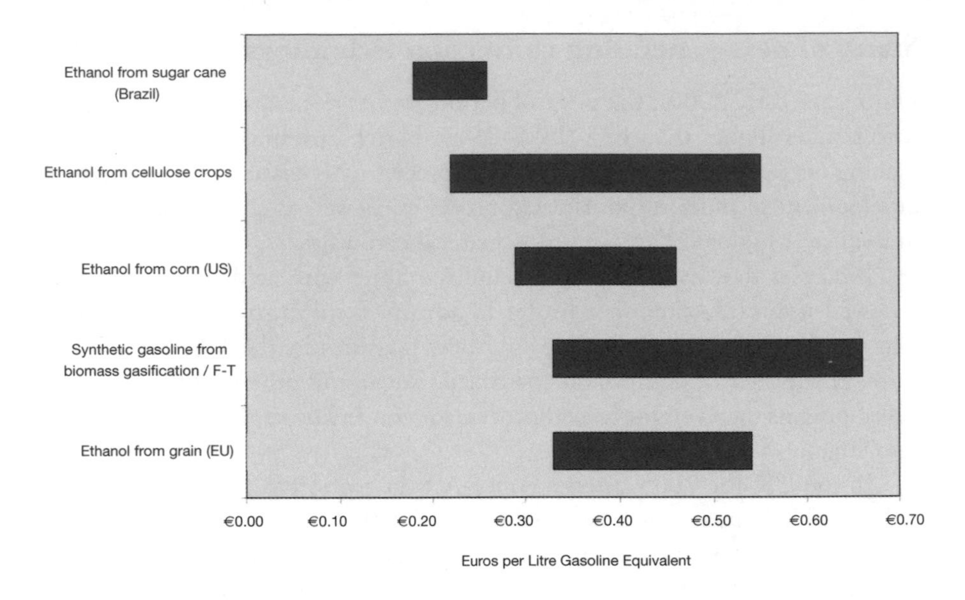

Figure 5.2 *Cost ranges for ethanol production after 2010*

Note: F-T stands for the Fischer-Tropsch process.

Source: Fulton et al (2004) pp85–86

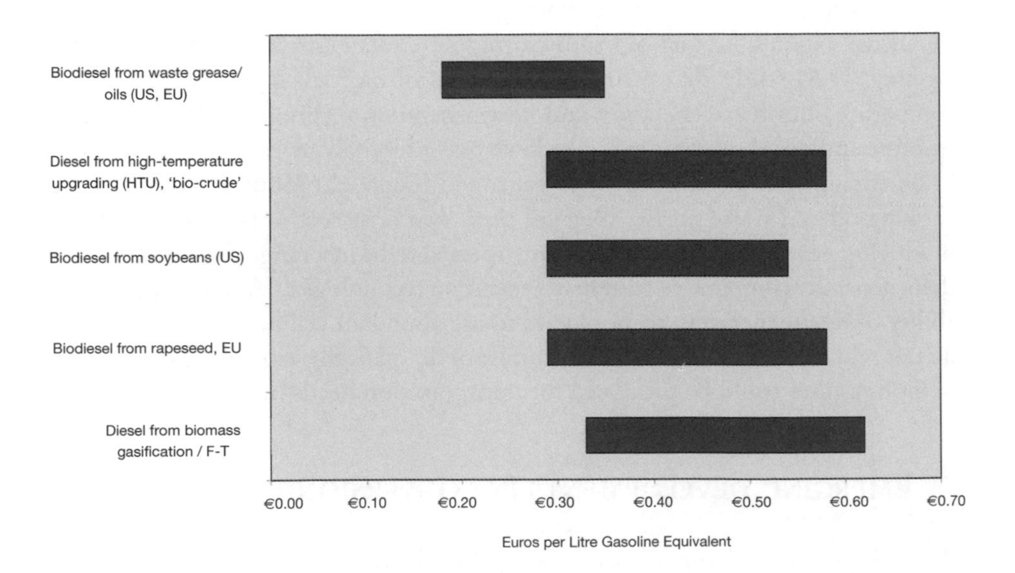

Figure 5.3 *Cost ranges for biodiesel and synthetic diesel production after 2010*

Note: F-T stands for the Fischer-Tropsch process.

Source: Fulton et al (2004) pp85–86

Biochemical technologies

The US has been the main driver of developments in biochemical conversion technologies. The US Department of Energy (DOE), through its National Renewable Energy Laboratory, has funded efforts to develop cheaper enzymes that hydrolyse cellulose into sugar, which are called cellulases, and cheaper micro-organisms that can ferment the unusual sugars in hemicellulose. The DOE has given two companies, Novozymes and Genencor, the largest grants towards this end, although other enzyme suppliers (e.g. Iogen and Dyadic) have also been developing biotechnologies for this conversion pathway. As a result of proprietary advances, Novozymes announced in early 2005 that it had reduced the cost of enzymes to between €0.22 to €0.44 per litre (US$0.10 to $0.20 per gallon), far less than the previous €1.09 per liter (US$5 per gallon).[32]

Further developments that could reduce the cost of the biological pathway are also considered likely. The processing of lignocellulose consists of four biologically mediated events: production of cellulose; cellulase-mediated hydrolysis of cellulose; fermentation of six-carbon sugars (hexoses); and fermentation of five-carbon sugars (pentoses). These events can be carried out with varying degrees of consolidation, as follows:

- In separate hydrolysis and fermentation (SHF), each of these four events is carried out in a separate reactor mediated by a separate biological catalyst.
- In simultaneous saccharification and fermentation (SSF), hydrolysis and conventional fermentation are combined into one process step.
- Saccharification and conventional fermentation are combined with the fermentation of five-carbon sugars in simultaneous saccharification and co-fermentation (SSCF). SHF, SSF and SSCF all feature a dedicated process step for cellulase production.
- In consolidated bioprocessing (CBP), all four biologically mediated events are combined into one process step, eliminating the need for dedicated cellulase production. CBP is widely regarded as the strategy offering the lowest cost in the long run. This kind of processing is, however, at a much earlier stage of development compared to processes featuring dedicated cellulase production.[33]

A variety of processes have been shown to be effective at pre-treating lignocellulose, including processes based on exposure to dilute acid, steam, or hot water, ammonia, lime and other agents. A variety of pre-treatment processes are under investigation, with none having yet emerged as the process of choice, and each offering some advantages.[34] In North America, the Consortium for Applied Fundamentals and Innovation (CAFI) has been focusing on the pre-treatment area, in particular; here, several leading pre-treatment experts are working collaboratively, led by Charles Wyman at Dartmouth University.[35]

In Europe, research on ethanol from cellulosic feedstocks is being conducted under a new EU-wide project called New Improvement for Lignocellulosic Ethanol (NILE), which covers a variety of conversion routes and aims to build demonstration plants in 2007.[36]

It is estimated that for the biochemical pathway of hydrolysing cellulosic biomass to ethanol, the cost will be about €0.52 (US$0.63) per litre for ethanol in the short term (the next 5–8 years); €0.31 (US$0.37) per litre in the mid term (8–12 years); and €0.21 (US$0.25) per litre in the longer term (13–20 years) (note that these are costs per litre of ethanol, which has two-thirds as much energy per litre as gasoline).[37]

Thermo-chemical technologies

Developments in the thermo-chemical pathway – that is, gasification and F-T synthesis – have occurred primarily in Europe. In particular, Choren, a company financed by DaimlerChrysler, Volkswagen and Royal Dutch Shell, has built a demonstration facility in Germany for converting wood wastes into synthetic diesel, which has operated successfully in Freiberg since mid 2003. The scale-up to a demonstration plant is now under way to further the technology to a pre-commercial state in the next three years.[38] The demonstration plant will have a capacity of 13,000 tonnes of biomass to liquid (BTL) per year, and will lay the base for the construction of commercial-sized systems in the order of 200,000 tonnes per year. Besides the Choren system, other routes with different gasifier designs are being tested as well (e.g. by Clausthaler Umwelttechnik Institut in Germany), and European research is quite active in that field.[39] Market introduction of gasification/F-T schemes is expected in the next decade.

Pilot and demonstration units producing electricity via biomass gasification are now under development in Scandinavia, the US, Brazil, and elsewhere in the EU, and both the US Department of Energy and the European Commission are actively involved in biomass power demonstration programmes.[40]

Several opportunities exist for reducing the cost of the gasification conversion pathway through technological advances, including by consolidating gasification, gas cleaning, and/or gas processing (i.e. steam methane reforming and water gas shift) in a single vessel, and by developing large-scale, pressurized oxygen-blown gasifiers.[41] Researchers in Europe and the US are also seeking more efficient catalysts and better methods of cleaning and preparing syngas for producing different products.[42] Gasifiers with self-contained gas cleaning have the potential to reduce capital costs by combining two unit operations. In order to produce co-products that can reduce the cost of fuels, researchers are also seeking markets for the many chemical by-products of gasification and F-T synthesis.[43]

Meanwhile, gasification using, not biomass, but fossil fuel feedstocks (primarily coal and petroleum) is already commercially established technology for the production of both electricity and liquid fuels. In 2004, there were 117 such plants operating worldwide, which generated a variety of outputs, including chemicals (produced in 37 per cent of plants), F-T liquids (36 per cent), power (19 per cent) and gaseous fuels (8 per cent). An additional 38 plants with a combined synthesis capacity of more than 25,000 megawatts-thermal are planned to come online by 2010.[44] This existing infrastructure could be adapted to utilize biomass feedstocks as well.

It is estimated that for the thermo-chemical pathway using F-T technology for converting cellulosic biomass to synthetic diesel fuel, the cost will be about €0.48 (US$0.59) per litre for diesel in the short term (the next 5–10 years), and €0.29 (US$0.35) per litre in the longer term (10–15 years). This analysis used the same methodology as that employed in estimating the costs for ethanol via enzymatic hydrolysis (see the previous section on 'Biotechnical technologies'). After adjusting for the lower energy content per litre for ethanol compared to diesel, this analysis indicates that F-T technology may be able to produce biofuels at a cost roughly 20 per cent lower than the enzymatic hydrolysis pathway in the long term (based on the cost per unit of biofuel energy produced).[45]

Pyrolysis

Other researchers have continued to develop pyrolysis pathways for producing biofuels, including new variations of 'hydrous' pyrolysis that can process wet biomass without pre-drying. Royal Dutch Shell re-evaluated its hydrothermal upgrading process in the early 2000s and has since seen more promising results with this technology.[46]

Changing World Technologies, a company that has adapted a pyrolysis technology first developed during the 1980s, has built a facility in Philadelphia where it is experimenting with processing food wastes, sludges, offal, rubber, animal manure, black liquor, plastics, coal, polychlorinated biphenyls (PCBs), dioxins and asphalt into pyrolysis oils, solids and gases. It has also built a full-scale facility for processing turkey carcasses from a Cargill slaughterhouse in Carthage, Missouri. This has since been shut down, however, because nearby residents blamed it for producing odours.[47]

FZK in Germany introduced a decentralized flash pyrolysis system to densify biomass for a more efficient transport to centralized gasification/FT-diesel plants, thus making direct use of the bio-oil for cogeneration, and the bio-coke would be transported as a feedstock to a larger-scale BTL plant. Currently, the design focuses on straw, but would also be suitable for other feedstocks.[48]

Implementation

Of the two pathways for cellulosic conversion, the biochemical one appears to be closer to large-scale implementation. It requires less capital-intensive facilities and can be economical on a smaller scale.[49] Moreover, because it is part of the rapidly developing biotechnology sector, the costs of this pathway are likely to decline faster than for other pathways, at least in the near to mid term, although not all experts agree on this.

Already, additional demonstration and commercial plants are being constructed that use advanced biological methods to refine ethanol from cellulose. In particular, the micro-organisms that can ferment the five-carbon sugars in hemicellulose are ready for utilization. Iogen is already using recombinant *Zymomona mobilis* bacteria to produce ethanol at its facility in Ottawa, with wheat residue as the main feedstock, and is currently raising €290 million (US$350 million) in capital to build a larger commercial facility in the US or Canada.[50] In Nebraska, the Spanish company Abengoa plans to use the cheap enzymes developed by Novozyme to break down corn stover.[51] In Louisiana, BC International is building a facility that will combine conventional ethanol production with conversion of sugar cane residues, using the *E. coli* bacteria developed at the University of Florida.[52] And Colusa Biomass Energy Corp is marketing its process for converting rice hulls into ethanol. While most of these efforts are concentrated in the US, SunOpta Inc plans to open the first commercial cellulosic ethanol plant in Spain in summer 2007, using technology from Abengoa.[53]

Although they are not as close to commercial success, gasification and F-T synthesis processes are progressing strongly as well. Gasification appears to be better able to accommodate a wide variety of feedstocks than enzymatic processes, and (as noted earlier) permits the conversion of a greater fraction of feedstock carbon to liquid fuels. It also has the theoretical advantage of being able to process batches of different feedstocks, and it can more easily be combined with coal gasification facilities.[54] Chinese planners have also shown interest in developing BTL technology, particularly by co-gasifying biomass with coal.[55]

Elsewhere, others are planning new pyrolysis biomass-conversion plants. Changing World Technologies aims to construct additional thermal de-polymerization plants in Europe. And in Latvia and Ukraine, Rika Ltd has licensed DynaMotive's 'fast pyrolysis' technology and has leased 10,000 hectares to grow energy crops for this conversion.[56]

Transitioning existing biofuel facilities into cellulosic refineries

There are several reasons to expect that cellulosic conversion facilities will initially be developed in an integrated manner with conventional biofuel production facilities. A key obstacle to advancing next-generation conversion technologies has been the expense of harvesting and collecting the biomass feedstock. Existing biofuel production facilities, however, already process large amounts of biomass. The high-fibre by-products of starch-based ethanol production processes – such as dried distillers grain (DDG) – are a pre-collected source of cellulosic mass. At sugar-based conversion facilities, bagasse and beet pulp are convenient sources of cellulose. With this convenient feedstock nearby, existing facilities are in a position to experiment with pilot plants that convert only a share of this cellulosic mass into fuels.

Previous models for producing liquid fuels from lignocellulosic matter assumed that both cellulose and hemicellulose would have to be utilized; otherwise, the expense of harvesting and collecting the feedstock would be too great. However, the availability of large quantities of low-value cellulosic biomass at integrated facilities could make it economical to convert the biomass into liquid fuel. Thus, existing conversion facilities could be a nursery for the evolution of cellulosic refineries.[57]

'MATURE' APPLICATIONS OF CELLULOSIC CONVERSION TECHNOLOGIES

The technologies for converting biomass into liquid fuels are developing along many different routes, and some are more viable today than others. As they mature, however, the requirements of specific applications will influence which technologies are most appropriate; it is also possible that combinations of several conversion technologies will be used to address the requirements of a particular feedstock or application.

For example, woody feedstock, including municipal solid waste, is probably more suitable for gasification and F-T synthesis because it contains relatively more lignin, which biological processing cannot convert. Compared to grassy feedstock, woody feedstock also contains less chlorine and alkali metals, which can foul up gasification operations. Grassy feedstock, including agricultural residues, is thus more suitable for the biological pathway because it contains less lignin. And hydrous pyrolysis is likely to remain a good solution for producing liquid fuels out of wet organic waste.

These processes could also work in tandem on the same feedstock. For example, pyrolysis has recently been considered a way of aggregating wet biomass and extracting some pyrolysis oil on a smaller scale, before sending the residual solids to a larger-scale gasification facility.[58]

The gasification pathway could also process the residual lignin left over after the cellulose and hemicellulose fractions of a feedstock are processed biologically. One detailed analysis of different conversion pathways concluded that this combination was the most economically and energetically efficient, as well as the most effective means for displacing petroleum, because it converts the lignin into liquid as well (although some analyses have found that gasification facilities will need to be larger than biological facilities, rather than smaller and secondary).[59] In this idealized scenario, ethanol would be produced through biological conversion of the feedstock's carbohydrate fraction. Then, the lignin-rich residue resulting from this bioprocessing, as well as the methane-rich biogas from wastewater treatment, would be thermo-chemically processed into F-T diesel and gasoline. This would result in very little waste (and energy loss), and permit a small amount of power export to the electricity grid (see Appendix 4 for a schematic diagram of this process). Figure 5.4 provides a comparison of its energy efficiency.[60]

The results of the analysis indicate that such mature biomass refineries can potentially produce fuels at costs competitive with gasoline. The ethanol, F-T fuels and power scenario, for example, produces ethanol at €0.14 per litre, or €0.20 per litre of gasoline equivalent (US$0.93 per gallon)[61] and F-T diesel at €0.22 per litre (US$1.04 per gallon) at a production scale of 5500 dry tonnes of feedstock per day. It also entails a 12 per cent internal rate of return (IRR), 60 per cent equity financing and electricity valued at €0.03 (US$0.04) per kilowatt hour. This scenario was most profitable at prices above roughly €6 (US$7.3) per gigajoule of gasoline equivalent (€0.19 per litre/US$0.89 per gallon wholesale), or about €21 per barrel (US$25) of crude oil. Below this price, dedicated power production was the most profitable configuration, achieving an IRR of just under 10 per cent.

THE BIOREFINERY CONCEPT

An oft-touted model for future biofuel production will be a kind of 'biorefinery' where both fuels and co-product materials are produced. This model would mimic petroleum refineries where fuels are produced simultaneously with chemicals and

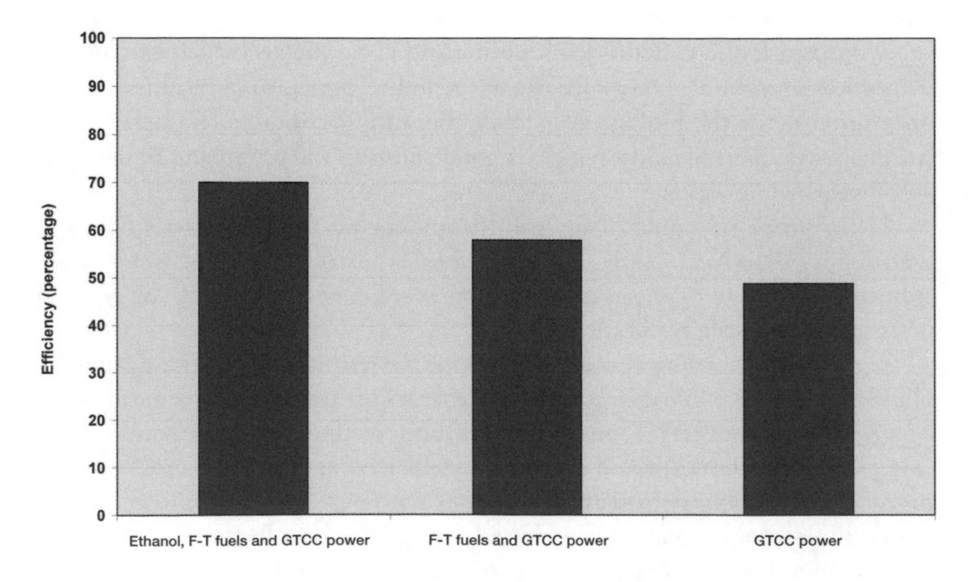

Figure 5.4 *Processing efficiency for three biorefining scenarios*

Note: GTCC stands for gas turbine combined cycle.

Source: see endnote 4 for this chapter

materials. In the past, conventional fuel production has been a stepping stone on the way to chemical production, as occurred with the petrochemical industry in the 1940s and with gasification-based coal processing in South Africa during the 1980s; similarly, commercial production of organic chemicals from biomass has been suggested as a stepping stone to commercial production of biofuels.[62]

Notwithstanding anticipated synergies between the production of chemicals and energy (fuel and power) from biomass, one complicating factor is the size of markets for organic chemicals, which tend to be relatively small compared to markets for liquid fuels. This will create a challenge in identifying chemical co-products for biorefinery production that will not exceed the market demand, since large quantities of liquid fuels are produced at these facilities.

Typically, fuels would represent the bulk of total biorefinery production, while chemicals and other materials would generate the bulk of the profits. This is similar to the situation with conventional fuels, where the flow of energy leaving the US oil-refining industry as liquid fuels is more than 25-fold greater than the flow leaving as petrochemicals. Chemical production makes a disproportionately large contribution to the profitability of oil refining and lowers the price of fuels because of its higher profit margins relative to fuel production; fuel production, meanwhile, lowers the price of chemicals by providing economies of scale.[63]

Biochemical conversion pathways may offer a more likely option for co-producing bio-based materials, particularly since the lignin fraction of the biomass feedstock is not converted to ethanol fuel in the conversion process. Although bio-based chemicals are currently 'speciality' chemicals produced in separate facilities, they could, instead, be economical co-products of biorefineries. Already, one conventional 'wet' mill for corn-ethanol in the US state of Nebraska generates a range of co-products, including high-fructose corn syrup, high-protein animal feedstock and speciality chemicals such as polylactic acid (a feedstock for corn-based plastics and fabrics).[64]

Even though most bio-based products are made via the biological pathway, gasification and F-T synthesis also has the potential to co-produce a wide range of materials. F-T synthesis could theoretically produce all the hydrocarbons produced by petroleum refineries, although the economics are unlikely to be the same for all products.

Future biorefineries will benefit from their ability to mimic the energy efficiency of modern oil refining as well. A key aspect of this refinery approach is extensive heat integration, whereby the heat available from some unit operations can be used to meet heat requirements of other operations within the process. Integrating heat flows in the most beneficial manner within a refinery with multiple heat sources and heat sinks is referred to as 'pinch analysis', and is routinely employed in designing and upgrading oil refineries. Without heat integration, which allows heat to be reused multiple times, auxiliary energy inputs to the process would be much higher and the efficiency would be lower.[65]

NEAR-TERM PROSPECTS FOR CELLULOSIC LIQUID FUELS

Despite their promise, cellulosic conversion technologies are still probably at least 8–15 years away from supplying a significant proportion of the world's liquid fuels. Biochemical pathways appear likely to begin significant commercial expansion in the next 8–10 years, and thermo-chemical F-T pathways are expected to begin such expansion in the next 10–15 years. This will depend partly upon the amount and quality of research and development funding that governments provide over this period; in particular, researchers will need to continue reducing the cost of producing valuable enzymes and cleaning F-T gasifiers and syngas.

The promise of cellulosic biofuels will also hinge in great part on how much support governments provide to help risk-averse investors build a large number of cellulosic conversion plants.[66] The cost and risk of building new conversion facilities is currently hampering development of these next-generation fuels. Building a new cellulosic biofuel plant requires a larger capital expenditure than building a conventional biofuel production plant, and investors are not assured that the price of petroleum will remain high enough for these operations to remain competitive.

In the future, cellulosic fuels could be cheaper than petroleum fuels should the latter prices remain high; but they are still likely to be more expensive than conventional first-generation biofuel conversion technologies for some time.

Moreover, even though the technologies appear to be approaching viability, it will take a long time for their application to ramp up to a level that can compare to the current petroleum-fuel infrastructure. Even with a dedicated effort to expand the production of cellulose-based fuels, it will probably be decades before these substantially rival petroleum fuels. Nevertheless, cellulosic fuels have great promise, and in the medium term they could begin to provide a growing share of the global fuel supply.

CONCLUSION

The technologies for converting lignocellulosic matter into biofuels exist and are becoming increasingly cost competitive as the technologies advance. Various analyses support the conclusion that mature conversion technology will be able to produce fuels from cellulosic biomass with high process efficiency and competitive costs – although, ultimately, the features of mature biomass conversion technology will be known only after such technology has been developed and widely commercialized. Moreover, the economic competitiveness of these fuels, and the development of the conversion pathways, will depend substantially upon the relative price of petroleum in the future.

The pathways for advancing today's evolving lignocellulosic conversion technology to maturity will need to be shaped by a clear vision of where we can expect this road to lead. Both technicians and policy-makers should continue to clarify which efforts at converting biomass into fuels will be most productive. Strategies to do this can be roughly categorized as falling into support for pre-commercial research and development (R&D), and support for commercialization *per se*. A combination of these approaches appears to be considerably more promising than doing either alone. Although a detailed commentary on R&D policy is outside the scope of this chapter, in general, most countries are underinvested in breakthrough-targeted R&D and applied fundamentals compared to other activities.

Efficient and lower-cost biomass conversion processes are necessary but not sufficient conditions for biomass to make a large contribution to providing energy services (e.g. heat, mobility, light and work). Societies will also need to be willing to invest in developing these technologies. As has been the experience with other promising renewable energy technologies, a period of development and governmental support is necessary to push them over the threshold into economical viability.

Since these conversion technologies are on the verge of viability, continued research and development could be helpful; but extensive deployment is perhaps more important. This will allow operators to streamline new facilities while also reducing the risk perceived by investors looking at an 'unproven' technology. Governments should thus offer supports such as loan guarantees and guaranteed markets for new cellulosic biofuel production facilities today, while also pursuing a vision of more integrated and efficient conversion processes in the future.

6

Long-Term Biofuel Production Potentials

INTRODUCTION

In the future, it may be possible to supply a substantial share of the world's energy needs using cellulosic biomass resources. But exactly how much remains uncertain. Estimates of the planet's biomass production potential vary considerably: in a highly optimistic scenario, biomass energy could well exceed total world energy requirements; in the most pessimistic scenario, it could provide few, if any, additional contributions.

Bioenergy is useful not only as a source of liquid transportation fuel, but also to meet energy needs in a wide range of other applications, including heat and power and the production of bio-based products such as construction materials, clothing and plastics. These different uses will probably combine to increase the harvesting of biomass residues and dedicated energy crops. In some cases, with more limited supplies of biomass, these different uses will compete. When production is more abundant, co-harvesting and biorefineries may permit lower-cost biofuels.

This chapter explores the potential contribution of cellulosic biomass in meeting future world energy needs. In particular, it rests on work done to consider various scenarios in which different levels of biomass energy can be sustainably harvested. Among the most significant variables in determining the future of biomass energy are the ability of agronomists to boost the food and feed yields of conventional crops and their ability to increase the biomass produced by dedicated energy crops. Humanity's collective demand for food and land is equally significant. Competing uses for land and biomass, as well as how we respond to climate change and ecological limitations, will also determine the quantity of biomass energy available for use as biofuels.

BIOENERGY IN THE WORLD ENERGY MIX

Currently, the world consumes roughly 430 exajoules (EJ) of energy per year. Of this, approximately 100EJ – 23 per cent – are used to meet transportation needs. Although it proves difficult to account for all uses of biomass resources in all regions, bio-based energy is estimated to account for roughly 9 per cent of the world's total energy, or nearly 40EJ.[1]

Yet, bioenergy's potential is enormous. Studies suggest that biomass could potentially supply anywhere between 0EJ to more than 1000EJ of energy by the year 2050. In the most optimistic scenarios, bioenergy could provide for more than two times the current global energy demand, without competing with food production, forest protection efforts and biodiversity. In the least favourable scenarios, however, bioenergy could supply only a fraction of current energy use, perhaps even less than it provides today.[2]

Figure 6.1 illustrates the results from 16 studies that have evaluated the potential to harvest energy from biomass.[3] Note that these studies have been mostly 'top-down' evaluations, derived from the anticipated demand for biomass or from extrapolations of the current supply. This book, however, relies primarily on a more 'bottom-up' analysis, which models different scenarios and, importantly, considers the prospects for increasing the yields of both food and energy crops.

Potential sources of biomass energy include energy crops grown on existing agricultural or marginal lands, agricultural residues, organic wastes and forest residues (see Chapter 4). Table 6.1 provides an overview of the potential contribution of each of these biomass types to the global energy supply until the year 2050.[4] Each of these categories is discussed in further detail in later sections of this chapter.

The potential for biomass to meet a substantial share of future transportation energy needs depends upon several key factors:

- *Developments in agronomy.* If farmers are able to increase their agricultural efficiencies significantly, they could not only produce more crop residue per hectare, but they may be able to free up prime agricultural land for the harvesting of dedicated energy crops. If plant breeders are able to develop highly productive energy crops that also grow on marginal lands, these crops could theoretically yield vast new quantities of biomass resources. Foresters might also increase forest yields (see Chapter 4).
- *The size of the human population and its collective appetite for food and land area.* If our ability to boost food crop yields per hectare continues to rise (e.g. through advanced biotechnology and crop-breeding improvements), some analysts surmise that humanity may be able to feed itself with less land than we use today – thus freeing up land for energy crop production. Food and related land requirements will also depend upon the level of demand for more calorie-intensive meat and dairy products, as well as upon the rate of urbanization, which threatens to pave over or subdivide lands otherwise available for biomass cultivation.
- *Energy conversion technologies.* Harvesting, processing and using this bioenergy more efficiently will reduce the need for additional acreage or higher yields. New technologies are making it possible to efficiently produce liquid transportation fuel from cellulosic biomass. Even if this conversion should prove comparably

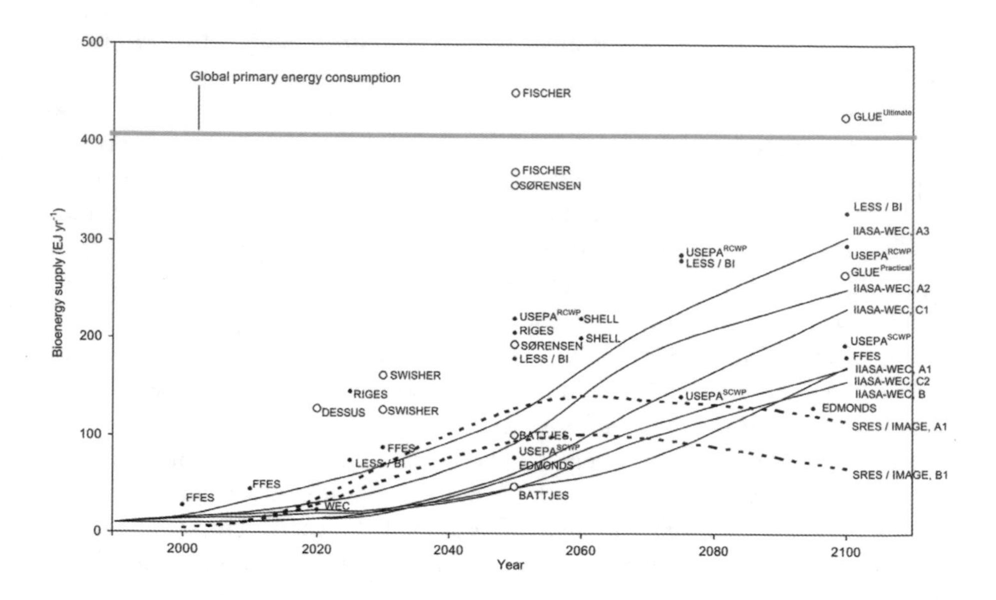

Figure 6.1 *A review of 16 studies on biomass potential*

Resource-focused studies are represented by hollow circles and demand-driven studies are represented by filled circles. IIASA-WEC and SRES/IMAGE are represented by solid and dashed lines respectively, with scenario variant names given without brackets at the right end of each line. The present approximate global primary energy consumption is included for comparison.

References for each of the study abbreviations are as follows:

WEC	WEC (1994)	SORENSEN	Sorensen (1999)
IIASA-WEC	Nakicenovic et al (1998)	RIGES	Johannson et al (1993)
FFES	Lazarus et al (1993)	LESS/BI	Willams (1995)
EDMONDS	Edmonds et al (1996)	LESS/IMAGE	Leemans et al (1996)
SWISHER	Swisher and Wilson (1993)	BATTJES	Battjes (1994)
USEPA	Lashof and Tirpak (1990)	DESSUS	Dessus et al (1992)
GLUE	Yamamoto et al (1999)	SHELL	Shell International (1995)
FISCHER	Fischer and Schrattenholyer (2000)	SRES/IMAGE	IPPC (2000)

Source: Berndes et al (2003)

Table 6.1 *Bioenergy production potentials for selected biomass types, 2050*

Biomass type	*Bioenergy potential (exajoules)*	*Main assumptions and remarks*
Agricultural residues	15–70EJ	• Based on estimates from various studies. • Potential depends upon yield/product ratios, total agricultural land area and type of production system. Extensive production systems require leaving of residues to maintain soil fertility; intensive systems allow for higher rates of residue energy use.
Organic wastes	5–50+EJ[b]	• Based on estimates from various studies. • Includes organic fraction of municipal solid waste (MSW) and waste wood. • Strongly dependent upon economic development and consumption, and as use for biomaterials. • Higher values possible through more intensive biomaterials use.
Dung	5–55EJ (or possibly 0)	• Use of dried dung. • Low value based on current global use; high value reflects technical potential. • Utilization (collection) over longer term is uncertain.
Forest residues	30–150EJ (or possibly 0)	• Figures include processing residues. • Part is natural forest (reserves). • Sustainable energy potential of world forests unclear. • Low value based on sustainable management; high value reflects technical potential.
Energy crop farming (current agricultural lands)	0–700EJ (100–300EJ is more average)	• Potential land availability of 0–4 gigahectares (Gha), though 1–2Gha is more average. • Based on productivity of 8–12 dry tonnes per hectare per year* (higher yields are likely with better soil quality). • If adaptation of intensive agricultural production is not feasible, bioenergy supply could be zero.
Energy crop farming (marginal lands)	60–150EJ (or possibly 0)	• Potential maximum land area of 1.7Gha. • Low productivity is 2–5 dry tonnes per hectare per year.* • Bioenergy supply could be low or zero due to poor economics or competition with food production.
Biomaterials	Minus 40–150EJ (or possibly 0)	• Provide an additional claim on biomass supplies. • Land area required to meet additional global demand is 0.2–0.8Gha. • Average productivity is 5 dry tonnes per hectare per year.† • Supply would come from energy crop farming if forests are unable to meet this demand.
Total	40–1100EJ (250 –500EJ is more average)	• Pessimistic scenario assumes no land for energy farming, only use of residues; optimistic scenario assumes intensive agriculture on better soils. • More average range is most realistic in a world aiming for large-scale bioenergy use.

Notes: * heating value: 19GJ per tonne dry matter; † the energy supply of biomaterials ending up as waste can vary between 20–55EJ (or 1100–2900 million tonnes of dry matter per year). Biomass lost during conversion, such as charcoal, is logically excluded from this range. This range excludes cascading and does not take into account the time delay between production of the material and its 'release' as (organic) waste.

Source: see endnote 4 for this chapter

inefficient, however, it will probably raise the economic potential for producing biomass by adding demand from the transportation energy sector.

These and other key elements determining bioenergy potential are laid out schematically in Appendix 5.

BIOMASS RESIDUES AND ORGANIC WASTES

As discussed in Chapter 4, vast amounts of biomass residue exist that could potentially be used as feedstock for biofuel production. Much of the plant matter produced by common crops is left on the fields after harvest, and a large portion of this decomposes into carbon dioxide rather than returning to the soil. Similarly, forestry practices leave behind large quantities of unharvested wood, and fire mitigation practices have allowed forest underbrush to accumulate. Of the biomass that is already being harvested, large amounts of residues are generated at agricultural processing facilities and forestry mills, including sugar cane bagasse, rice hulls, nut shells, sawdust and black liquor (at paper mills). In urban areas, cellulosic residues include portions of municipal solid waste, grass clippings, and wood from tree trimmings and land clearing activities.

Studies of biomass potential estimate that together these residues and organic wastes could supply between 40EJ and 170EJ of energy per year on a global basis. Although additional energy would be required to process the biomass into liquid fuels, this range of potential biomass energy is roughly comparable to the 100EJ used worldwide to meet transportation needs today. Achieving the high end of the range would require boosting the productivity of crops and forests (to increase residue availability) and developing economical ways to simultaneously harvest conventional food crops and crop residues, as well as timber and forest residues.[5]

Residues and wastes have several advantages over dedicated energy crops. Most of them would require no additional land acreage since they are typically pre-collected into piles at large agricultural and forestry facilities, and often represent 'waste' that must otherwise be disposed of. As a result, this feedstock is cheaper and is likely to be the first source of biomass to be tapped. Already, the wood products industry uses most lumber residues and much of the forestry residue in Europe and the US for processing purposes and to generate co-products such as wood chips and fibreboard.[6]

For sustainability reasons, however, estimates of potential bioenergy from waste and residues would probably be lower than those suggested above. In general, it is a good idea to retain some portion of biomass residue in the field or forest to hold carbon, water and other nutrients in the soil, and to provide habitats for various species. Leaving a protective amount of residue behind is especially important on steep slopes or on ecologically sensitive sites that have particularly erodible soils, or are near riparian areas.[7]

Even taking these considerations into account, studies suggest that large quantities of residue and waste biomass would still be available for harvesting. One study exploring the biomass potential in the US found that harvesting residues from agriculture and forestry alone could provide more than 700 million tonnes of biomass annually, or enough energy to displace more than 20 per cent of the country's current petroleum fuel use.[8] Global estimates of this potential energy have ranged from 15EJ to more than 200EJ.[9]

Agricultural residues

The production potential for agricultural residues depends upon the various yields of different agricultural products, the total agricultural land area and the type of production system. Less-intensive management systems require the reuse of residues for maintaining soil fertility, reducing the total amount that can be sustainably removed. More intensively managed systems, meanwhile, allow for higher use rates of residues, but also typically rely on crops with lower crop-to-residue ratios, such as corn.[10]

Estimates of the energy potentially supplied by agricultural residues vary from around 15–70EJ per year. The latter figure is based on the regional production of food multiplied by the co-production of residue and the amount of residue that can be sustainably harvested.[11] These figures do not take into account potential alternative uses for agricultural residues. Hall et al (1993) estimate that just by harvesting residues from the world's major agricultural crops (e.g. wheat, rice, corn, barley and sugar cane), a 25 per cent recovery rate could generate 38EJ of bioenergy.[12] Worldwide, large stores of rice hulls, coconut husks and sugar cane bagasse currently exist, and in some instances this is already used to provide a local source of power (see Table 6.2).[13]

Organic wastes

Organic wastes, such as the organic fraction of municipal solid waste (MSW) and waste wood (e.g. demolition wood), are a particularly attractive source of biomass energy because they can have a 'negative' price. In other words, collecting and utilizing these can result in savings from landfill tipping fees.[14]

Estimates of the bioenergy potential of wastes depend strongly upon assumptions about economic development, consumption and the use of biomaterials; nevertheless, the ranges projected for MSW in the longer term (beyond 2040) are between 5EJ and 50EJ (higher values are possible when biomaterials are more intensively used and then made available for recycling). Translated into biofuel production, a city of 1 million people could theoretically provide enough feedstock to produce about 430,000 litres of ethanol per day (see Box 6.1).[15]

Table 6.2 *Annual bagasse availability in the top ten sugar cane-producing countries*

Country	*Bagasse availability (million tonnes)*	*Potential bioenergy (exajoules)*
Brazil	67.3	0.521
India	56.8	0.439
China	28.0	0.216
Australia	18.0	0.139
Thailand	17.8	0.138
Mexico	16.4	0.127
Cuba	12.6	0.098
US	12.2	0.095
Pakistan	12.1	0.093
South Africa	8.3	0.064

Note: Assumes a yield of 3.26 tonnes of fuel bagasse per tonne of cane sugar produced, at 50 per cent humidity.

Source: see endnote 13 for this chapter

Box 6.1 How much ethanol could the municipal solid waste from a city with 1 million people produce?

The average person in the US generates approximately 1.8kg of municipal solid waste (MSW) per day. This typically contains about 75 per cent of mostly cellulosic organic material, including waste paper, wood wastes, cardboard and waste food scraps. Thus, a city with 1 million people produces around 1800 tonnes of MSW in total, or about 1300 tonnes per day of organic material.

With technology that could convert organic waste to ethanol, roughly 330 litres of ethanol could be produced per tonne of organic waste. Thus, 1300 tonnes per day of organic waste from a city with 1 million people would be enough feedstock to produce about 430,000 litres of ethanol per day, or approximately 150 million litres per year. This is enough fuel to meet the needs of more than 58,000 people in the US, 360,000 people in France or nearly 2.6 million people in China at current rates of per capita fuel use.

Source: see endnote 15 for this chapter

Animal excrement, too, could supply a significant amount of bioenergy, perhaps between 5EJ and 50EJ per year. Yet, even more so than crop residues, it is valuable for improving and fertilizing soils. It can also be harder to recover and can cause health problems among those who work with it.[16]

Forest residues

The sustainable energy potential of the world's forests is also uncertain. However, a recent evaluation of forest reserves and trends in global wood demand concluded that even the highest projected demand for wood products could be met without further deforestation, suggesting that forest residues can be an ecologically benign feedstock for energy production.[17]

In terms of potential bioenergy provision, studies show that forest residues can contribute between 30EJ and 150EJ per year in 2050. The most promising producer regions are Latin America and the Caribbean, the former Soviet Union and, to some extent, North America. Key variables include the demand for industrial roundwood and fuelwood (obtained both legally and illegally), plantation establishment rates, natural forest growth and the impact of technology and recycling.[18]

Despite this potential, the amount of energy that can be obtained from forest residues and other waste biomass resources will be limited in comparison to energy crops; moreover, these reserves will probably be depleted first as demand for bioenergy grows. Finland, which has focused on harnessing biomass energy for many years, has already used all of its accessible residues and wastes and is now importing wood for producing energy.[19]

ENERGY CROPS AND LAND AVAILABILITY

Energy crops, bred and cultivated to produce the maximum amount of biomass energy per hectare, hold the greatest promise for increasing the availability of bioenergy. As discussed in Chapter 4, fast-growing grasses and trees cultivated on plantations can be highly productive. They can grow on pastureland, degraded land and lands otherwise considered undesirable for agriculture. Agronomists could potentially increase energy crop yields by twice or more.[20]

Depending upon yield improvements and the quality of land available for their production, high-yielding energy crops could provide the bulk of future bioenergy supplies, ranging from 0 to 850EJ per year (for farming on current agricultural as well as marginal lands). At the high end, this would represent as much as twice the current global energy use (430EJ) and about eight times current transportation energy use (100EJ).[21] At the low end, increased demand for food, coupled with the failure to increase agricultural efficiencies quickly enough, could eliminate the availability of even marginal lands for energy production. It is also likely that harvesting bioenergy crops on particularly unproductive lands will prove uneconomical.

Growing energy crops on marginal lands

So-called 'marginal' lands cover an estimated 1 billion to 3 billion hectares, or about 7 to 20 per cent of the Earth's land surface.[22] They may be considered undesirable for agriculture for a variety of reasons, including high acidity; salinity; high levels of phosphorus, aluminium or ferric oxides; poor drainage; erodability; shallowness; or tendency to expand and contract.[23] In many cases, land that was once highly productive has been made marginal by over-harvesting or over-grazing.[24]

Studies estimate that the marginal land area available for cultivating energy crops is between 100 million and 1 billion hectares.[25] The remainder lacks sufficient water or soil quality to be economical for energy crop harvesting. Other sites, such as farmland set aside to protect water quality, may have yields equivalent to those possible on high-quality agricultural land. The range of these yields will depend upon the price of biomass and the ability of agronomists to develop crops particularly suited to specific conditions.

Cultivating energy crops on marginal lands could be either detrimental or beneficial to these areas, depending upon what the crops replace and how they are managed. For instance, if energy crops replace more varied ecosystems with monocultures or are grown using fertilizers and pesticides, these practices could cause damage to biodiversity or ecological systems. Large-scale expansion of energy crop production would also lead to an increase in water use for irrigation, which in some countries would further strain stressed water resources. Issues of water supply and demand need to be better incorporated within future assessments of bioenergy potentials.[26]

At the same time, energy crops could be a means of restoring degraded lands, improving soil quality and restoring water cycles in regions affected by desertification. Virtually all of the promising new energy crops are perennials, with roots that hold soil in place all year round. Such crops could be planted as part of efforts to protect riparian areas or reduce erosion. Energy cropping could also serve as an ecological intermediary between dedicated agricultural land and dedicated conservation land (for more on the environmental risks and benefits associated with energy crop cultivation, see Chapter 12).[27]

Growing energy crops on agricultural land

The greatest theoretical potential for producing bioenergy lies on land that is useful for agriculture as well. Like conventional food and feed crops, energy crops grow best on prime farmland. While some analysts assume that the growing human demand for food will require the use of all available cropland, this is not necessarily the case. Whether and how much agricultural land could be made available for energy cultivation is a central uncertainty in the future of next-generation biofuels (see Chapter 8 for a discussion of food versus fuel).

Many assessments of biomass potential have failed to consider the extent to which increasing agricultural efficiencies could contribute to bioenergy production. Agricultural yields per hectare have been improving by around 1 to 2 per cent annually for decades, and the meat industry has increased yields even faster by feeding cattle with grains, such as corn, rather than wild grasses.[28] So far, agriculture's ability to outpace population growth and the demand for food has led to regional overproduction, one of the primary reasons governments have launched conventional biofuel programmes to begin with (see Chapter 1). Although the increase in yields is expected to slow, many experts still expect output to keep rising, particularly with wider use of hybrid crops, chemical inputs, irrigation and mechanization.[29]

Unfortunately, many of the key ways of improving agricultural efficiency can also be socially and environmentally destructive. Growing crops where they are most productive can lead to monocultures and the loss of genetic and biological diversity. Mechanization can reduce the employment opportunities in rural communities. The use of fertilizers and pesticides can contaminate soils and waterways. Transitioning from pastures to feedlots can be unhealthy for the animals. And genetically modified organisms represent a new degree of tampering with both genes and ecosystems. Nonetheless, such techniques are often critical for improving the efficiency of human land use, and may be essential if the goal is to increase agricultural output per hectare of land.[30]

Improving yields in sub-Saharan Africa

The opportunity for increasing agricultural efficiencies is greatest in developing countries, where yields can be ten times lower than those on comparable land in industrialized countries.[31] African countries stand out for their potential to improve land productivity. Mozambique, for example, has abundant rainfall and fertile soils; but the small-scale, low-input nature of its agriculture has typically meant lower yields. Average cereal yields in the country are just 1 tonne per hectare, well below the 8–12 tonnes per hectare yields attainable in high-input agricultural systems in industrialized countries.[32] Batidzirai et al (2006) estimate that farmers in Mozambique could increase their productivity by seven times with just moderate use of agricultural technologies, such as fertilizers, pesticides, selected seeds and large-scale harvesting practices. Similar yield increases are a potential elsewhere in Africa as well (see Table 6.3).[33]

Implementing more efficient agricultural practices across sub-Saharan Africa could free up as much as 700 million hectares of surplus agricultural land. By planting highly productive plantations of eucalyptus and other energy crops on this land, sub-Saharan Africans could harvest as much as 347EJ of bioenergy.[34]

It should be noted that both local agencies and international development organizations have sought to improve agricultural yields in Africa for decades; however, due in large part to continuing poverty, these efforts have met with only limited success.[35] Moreover, unless such agricultural changes are managed carefully, they could cause considerable social and ecological disruption. Nevertheless, by generating additional revenue for rural communities, bioenergy crops could theoretically help to improve conventional agricultural yields by enabling

Table 6.3 *Potential yield increases by 2050 for four agricultural scenarios*

Region	*Scenario 1*	*Scenario 2*	*Scenario 3*	*Scenario 4*
		(factor of increase, 1998 = 1)		
North America	1.6	2.3	2.3	3.2
Oceania	2.4	3.7	3.7	4.6
Japan	2.7	2.8	2.4	3.0
West Europe	0.9	1.5	1.3	1.9
East Europe	2.1	3.3	3.3	4.1
Former Soviet Union and Baltic States	3.2	5.4	5.3	6.7
Sub-Saharan Africa	5.6	6.2	6.2	7.7
Caribbean and Latin America	2.8	3.6	3.5	4.5
Middle East and North Africa	1.4	2.3	2.3	2.9
East Asia	2.3	2.7	2.5	3.2
South Asia	3.7	4.5	4.5	5.6
World	2.9	3.6	3.6	4.6

Notes: Table 6.3 shows the potential increase in crop yields from 1998 to 2050, based on the average for all crops analysed. Scenarios 1 to 4 represent four different but relatively optimistic possibilities: scenario 4 assumes that animals do not use grazing land and utilize grain foods efficiently, while crops are irrigated and have 'extremely' high yields; scenarios 2 and 3 assume that crops are irrigated but have only 'very high' yields, and scenario 2 assumes that some meat will be raised via grazing. Scenario 1 also assumes some animal grazing and 'very high' yields, but no irrigation.

Source: see endnote 33 for this chapter

greater investment in agricultural technologies. Since capital and investments in infrastructure are critical to improving agricultural yields, such schemes warrant further analysis (see Chapter 8 for more on the benefits of biofuel production for rural communities).[36]

Improving yields in Central and Eastern Europe

There are also opportunities for improving agricultural efficiencies in Central and Eastern Europe. Following the collapse of the Soviet Union, the farm acreage in many of these historically agricultural countries – including Ukraine, Poland and Romania – declined. Most of these countries also continue to use agricultural practices that are inefficient compared to those in Western Europe. As these countries enter the European Union and adopt the EU's Common Agricultural Policy (CAP), efficiency levels are likely to improve, leading to greater mechanization and use of inputs, and higher yields. Bioenergy crops may provide a new market for European farmers displaced by this transition.[37]

Shrinking grazing lands

Grazing lands are considered key sites for future bioenergy crops. Studies suggest that transitioning livestock from pastures to more concentrated feedlots could free up many hectares of semi-agricultural and marginal land, either for production of energy grasses or partial conversion to woodlands.[38] However, such shifts are controversial. Confining animals in smaller areas can increase the risk of both animal and human disease, create enormous pollution issues and be unhealthy for the animals.[39] It also entails using a larger share of human food crops for animal feed. On the other hand, the transition from grazing land to perennial energy crop cultivation could permit the restoration of healthier grassland ecosystems (see Chapter 12).

Rising human demand for food

To free up agricultural land for the cultivation of energy crops, agricultural efficiency will need to rise faster than the human appetite for food. Population growth is the most significant driver behind increased food demand. While human numbers are expected to stabilize by mid century, the 'medium-range' forecast of the United Nations still puts world population at around 8.9 billion people in 2050, up from 6.5 billion in early 2006.[40] Over the next three decades, food demand is projected to increase by 1.5 per cent per year – with 1 per cent of this annual growth due to continued population increase, according to the UN Food and Agriculture Organization (FAO).[41]

The rest of the growth in projected food demand comes from changes in diet as more of the world's people are able to afford calorie-intensive meat and dairy products. Producing these items can require large resource inputs, including additional land to grow crops for animal feed.[42]

All of the scenarios presented in this chapter have accounted for the increasing human demand for food, considering both the forecasted size of the global population and the expected dietary preference for meat.

Climate change

Projected climatic changes in the decades to come only add to the uncertainty about humanity's potential to boost agricultural yields. Shifts in the hydrological cycle may cause droughts in some areas, while bringing excessive or unseasonal rains to others. Temperature increases could eventually shift bands of vegetative growth away from the Equator, requiring regional adaptations in agricultural practices. Some areas might even benefit as increased levels of carbon dioxide stimulate faster plant growth: Canada's plains, in particular, could benefit from longer growing seasons and more plentiful rain. However, other regions – for example, in Southern Europe or Africa – would probably suffer.

In tropical areas, in particular, higher temperatures could contribute to droughts and interfere with the ability of plants to pollinate. Recent studies show that rising temperatures associated with climate change may be detrimental to crop production. A study that evaluated data collected by the International Rice Research Institute between 1979 and 2003 found that increasing the mean minimum temperature by just 1°C during the dry cropping season was associated with a 10 per cent reduction in grain production.[43] A second study in the US, relying on data between 1982 and 1998, came to similar conclusions – with a 1°C temperature increase corresponding to a decrease of roughly 17 per cent in both corn and soybean yield.[44]

Despite the huge risks of climate change – and in particular shifts that occur more rapidly than most species and ecosystems can adapt – bioenergy crops may offer a strategy for mitigating the damage. Perennial energy crops, with their semi-permanent roots, can help soils to survive both floods and droughts, and hardy energy crops can be an alternative for farmers in areas where the climate is no longer appropriate for their traditional crops.[45]

GLOBAL SCENARIOS FOR BIOMASS PRODUCTION

Several long-term models have attempted to assess the potential of biomass to meet world energy needs, and provide interesting results. As discussed, bioenergy potential can vary greatly depending upon such factors as the availability of marginal land, the relationship between increasing agricultural productivity and increasing food demand, the extent of international food trade, and the effects of climate change.

In 2004, Hoogwijk detailed four scenarios for bioenergy production based on models published in the Intergovernmental Panel on Climate Change's (IPCC's) *Special Report on Emissions Scenarios* (SRES) (see Table 6.4).[46] The models describe futures with very different social, economic, technological, environmental and policy developments, constructed along two dimensions: the degree of globalization versus regionalization, and the degree of orientation towards material versus social and ecological values. In the more globalized scenarios, population growth is lower (due to declining birth rates) and agricultural efficiencies increase faster (reflecting the fact that crops are more often grown with 'modern' technologies, in their best environments, and then traded internationally). In scenarios that place greater emphasis on social and environmental values, people east less meat and larger portions of land are set aside for conservation purposes.

In the so-called B1 scenario (the environmentally and globally oriented combination), the estimated potential for bioenergy production exceeds the total primary energy demand for that particular future. In contrast, the A2 (materialistic and regionally oriented) scenario results in the highest total energy demand and the lowest biomass potential. Thus, while the share of biomass in the total energy

Table 6.4 *Schematic description of the four scenarios used by Hoogwijk*

MATERIAL/ ECONOMIC

GLOBAL ORIENTED | **REGIONAL ORIENTED**

A1

Food trade	Maximal
Consumption of meat	High
Technology development	High
Average management factor for food crops	2050: 0.82 2010: 0.89
Fertilization of food crops	Very high
Crop intensity growth	High
Population	2050: 8.7 billion 2100: 7.1 billion
Gross domestic product (GDP)	2050: US$24,200 billion per year 2100: US$86,200 billion per year

A2

Food trade	Low
Consumption of meat	High
Technology development	Low
Average management factor for food crops	2050: 0.78 2100: 0.86
Fertilization of food crops	High
Crop intensity growth	Low
Population	2050: 11.3 billion 2100: 15.1 billion
GDP	2050: US$8000 billion per year 2100: US$17,900 billion per year

B1

Food trade	High
Consumption of meat	Low
Technology development	High
Average management factor for food crops	2050: 0.82 2100: 0.89
Fertilization of food crops	Low
Crop intensity growth	High
Population	2050: 8.7 billion 2100: 7.1 billion
GDP	2050: US$18,400 billion per year 2100: US$53,900 billion per year

B2

Food trade	Very low
Consumption of meat	Low
Technology development	Low
Average management factor for food crops	2050: 0.78 2100: 0.89
Fertilization of food crops	Low
Crop intensity growth	Low
Population	2050: 9.4 billion 2100: 10.4 billion
GDP	2050: US$9400 billion per year 2100: US$27,700 billion per year

ENVIRONMENT/ SOCIAL

Source: see endnote 46 for this chapter

Note: GDP expressed in 1995 US$ values

mixture could reach 100 per cent in an environmental global scenario, it would always be limited in a material regional world (contributing only around 22 per cent of total energy).[47]

The potential for using abandoned agricultural land is the largest in the two globally oriented scenarios (A1, B1). In these cases, the potentials are comparable to the present world energy consumption of about 430EJ per year, and the regions with the most significant potentials are the former Soviet Union, East Asia and South America. One reason for higher use of marginal land in these scenarios is that both describe a world where population growth is low and technical development is high, thus avoiding issues of food competition or poor agricultural efficiency. In the global environmental scenario (B1), competing land-use options such as conservation restrict biomass potential more than in the global material scenario (A1), although large quantities of land are still potentially available.[48]

Of the four scenarios, the regional materialistic world (A2) has the lowest bioenergy potential. It is a world with rapid population growth, where agricultural trade does little to improve inefficiencies in global land use.

Geographic potentials for harvesting biomass energy

In a different study, Smeets et al (2004) also modelled four scenarios for 2050 and came to several useful conclusions about the potential for bioenergy production in different geographic regions. The results are presented in Appendix 6 and are summarized as follows.

In terms of overall biomass production, Latin America, sub-Saharan Africa and Eastern Europe are particularly promising regions. Oceania and East and Northeast Asia also have potential over the longer term, as population growth in these regions slows by 2030 and as rapid technological progress in agriculture leads to substantial productivity increases.

The largest potential surplus cropland is in Latin America and the Caribbean and in sub-Saharan Africa. Potential cropland in Latin America and the Caribbean could provide for 87–279EJ of bioenergy per year, and potential cropland in sub-Saharan Africa could provide for 49–347EJ of bioenergy per year. The high bioenergy potential in these regions stems mainly from the large areas of surplus pastureland currently in use, as well as currently inefficient production systems and land use.

The most land-stressed regions are the Near East and North Africa, South Asia and parts of East Asia. These regions would need to meet their energy needs through imports from other regions. However, they have some bioenergy production potential in areas currently classified as not suitable for crop production.

The most robust of all regions is the Commonwealth of Independent States (CIS) and the Baltic States. It has a considerable biomass production potential equal to one fifth to one third of the total agricultural land use, with a bioenergy potential of 83–269EJ per year. Due to the collapse of communism and subsequent economic restructuring, income, consumption, production and agricultural yields in the region have all decreased, and it will take several decades before consumption levels again reach levels common in the Soviet period. In addition, the population is projected to decrease by 2050. Consequently, the agricultural land area is relatively large compared to the projected demand for food.

The industrialized region with the largest potential to boost yields and reduce the area of conventional agricultural land is Oceania. Between 42 to 84 per cent of the total agricultural land in use in this region in 1998 could be abandoned and used for bioenergy production, equal to 40–114EJ per year.

North America, meanwhile, has the potential to produce 39–204EJ of bioenergy per year, despite a projected increase in population. Processing forest and agricultural residues accounts for 19–24EJ of this annual potential.

Worldwide, the surplus production potential of wood from natural forests is estimated at 20–36EJ per year. However, various limiting factors, such as the exclusion of undisturbed forests or the possibility that forest use is economically unattractive, may reduce or eliminate this potential.

Exploiting the bioenergy potential described here would require major efforts, particularly significant improvement in agricultural efficiency in developing countries. It is uncertain to what extent and how rapidly such transitions can occur. Under less favourable conditions, the regional bioenergy potential could be quite low. It should be noted that technological developments in the conversion of biomass into transportable pellets and liquids, as well as long-distance biomass supply chains, dramatically improve the competitiveness and efficiency of bioenergy.[49]

Cost–supply curves

The relative costs of producing energy crops will develop over time. They may increase due to the rising cost of labour or decrease due to improvements in yield. Based on the four scenarios by Hoogwijk (2004) (see Table 6.4), global cost–supply curves have been constructed for the year 2050. They show that in 2050, a significant share (130–270EJ per year) of the biomass production potential may theoretically fall below €1.65 (US$2) per gigajoule (GJ). This price is comparable to the upper level of the price for coal. In large areas in the former Soviet Union, energy crops could potentially be produced at even lower costs. The lowest costs are in East Africa, at only €0.66 (US$0.8) per gigajoule (achieved in the A1 materialistic–globally oriented scenario).

Improving the models

While the literature examining the global potential for biomass energy has improved in recent years, it needs further refinement. In particular, a next step in these assessments is moving beyond more abstract analyses and making specific estimates of the economic and implementation potential for bioenergy. In particular, data is lacking on the use and sources of fuelwood; feed composition and feed conversion efficiencies; production capacities of natural pastures and the impact of various management systems; the extent and severity of environmental degradation and the impact of various management systems; sustainable forest management; and the impact of wood harvests.[50]

There is also a lack of information on the relationship between socio-economic systems and land-use patterns and yields – in particular, the impact of large-scale energy crop production on the cost of land and other production costs. Furthermore, substitution correlations between various production factors (substitution elasticities) are relatively unknown and should be addressed in national and sub-national case studies. Estimates of biomass potential need to be presented carefully, emphasizing the range of variables involved and the significant risks that large-scale biomass cultivation present to both environmental quality and social stability.[51]

The next step in assessing the realizable potential of biomass energy is to begin implementing actual projects for harvesting sustainable biomass. Projects that harvest crop residues or attempt to remediate degraded land with bioenergy crops can help producers to gain practical experience and results. This is a key point: the true potential for biomass energy will become clearer as it is cultivated in more situations and on a larger scale.

COMPETING USES FOR BIOMASS

As the above studies have shown, the potential to harness large quantities of global biomass for energy is enormous. However, not all of this will be converted into liquid biofuels for transportation. Among the important competing uses for biomass are heat, electricity and materials. Indeed, refining bioenergy into transportation fuels may be a less energetically efficient option than using bioenergy for heat and power. Likewise, biofuels may be less valuable than other biomaterials, such as bio-based plastics, fabrics and chemicals, whose production is likely to take precedence.[52]

Nonetheless, biofuels are likely to draw on a significant portion of future biomass supplies. While there exist a range of other ways to produce heat and power, including wind, solar and nuclear power, few alternatives to petroleum exist for producing liquid fuels. And although the production of biomaterials will probably continue to increase in the future, these co-products create less of a rivalry with biofuels. Not only are they likely to use only a fraction of total biomass supplies, but theoretically they can also be recycled into biofuels after the end of their productive life.[53]

Heat and power

When considering the use of biomass for heat and power versus fuel, it is important to note that, unlike the burgeoning options for producing biofuels, there are already well-established and widely used technologies for converting lignocellulosic biomass into heat and power. Moreover, just as new biofuel conversion technologies are becoming economical, gasification technologies such as integrated gasification/combined cycle systems are simultaneously making the conversion of biomass into electricity more competitive.[54]

Nonetheless, there is not a considerable difference in the cost of using biomass for electricity versus using it to produce the next generation of biofuels. Indeed, both routes can achieve competitive cost levels compared to fossil fuel-based power and fuel production when primary biomass resources are available at about €1.65 to €2.48 (US$2 to $3) per gigajoule. Nevertheless, fluctuations in the price of primary fuels (coal, natural gas and oil) can strongly affect the attractiveness of using biomass for one application or the other. A high price for petroleum and a low price for coal would combine to make biofuel production more desirable.[55]

In the longer term (beyond 2020), a scarcity of cheap conventional oil, combined with rising demand for transport fuels, will make next-generation biofuels an increasingly attractive use of biomass compared to power and heat. For strategic reasons as well, developing an alternative fuel to petroleum may be the preferred route for use of biomass resources. As an energy resource, oil has a far more constrained supply than coal, the dominant resource for power generation (see Chapter 7 for a discussion of oil security).

If the primary goal is to reduce the threat of climate change, however, biomass can reduce carbon emissions more by displacing coal (for electricity) than by displacing petroleum (for fuel). Co-firing biomass in coal-fired power stations has a higher avoided emission per unit of biomass than does using biomass to displace diesel or gasoline. This is true even when next-generation biofuels are concerned (see Chapter 11 for more on climate mitigation issues).[56]

Even so, there are far fewer alternatives for reducing the greenhouse gas emissions associated with liquid transportation fuels than for reducing those associated with electricity conversion. Alternative energy options for power generation include wind energy, photovoltaics and other solar technologies, geothermal power, small-scale hydropower, natural gas, nuclear power, and potentially carbon capture and storage. It is thus likely that using biomass for transport fuels will gradually become more attractive from a carbon-mitigation perspective. In the shorter term, however, careful strategies and policies are needed to avoid brisk allocation of biomass resources away from efficient and effective utilization in power and heat production.

Policy supports currently favour the production of biofuels. As of early 2006, tax exemptions in Europe were supporting biofuel production more than feed-in tariffs and carbon taxes support biomass for heat and power generation. Although these policies currently favour conventional liquid biofuels, in the future they could support next-generation biofuels converted from lignocellulosic biomass.[57]

It should be noted that all next-generation biofuel production facilities can be equipped to generate all or a portion of their own heat and power from biomass, in addition to producing biofuels. For gasification facilities, which can produce both electricity and liquid fuels, the output of power and fuel can vary over time, possibly leading to operational and economic advantages for plant owners. New large-scale conversion of biomass can therefore add to the flexibility of the energy system as a whole, bringing an important strategic benefit for the coming decades.[58]

Biomaterials

The use of biomass for materials ('biomaterials') can be an important competitor to applications of biomass for energy. Furniture, building frames, packaging, clothing and paper are already significant biomaterials, and, as discussed in Chapter 5, bio-based plastics and fabrics are likely to become more important in the future. Trends indicate that the future demand for biomass as a feedstock for materials will surpass the historic demand for biomass as a source of energy. However, this demand does not necessarily rival the supply of biofuels.[59]

The use of biomass for biomaterials will increase both in well-established markets (such as paper and construction) and in large new markets (such as biochemicals and plastics, and use of charcoal for steel-making). Given that many biomaterial applications have a high economic value, this adds to the competition for biomass (particularly forest biomass) and land resources for producing woody biomass and other crops. On the other hand, increased demand for biomass for energy will increase feedstock prices for material applications. Such price effects are already observed with forestry and paper pulp products – in particular, triggered by financial support of bioenergy applications in places such as Europe.[60]

However, biomaterials can be recycled and eventually down-cycled into feedstocks for biofuels. Construction wood ends up as waste wood, paper as waste paper, and bioplastics as municipal solid waste. Such waste streams are also biomass feedstocks and are often available at very low or even negative costs. Dornburg and and Faaij (2005) have investigated the climate and economic impacts of 'cascading' biomass in this way. They conclude that this strategy for using biomass could provide very large carbon dioxide (CO_2) reductions when compared to using biomass directly for energy. Although this is not true for every combination of biomaterial and biofuel, such recycling is a promising strategy.[61]

CONCLUSION

The potential for harnessing large quantities of biomass energy is very large. Biomass can theoretically rival fossil fuel supplies, although the longer-term potential for biomass resources varies widely and depends upon factors that can be hard to predict. If the human population increases only slightly, vegetarianism remains a prevalent diet in much of the world and agricultural yields continue to increase, then there could be a large reservoir of biomass energy to tap. If, however, the human population doubles and the demand for meat and dairy products continues its rapid rise, while climate change and limited investment in poor rural areas impede growth in food crop yields, then there may be only very limited supplies of biomass energy.

The large-scale cultivation and harvesting of photosynthetic energy brings with it a different set of challenges and concerns. While fossil fuels pose a greater threat to greenhouse gas concentrations, biomass fuels potentially pose a larger threat to wild ecosystems, soil quality and water use. At the same time, should biofuels be cultivated carefully, they might also bring net ecological benefits. Perennial crops, in particular, can bring advantages: they have the potential to sequester carbon, diversify the habitat of old farmland, and maintain or restore the soils of degraded or marginal lands. At best, the value of biomass will provide an economic motive for prudent ecosystem management.

The advancement of agriculture is one key to achieving the high end of biomass's potential.

Policy-makers will want to evaluate the best ways to sustainably improve the productivity of agricultural systems and to encourage their adoption.

Freer international trade in both food and biomass could help to expedite the development of bioenergy resources, as both food and energy crops are grown in their best conditions and then traded. However, domestic environmental laws and international certification standards would be necessary to ensure that biomass is not cultivated recklessly (see Chapter 18).

Biomass should also be encouraged as part of a strategy to aid rural communities. Should biomass be harvested for export, it may be cultivated at the expense of rural residents, who may be removed from their land by large-scale plantation efforts. At the same time, a new market for agricultural goods could bring additional income to farmers and provide them with the capital they need to increase their crop efficiencies and yields (see Chapter 8).

There are considerable volumes of forest and agricultural residues and organic wastes available; but large-scale harvesting of dedicated energy crops is still minimal. The rapid development of global bioenergy will mobilize residue and waste resources quickly since these are often cheaper and more readily available. But over the next 10–20 years, the active production of energy crops will probably become more widespread. It is important that the critical lessons about sustainable harvesting of bioenergy crops be learned during this early expansion, which, notably, will occur even before the next generation of biofuel conversion technologies permits significant production of cellulosic biofuels.

In order to refine assessments of the true potential for cultivating biomass energy, and in order to learn how best to harvest biomass sustainably, it is important to begin accumulating and evaluating practical experience. Analyses could then extrapolate from real experience with introducing biomass production systems into the agricultural sector and onto marginal lands, while also testing the socio-economic effects of different harvesting schemes on the economic health of rural communities in different contexts. As biofuel production technologies mature, so should agronomic techniques for harvesting cellulosic biomass.

As a feedstock for producing liquid transportation fuels, biomass is perhaps the best candidate for replacing petroleum fuels. The technologies available to convert lignocellulosic fibres into liquid fuels have so far produced only negligible amounts of fuel for transportation, and it will take some time to ramp up the production of such fuels. In the meantime, biomass is already used widely as a fuel for producing heat and generating electricity, and ramping up the available supply of bioenergy will probably improve the prospects for the future conversion of this biomass into liquid fuels for transportation. Policy-makers might thus expedite the use of biomass for biofuels by increasing the use of biomass for all purposes.

Part III

Key Economic and Social Issues

7

Economic and Energy Security

Introduction

Mobility is critical to the supply of goods, food and labour. Yet, the transportation infrastructure that underpins today's economy is vulnerable, due in part to its overwhelming dependence upon a single fuel source. Petroleum fuels provide an estimated 96 per cent of global energy for transport.[1] Oil reserves are concentrated in a relatively small number of countries, many of them beset by economic and political problems that threaten their stability. Myriad political, environmental and economic factors are making trade between oil exporters and highly dependent importing nations increasingly tense and vulnerable to disruption.

Biofuels offer a large-scale alternative to petroleum-based fuels for transportation. They have the ability to diversify fuel supplies and thereby alleviate pressures on the oil market. Biofuels can reduce the level of oil import dependence in many nations and can strengthen their rural economies by redirecting spending that otherwise would have been sent abroad. However, biofuels alone cannot meet increasing global energy demands for transport, and must be paired with greater efficiency and other demand-reduction strategies if a more sustainable transportation economy is to be achieved.

This chapter discusses the role of oil in the global economy and explores ways in which expanded biofuels development could benefit economic and political relations. It describes how changing oil prices and increased petroleum demand contribute to economic and civil insecurity, and points to some of the potential advantages biofuels bring in these areas. The chapter also provides a comparison of different government supports for oil versus biofuel production and examines how biofuels may affect global prices for food and other agricultural commodities.

Rising demand for liquid fuels

Over the past 50 years, world oil consumption has surged steadily upward, and in recent years, oil producers have begun to have difficulty keeping up with the pace of demand, which is now roughly 85 million barrels per day (see Figure 7.1).[2] This has pushed real oil prices to their highest levels in two decades, causing increasing economic hardship, particularly for poorer oil-importing countries. Unlike in the past, when oil shortages were met with increased production from Persian Gulf producers, oil producers are now operating with little excess capacity to release when supplies become tight.[3]

Rising demand for liquid fuels has led to increased investment by oil companies and is projected to add 3.6 million barrels of oil production capacity in 2006, 3.7 million in 2007 and 3.1 million barrels of capacity in 2008; however, this is far short of rising energy demand, and ExxonMobil reports that more than half of the world's hydrocarbon needs over the next 15 years have yet to be found.[4] Notwithstanding, oil price forecasts vary, and some predict the return of lower prices in the coming years, a trend that could have serious implications for investments in biofuels and other liquid fuel alternatives. Factors including climate priorities and dependencies upon oil, as well as security, will all come to bear on the oil market of the future.

The implications of this tight oil market were evident in late 2005 and early 2006 as oil prices ranged between US$60 and $70 a barrel (€50–58 a barrel). The price spike that followed two major hurricanes in the US Gulf Coast region demonstrated the additional vulnerability of today's oil-based fuels market (see Figure 7.2).[5] The effects of skyrocketing oil prices were apparent in Latin America, where importing countries witnessed rising inflation and a slowdown in consumer demand.[6] And in parts of Asia and Africa, public unrest ensued as governments

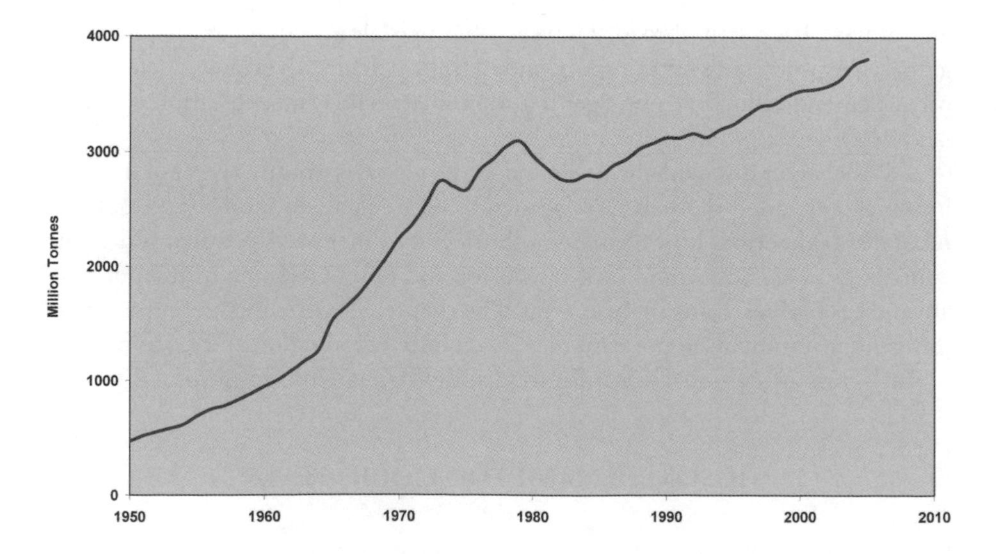

Figure 7.1 *World oil consumption, 1950–2005*

Source: see endnote 2 for this chapter

tried to lower oil subsidies – traditionally kept high to maintain low prices for consumers – in response to upwardly spiralling oil costs.[7]

In 2006, the increased demand for oil worldwide has kept prices high, and the situation is not expected to change anytime soon. The rapidly industrializing economies of China and India, in particular, are projected to increase their consumption of petroleum fuels dramatically in the coming decades as levels of consumer spending and car ownership rise. In China alone, car ownership is projected to rise by 15 per cent per year; with an estimated 28 million private cars on the road in 2005, petroleum consumption already accounted for 81.2 million tonnes, or 594 million barrels, nearly one third of the Chinese total petroleum requirement.[8] Along with other developing countries, China is projected to account for more than two-thirds of global energy demand by 2030.[9] Chinese demand for oil alone jumped by 36 per cent between 2002 and 2005 to represent more than 8 per cent of global demand, and China is now the world's second largest petroleum consumer after the US.[10]

As conventional crude oil supplies appear less able to satisfy growing energy demand, non conventional oils, such as tar sands, shale oil and synthetic petroleum that must be extracted under difficult geographical or climatic conditions, are increasingly being developed as substitutes.[11] However, these supplies are more costly than conventional petroleum in both economic and environmental terms as they tend to involve more disruptive extraction techniques and are more greenhouse-gas intensive to refine (see Chapters 11 and 13).

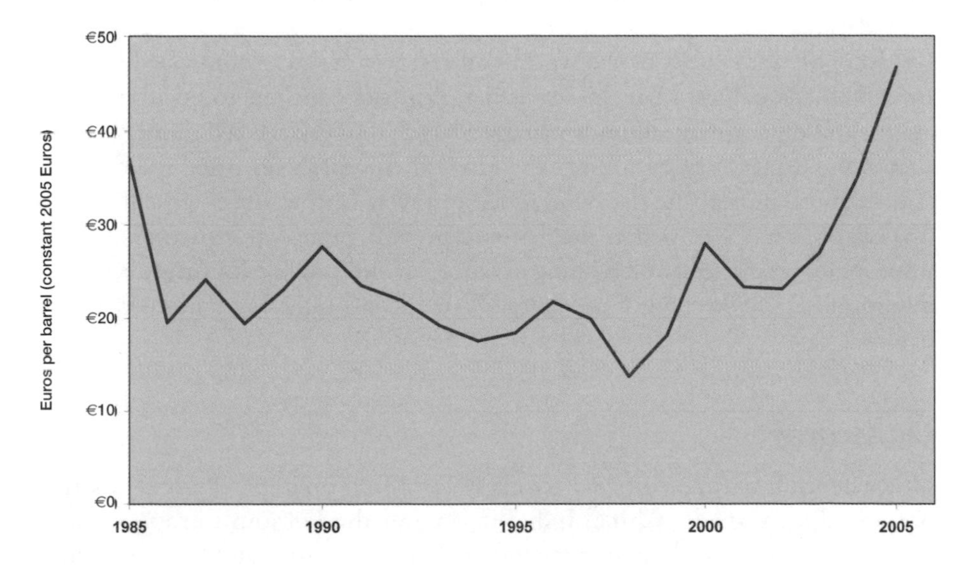

Figure 7.2 *Spot crude oil prices (West Texas Intermediate price), 1985–2005*

Source: US DOE and EIA (2006d)

Note: €1 ≈ $1.20

The power of oil-exporting nations

In the past, oil-producing nations have sometimes used their power over supply and distribution to achieve specific political or economic ends. For instance, both the US (when it was a net oil exporter) and the Arab members of the Organization of Petroleum Exporting Countries (OPEC) – home to 65 per cent of the world's proven oil reserves – have imposed oil embargoes for political or economic reasons.[12]

International oil trade is a major contributor to the flow of capital worldwide and has considerable influence over the global economic and political system.[13] During the oil shocks of the 1970s and 1980s, industrialized country banks were flooded with 'petrodollars', which in turn encouraged the widespread lending that precipitated the debt crisis in Latin America.[14] In 2005, during a period that has seen the highest sustained oil prices since the 1970s, oil-exporting countries earned record profits. Compared with the oft-touted account surpluses of the rising Chinese economy (6 per cent in 2005), Middle Eastern oil exporters earned surpluses averaging 25 per cent of gross domestic product (GDP) in 2005.[15]

Oil prices are a key driver of the global economy, and oil price fluctuations determine the direction of large-scale currency flows in the global market. When prices are high, oil-exporting countries earn huge profits and see their accounts expand, while the accounts of oil-importing countries swing towards deficit. Oil revenues have brought some nations, including Norway and Kuwait, great economic success; others, however, have been plagued by corruption and environmental degradation.

Roughly four-fifths of the world's oil reserves are now controlled by state-owned oil companies. Many oil-exporting countries continue to invest in private equity and other financial funds managed largely in Western economies (a factor that some experts say cushioned the blow of the 2005 oil price spikes to the American economy). At the same time, however, further development of local financial markets – as well as the desire to provide more infrastructure at home, improve oil production and refining capabilities, and provide for future economic development – is keeping more petrodollars in oil-producing nations than ever before.[16]

Oil security

Heavy dependence upon oil can have negative impacts on importing and exporting nations alike. Canada, China, India, Japan and the US join Europe in facing significant energy and environmental challenges, most notably from climate change and increasing security costs that result from oil dependency.[17] The ease of transporting oil and its utility and relatively low cost over the last century gave rise to the oil-dependent global industrialized economy. Oil dependency has, at times, led to military intervention to ensure the steady supply of oil. In addition, climatic changes, including more severe storms, resulting from rising fossil fuel use are increasingly apparent (see Chapter 11 for more on climate). Also, as indicated above, lack of spare production capacity precipitates dramatic oil price increases, leaving nations with oil-centric energy portfolios in dire financial straits.

Furthermore, as oil supplies in many parts of the world begin to dwindle in the years ahead, dependence upon Middle Eastern oil is expected to grow, leaving the entire world more vulnerable to social and political developments in one of the world's least stable regions. In fact, of the world's known potential of conventional petroleum (364 billion tonnes), more than 70 per cent is located in the so called 'strategic ellipse', an area spanning much of the Middle East and Central Asia that is also home to 69 per cent of known natural gas reserves.[18] The European Commission estimates that European energy import dependency will rise to approximately 70 per cent of the European Union's energy requirements in the next 20–30 years, compared to 50 per cent today, from regions threatened by insecurity.[19]

Any significant interruption of oil flowing through the Persian Gulf in today's market conditions would probably send oil prices over €83 (US$100) per barrel. And because the price of oil is set in a single integrated market, consumers around the world would be affected by such a development, even if their own supplies came from nearby fields. Outside of challenges resulting from concentration of energy resources in OPEC and the Persian Gulf, other barriers to increased supply present themselves. For example, in Russia the renationalization of the oil and gas industry through Gazprom and other state companies has given Moscow increasing power over its Western European counterparts, as evidenced by the Ukraine incident in the autumn of 2005.

All told, 80 per cent of the world's conventional oil reserves are under state control and off limits to private investment. Widespread state ownership and declining existing reserves leaves the world's six largest publicly traded oil firms projecting falling production over the next two years and global consumers with rising prices due to more limited supply unless affordable liquid fuel alternatives are quickly developed.[20] In sum, oil is unlikely to be a stable or secure supply of energy in the years ahead – and most nations are in urgent need of a more diverse energy supply.

Large concentrations of easily extractable resources can allow elite groups in oil-exporting nations to assert economic and political control and reduce the need to implement reforms that would allow a more diversified economy to develop.[21] This concentration of power, as well as the lack of economic opportunities for a majority share of the population, have contributed to the rise of insurgent groups and religious fundamentalism in some regions. The so-called 'resource curse' – the tendency for nations endowed with oil to be plagued by corruption and conflict, rather than being able to support growth and development – is exemplified by well-known tragedies associated with oil extraction in Angola, Congo, Sudan and elsewhere.

Effects on oil-importing countries

Even without disruptions in supply due to political or natural crises, today's concentrated oil infrastructure poses challenges for many importing countries. In general, these economies are prone to endemic problems related to their lack of domestic oil resources: high imports contribute to growing trade imbalances and cause short-term shortages in foreign currency reserves, which, coupled with inflation and currency convertibility issues, can slow economic growth.

Rising oil prices only worsen these problems. In developing countries, in particular, the landed prices of crude oil and petroleum products are often considerably higher than prices on the international spot market. This is especially true in landlocked African nations, where transport costs can add up to 50 per cent to the delivered price of fuel. High fuel costs put industries in countries with more rigid currencies at a considerable disadvantage: as energy prices rise, they face foreign exchange shortfalls and must bear the costs of paying more for imported fuel.[22]

Another problem for oil-importing countries is that domestic fuel subsidies, designed to lower the price of fuel for consumers, can stress already weak economies and divert scarce resources from social and human development. As oil prices continue to rise, governments must pay even more to keep energy affordable. According to some estimates, at oil prices of €50 to €57 (US$60 to $70) a barrel, weaker African countries are forced to spend a significant share of their foreign exchange earnings on petroleum imports.[23] In India, the International Monetary Fund estimates that every €7 (US$10) increase in the price of oil will result in a 1 per cent decrease in GDP and a 1.2 per cent point deterioration in the balance of trade.[24] And in the world's very poorest countries – the so-called highly indebted poor countries (HIPCs), many of which are in Africa – sustained crude oil prices would cancel all the benefits of debt cancellation proposed at the G8 Summit in 2005.[25]

THE BIOFUELS ALTERNATIVE

Compared with oil, biofuels can reduce many of the vulnerabilities associated with today's highly concentrated energy economy. Petroleum fuels rely on a narrow concentrated network of extraction, refining and distribution, with most transported by ship to a limited number of refineries, and then by ground or pipeline to market. Biofuel production, in contrast, is considerably less concentrated because of the large land area needed to cultivate the feedstock and the low energy density of this feedstock, which makes it less economical to transport long distances. As a result, biofuel processing facilities are more numerous and spread over a wider geographical area, contributing to a liquid fuel supply that is less vulnerable to disruption. Biofuels also offer an opportunity for a more dispersed and equitably distributed revenue stream.

In terms of global security, biofuels are not likely to produce the same powerful or potentially dangerous alliances that oil has generated. Revenues from Brazilian ethanol, which has thus far received the largest export profits of any biofuel, are unlikely to finance groups that are politically destabilizing. Other countries with large biofuel potential – such as Australia, China, Germany, Guatemala, India, Indonesia, Malaysia, Thailand, the US and Zimbabwe – will probably use biofuels to meet domestic energy needs, substituting for costly oil imports, in the near to medium term.

The overall scale of biofuel production, however, can greatly affect the degree of vulnerability associated with these fuels. Small-scale production, in particular, can help to circumvent supply and distribution concerns by bringing both energy and profits directly to rural consumers. Locally produced fuels can also help to avoid the environmental impacts that can occur with long-distance shipping and other transport. Larger-scale biofuel production, on the other hand, may result in greater industry concentration, be of less benefit to local communities and will probably require more complex infrastructure, such as the use of pipelines and large processing refineries. This could lead to political, economic, social and environmental effects more akin to those with fossil fuels (see Chapters 11 and 13).

Although they may not face the same logistical and political vulnerabilities as oil, biofuels still remain susceptible to natural and human-caused disasters, including crop failures, irregular weather patterns and droughts, which could increase with climate change. This vulnerability is of particular concern with biofuels produced from first-generation agricultural feedstocks; next-generation feedstocks derived from residues and wastes, such as municipal waste streams, are less affected by weather-related disruption.

The economics of biofuels

Biofuels also offer potential economic advantages over fossil fuels, although direct cost comparisons can be difficult. This is because varying biofuel feedstock and processing options have different costs and benefits, and because the negative externalities associated with fossil fuels – in terms of military expenditures and health and environmental costs – tend to be poorly quantified. Although many studies have endeavoured to assign prices to such factors as climate benefits, air quality improvements, human health, sustainability and increased security, accurate quantification of these variables still proves challenging.

Biofuels can look uncompetitive if measured on a direct-cost basis. For example, the market price for biomass pellets in The Netherlands was about €7 to €7.5 ($8.5–9.1) per gigajoule (GJ) in 2004 and is expected to stabilize around €5.6 to €6.4 ($6.8–7.8) per GJ in the short term, while the cost of coal generally remains around €1.2 ($1.5) per GJ. Brazilian ethanol costs, however, are around €6

(US$7.2) per GJ, suggesting that with continued gains in technology and efficiency, biofuels may eventually be able to compete outright with fossil fuels.[26] Biofuels also have the potential to generate many positive externalities, such as reduced greenhouse gas emissions, decreased air pollution and job creation, making them a more socially and environmentally desirable liquid fuel.

Some developing countries, in particular, may be able to develop significant biofuel industries, based on the comparatively low land and labour costs and favourable growing conditions in these regions. Because feedstock costs generally represent about 80 per cent of the total production cost of current-generation biofuels, overall production costs in tropical countries would be low relative to more temperate countries – as is the case with sugar production (see Table 7.1).[27] Increased trade in biofuels would generate more income for producing countries, reduce vulnerabilities (in part, by increasing the number of producers) and result in a more diversified global supply of liquid fuels for transport (see Chapters 2 and 9).[28]

When comparing the environmental costs of biofuels versus fossil fuels, a biofuel market may actually provide long-term economic benefits, provided it develops in a sustainable manner. Carbon dioxide (CO_2) is one negative externality of fossil fuel use that biofuels can reduce, with benefits increasing even more dramatically with next-generation feedstocks and fuel technologies (see Chapter 11). Through

Table 7.1 *Cost ranges for various sugars and sweeteners, selected regions, 1997/1998–2001/2002*

Category	*Cost of production, ex-mill, factory basis (Eurocents per kilogram)*
Raw cane sugar	
Low-cost producers[a]	9.6–20.0
Major exporters[b]	11.2–31.6
Weighted world average	20.4–22.9
Cane sugar, white equivalent	
Low-cost producers[a]	15.8–27.2
Major exporters[b]	17.6–39.7
Weighted world average	27.6–30.2
Beet sugar, refined value	
Low-cost producers[c]	29.6–53.1
Major exporters[d]	40.4–57.8
Weighted world average	54.2– 56.9

Notes: [a] Australia, Brazil (centre-south), Guatemala, Zambia and Zimbabwe; [b] Australia, Brazil, Colombia, Cuba, Guatemala, Zambia and Thailand; [c] Belgium, Canada, Chile, France, Turkey, UK and US; [d] Belgium, France, Germany and Turkey.

Source: see endnote 27 for this chapter

emerging carbon markets such as the European Union (EU) Emissions Trading System and the Chicago Climate Exchange, biofuels will benefit and fossil fuels will lose out as a result of their differing carbon intensities (see Chapter 12).[29] Although transport is not currently covered by emissions trading schemes, biofuels could be a more valuable asset in future systems that will probably include the transport sector. The Kyoto Protocol's Clean Development Mechanism may provide widespread opportunities for rural development while contributing to the reduction in global greenhouse gases if existing hurdles to implementation can be overcome. According to a recent World Bank study, at carbon prices of €2.5 to €16.5 (US$3 to $20) per tonne of CO_2 equivalent, biofuels could earn €0.005 to €0.06 (US$0.01 to $0.07) per litre on the carbon market (see Chapter 18).[30]

Savings from avoided oil imports and positive externalities

Biofuels can be an especially important energy alternative in oil-importing developing countries where landed petroleum costs are high due to poor distribution infrastructure.[31] Investing in a domestic biofuels industry could not only provide increased employment opportunities in rural areas, but it would allow developing nations to internalize a share of the economic value of the locally produced fuels.[32] However, this process is neither immediate nor guaranteed, and must be supported by appropriate policy measures.

The economic savings from avoided oil imports can be considerable, but must be taken in the context of government spending required to support biofuel development. In Brazil, years of domestic support for the ethanol programme allowed the country to reduce oil imports and to produce ethanol at a lower domestic cost than gasoline.[33] Between 1975 and 1987, ethanol saved Brazil €8.6 billion (US$10.4 billion) in foreign exchange and cost the government €7.44 (US$9 billion) in subsidies.[34] More recent studies show that from 1976 to 2004, Brazil's ethanol production substituted for oil imports worth €50.2 billion (US$60.7 billion), or €100.3 billion (US$121.3 billion) including interest on the foreign debt previously incurred when financing oil imports.[35] Many countries are eager to follow Brazil's example, saving foreign exchange and retaining more money in the domestic economy.

In India, it is estimated that gradually substituting 15 per cent of the current transport-sector fuel with biofuels would save some €2.1 billion (US$2.5 billion) worth of foreign exchange by 2012 to 2013.[36] And in sugar-producing countries in Central America, experts estimate that, on average, domestic biofuels industries could generate more than €62 million (US$75 million) annually in savings from avoided oil imports.[37] This is especially significant for countries where the rural poor are hit hard by rising energy prices: in 2005, inflation rates reached 8 per cent in Guatemala and 14 per cent in Costa Rica in response to 30 per cent rises in fuel prices – increasing the number of people living on less than US$1 a day.[38]

Some countries are already investing in biofuels development to offset oil imports. In Jamaica, where oil imports eclipsed US$1 billion in 2005 and officials have declared the energy consumption-to-production ratio 'economically challenging', the government recently formed a partnership with the Brazilian company Coimex to boost the ethanol content of gasoline from 5 per cent to 10 per cent to reduce oil imports.[39] Nigeria, meanwhile, expects to save €202.6 million (US$245 million) in avoided refined petroleum product imports per year through recent biofuel investments and ethanol-blending mandates.[40]

SUBSIDIES

A variety of economic and policy instruments, including subsidies, blending mandates and tax incentives, have been critical in spurring the development of biofuels worldwide. However, this support must be considered in the context of the support enjoyed by the global oil industry, which – despite being a mature industry – continues to receive massive subsidies to secure supply, including depletion tax credits and indirect subsidies from health-care systems and international military presence.

While reliable quantification of global oil subsidies is nearly impossible, it is fair to say that direct petroleum subsidies are considerably larger than direct biofuels subsidies in absolute terms; however, the much larger size of the oil and gas industry must be considered. The US General Accounting Office estimated that in the 32 years prior to 2001, the oil industry received more than US$130 billion in tax incentives, compared with roughly US$11 billion given to the ethanol industry in the 21 years prior to 2001 (see Table 7.2).[41] Although this does not include the military and health costs of fossil fuels, US subsidies to the petroleum industry, equal to roughly €0.002 per litre (US$0.003 per litre), are far lower than the per litre subsidy to biofuels in the US.[42]

In terms of indirect (or implicit) expenditures for fossil fuels, in 2003 the US National Defense Council Foundation estimated that some €40.6 billion (US$49.1 billion) in annual defence outlays is required to defend the flow of Persian Gulf oil to importing countries – the equivalent of adding €0.26 (US$0.30) to the price of a litre of gasoline.[43] Health expenditures related to fossil fuels add up as well. The US Congressional Research Service estimated that the additional cost due to ozone-related respiratory health problems was €3.31 billion (US$4 billion), or €0.01 per litre (US$0.05 per gallon) of gasoline, while the additional cost due to morbidity and premature mortality caused by particulates and acidic aerosols was tens of billions of dollars, or €0.13 per litre (US$0.59 per gallon) of diesel.[44] Estimates of environmental damage from automotive diesel, based on 1993 data, averaged €0.26 (US$0.31) per litre of gasoline.[45]

Financial support for biofuels is considerably lower. In 2005, the US government provided direct subsidies of more than €82.7 million (US$100 million) to support

Table 7.2 *Tax incentives for petroleum versus ethanol in the US, 1968–2000*

Tax incentive	Time period	Government revenue losses[a] (US$ million, adjusted for 2000)
Petroleum industry		
Excess of percentage over cost depletion[b]	1968–2000	81,679–82,085[c]
Expensing of exploration and development costs[b]	1968–2000	42,855–54,580
Alternative (non-conventional) fuel production credit	1980–2000	8411–10,542
Oil and gas exemption from passive loss limitation	1988–2000	1065
Credit for enhanced oil recovery costs	1994–2000	482–1002
Expensing of tertiary injectants	1980–2000	330
All incentives	1968–2000	134,822–149,602[d]
Ethanol industry		
Partial exemption from the excise tax for alcohol fuels	1979–2000	7523–11,183
Income tax credits for alcohol fuels	1980–2000	198–478
All incentives	1979–2000	7721–11,661[d]

Notes: [a] Estimates include both corporate and individual income tax revenue losses except for the partial exemption from the excise tax for alcohol fuels, which represents revenue losses from the federal excise tax on gasoline. [b] Ranges are based on varying estimates given by the US Department of Treasury and the Joint Committee on Taxation. [c] In some years, revenue losses associated with other fuels and non-fuel minerals were included with revenue losses from oil and gas. [d] Estimates of total revenue losses are very rough – the sum of two or more incentives could result in a total change in tax liability that might have a lesser or a greater effect on revenue than the amounts shown for each item separately.

Source: see endnote 41 for this chapter

domestic ethanol, one of the world's largest biofuel industries. Ethanol receives about twice as much funding in the US as biodiesel; however, per litre, biodiesel is more subsidized because it represents a much smaller share of the US market.[46] Federal tax credits for the fuels were worth an additional €1.65 billion (US$2 billion).[47] Additionally, many states provide assistance for the construction of new plants, as well as exempting biofuels from state excise taxes normally applied to transport fuels. US ethanol benefits from one additional support as well: a tariff on competing imports (mainly from Brazil) of €0.12 per litre (US$0.54 per gallon).

In Brazil, ethanol production was heavily subsidized from the 1930s leading up to and during the *Proálcool* programme, launched during the 1970s. Price guarantees and subsidies, public loans and state-guaranteed private bank loans were all used to support the ethanol industry during its development, and at one point the interest on unpaid debt from this industry alone was equivalent to €0.41 (US$0.49) per litre.[48] Today, ethanol production from sugar cane in the centre-south does not receive any direct government subsidies; however, Brazil employs a series of policies that secure ethanol's place in the country's energy matrix, including:

- a mandate requiring that all gasoline be blended with a minimum of 20–25 per cent ethanol (flexible with respect to changing sugar and ethanol prices on the world market);
- an import tariff on gasoline that is one of the highest in the world;
- a ban on diesel-powered personal vehicles to boost the demand for ethanol-powered vehicles;
- a requirement that all government entities purchase 100 per cent hydrated alcohol-fuelled vehicles; and
- low interest loans for financing producer-owned stocks.[49]

Production costs in Brazil make ethanol competitive at average crude oil prices in 2005. While many studies assessing the long-term feasibility of biofuel production are reluctant to assume that oil prices will stay high enough to keep these fuels attractive, most signs point to high oil prices in the future – including the rising cost of extracting petroleum from deeper and more remote deposits, government policies to internalize externalities and accelerating demand. Even the fairly conservative US Energy Information Administration has projected that oil prices will remain near US$50 per barrel for the next few decades.[50] Biofuels will not be immune to high oil prices, and until they represent a significant share of the global market, they will continue to be price takers and will shadow increases in gasoline prices.

Technological improvements can bring significant cost reductions for competitive biofuels producers, as well. During the early years of Brazil's ethanol programme, prices were tightly controlled by the government; but today, as a result of increasing economies of scale, technological efficiency and biotechnology, direct subsidies are no longer required, and prices shadow the oil market and the global sugar market. Many experts feel that with increased research and development, improvements in crop yields due to genetic engineering and enzyme development for biofuels processing can further reduce the costs of ethanol production (see Chapters 4 to 6).[51]

BIOFUELS AND THE AGRICULTURAL MARKET

Biofuels could also have significant impacts on agricultural markets. Global prices for agricultural commodities, including crops such as corn, wheat and cotton, often fall below the costs of production because government subsidies and policies in industrialized countries favour urban consumers over farmers, resulting in excess supply.[52]. Low agricultural prices have the greatest impact on small-scale grain and oilseed producers in developing countries, which are often unable to grow alternative crops or find other work.[53] A move towards agricultural-based energy production via biofuels could absorb excess supply, transition land away from traditional tradable corps and help to maintain higher commodity prices.

Higher prices have already been recorded for the most commonly used biofuel feedstocks – corn, sugar cane, rapeseed and palm oil – in the wake of high oil prices in early 2006.[54] According to 2006 Organisation for Economic Co-operation and Development (OECD) estimates, additional demand for agricultural commodities due to increased biofuel use will have the strongest impact on sugar markets, with up to 60 per cent increases in price by 2014. In a conservative scenario that assumes current levels of biofuels use will continue, vegetable oil prices are projected to increase by up to 20 per cent and cereals will increase by 4 per cent.[55] In a scenario that assumes sustained oil prices around €50 (US$60) per barrel, the impact of additional biofuel production would increase the sugar price an additional 4.2 per cent and vegetable oils an additional 4.3 per cent.[56] These higher prices have the potential to bring increasing profits to rural areas worldwide, increase employment and supply a less-harmful biomass source for developing country energy needs (see Chapter 8)

While long-term commodity price increases could boost farmer incomes worldwide, in the short term such price increases may hurt consumers in countries that are net food importers. This is especially true in the world's poorest countries where, faced with higher food prices, people tend to consume fewer high-valued goods such as meat and dairy. Although food demand is relatively inelastic, studies show that for every 1 per cent increase in the price of food, consumers in developing countries decrease their consumption by three-quarters of a per cent, compared to only one third of a per cent in industrialized countries.[57] At times when agricultural feedstock is more valuable for food production than for energy production, this may also result in shortages in biofuels, as happened during the 1990s when Brazil was forced to import ethanol from the US.[58]

Adding biofuel production capacity as a way to take advantage of the changing availability of different feedstock sources is a good short-term response to fluctuations in commodity output and prices. However, as the markets for food, fuel and energy become increasingly intertwined, there may be risks as well. Agricultural surpluses could turn into regional shortages, pushing prices for biofuel feedstocks and related commodities, including food crops, even higher. A recent study by the Centre for International Economics suggests that ethanol substitution could harm some sectors of the rural economy. Using research from Australia, it concludes that mandating a 10–15 per cent ethanol blend in gasoline would increase grain prices by up to 25 per cent, adversely affecting the domestic livestock industry and weakening its export position.[59] Potential savings from averted petroleum and diesel imports, valued at around €1.08 billion (US$1.3 billion), would be offset by losses in livestock exports valued at around €1.74 billion (US$2.1 billion), as well as by the additional cost of importing grain to make up for diverted feedstock, estimated at €314.2 million (US$380 million).[60]

Continued research and development into the most integrative technologies can provide increasingly stable demand for agricultural commodities and their co-products (see Chapter 6). In this regard, the development of next-generation biofuels is more desirable because these make use of lower-value agricultural products, such as surplus and waste feedstocks, which do not compete directly with food commodities.

In the short term, biorefining capacity can provide a useful way of eliminating surplus when a bumper crop is harvested, or when global demand for a commodity falls. In fact, the availability of surplus sugar cane feedstock spurred India's ambitious blending mandates in 2000 (which were subsequently placed on hold as a result of drought).[61] And China's excess of grain encouraged renewed investment in ethanol in 2001, although these surpluses have now largely been exhausted. A report for the US Department of Agriculture by the Energy and Environmental Study Institute found that an abundance of 'centralized biomass' created the most favourable conditions for the development of biofuels and bio-based products.[62] However, regular surpluses may be indicative of a larger market failure and may not be sustainable.

Currently, regular surpluses of molasses exist in many African and Latin American countries and could be used to produce ethanol at very low prices. A large quantity of surplus molasses in close proximity to ethanol production facilities allows for the economies of scale necessary to make the ethanol price competitive with gasoline.[63] If trade in agricultural goods liberalizes in response to agreements under the World Trade Organization (WTO), surplus crops may not be as abundant in industrialized countries because domestic support measures must be drastically reduced (see Chapter 9). This could provide an increased opportunity for more efficient producers of energy crops in developing countries to earn income, especially in Latin America and the Caribbean, where agricultural products still contribute 20 to 30 per cent of GDP.[64]

THE ECONOMIC PROMISE OF NEXT-GENERATION BIOFUELS

Feedstock costs for emerging biofuel technologies – such as cellulosic ethanol or the expansion of biomass-to-fuel production – are projected to be much lower than those assumed for current or conventional biofuels (where feedstock costs account for a large share of the overall production cost; see Chapter 3). This is because next-generation feedstocks – including municipal solid waste, crop residues, animal waste, wood wastes and residues, and other waste materials – can be converted to liquid fuels at little or no cost for the feedstock, and could even result in negative costs if landfill tipping expenditures are considered.[65] Similarly, waste-to-fuel projects can prevent the destructive environmental effects that agriculture and municipal wastes have on streams and aquifers, and could lead to savings from avoided human health and environmental clean-up (see Chapters

12 and 13).[66] As wastes become more important for biofuel production, their costs may increase. Further analysis is necessary to determine how the use of crop residues and other waste products will affect farmers' incomes and harvesting and production techniques.

At the same time, these next-generation biofuels have the capability of being produced without the need for long-distance transport because of the diversity of potential feedstocks and the possibility of building biorefineries near centres of demand. A recent study in the US estimates that if all available crop residues in the country were converted to fuel ethanol, approximately 76 billion litres (20 billion gallons) of ethanol could be produced every year, corresponding to ten times the output of the existing corn ethanol industry or the equivalent of 14 billion gallons of gasoline – more than 10 per cent of current consumption (see Chapter 6 for more on future feedstock potentials).[67]

Perennial energy crops, such as switchgrass, miscanthus and short-rotation forestry, represent a far more positive economic picture for farmers and fuel producers than current feedstock crops such as rapeseed, corn and soybeans. A recent University of Illinois study calculated that if the US state of Illinois expanded cultivation of miscanthus as a dedicated energy crop, as the EU has done on a small scale, it is expected to be competitive with corn and soybeans, without subsidies, within ten years.[68] Another study by the Ceres network predicts that dedicated energy crops can earn farmers more than corn and wheat due to low fixed costs and high market prices.[69] Studies projecting the potential for this feedstock in Germany have not yielded such favourable cost projections, and national conditions with regard to agricultural production must be considered in determining future feedstock uses.

Increasing Efficiency and Demand-Reduction Strategies

When considering the economics of biofuels, it is important to note that countries and industries vary greatly in the energy intensities of their production (the output produced by each unit of energy input), leading to differing sensitivities to changing energy prices. Developing countries that are oil importers, for example, require twice the amount of oil to produce the same amount of output as countries in the industrialized world due to less-efficient industrial production processes that use older technologies.[70] The greater the energy intensity of an economy, the more detrimental the effects of oil price hikes will be. As a result, high oil prices are significantly more crippling for oil-importing countries than for producers. India, for example, spent 16 per cent more on oil imports in 2003 than it did in 2001 – equivalent to 3 per cent of GDP.[71]

Biofuels could provide an important leapfrogging opportunity for many developing nations, enabling them to bypass many of the economic, environmental and social costs of petroleum fuels that industrialized countries face. However, biofuels development will not contribute to sustainable economic development in the absence of increasing energy efficiency. Those countries with high energy intensities in their agricultural sectors will therefore need to take steps to improve the overall efficiency of production.

Agriculture has become one of the world's most energy-intensive industries, with the US agricultural sector alone consuming 10 quadrillion Btu (10.55 exajoules) of energy per year, equivalent to France's entire annual energy consumption.[72] Globally, 28 per cent of the agricultural energy used goes to manufacturing fertilizer, and 7 per cent is for irrigation, while 34 per cent is consumed as diesel and gasoline by farm vehicles used to plant, till and harvest crops.[73] The balance is absorbed by the agricultural distribution chain, which requires energy-intensive packaging, refrigeration and transportation. This high level of energy intensity is unsustainable and has caused producers in poorer oil-importing nations to limit agricultural production when fuel prices are high.[74]

In the future, if crude import prices remain around 2006 levels of €50 (US$60) per barrel, agricultural outputs are expected to fall 1.5 per cent for wheat and 2.8 per cent for oilseeds, all else being equal. However, OECD models indicate that this price of oil could stimulate increased biofuel production of up to 8 per cent for ethanol in the US and Brazil and up to 16 per cent for biodiesel, dominated by the EU.[75]

A biofuels industry that is developed to minimize energy inputs and improve efficiency will result in significantly higher economic benefits, and can help to alleviate the pressure on oil supplies and increase countries' competitiveness. First-generation feedstocks that require fewer fossil energy inputs for fertilizing, irrigating and harvesting (such as Brazilian sugar cane and jatropha) will be more profitable and sustainable than more energy-intensive feedstocks such as corn and wheat. Next-generation feedstocks, such as municipal solid waste and short-rotation forestry crops, will also need to optimize energy intensity (among other factors) as well in order to make these fuels as sustainable and cost-competitive as possible. Gains in agricultural efficiency from no-till technology and more-efficient energy uses for industrial production can also bring benefits.[76]

Efficiency choices made in the transportation sector will be decisive in determining the future prices of liquid fuels. Without improvements in vehicle fuel economy and the development of alternative vehicle technologies and public transport, biofuels and petroleum fuels alone will be unable to meet the projected growth in fuel demand, particularly as car ownership increases in developing countries. Fortunately, high fuel prices are beginning to drive consumers towards more efficient vehicles, and many governments are offering incentives to those who purchase alternative-fuel vehicles.[77]

Failure to respond to efficiency needs will result in continued environmental degradation from fossil fuel emissions and, ultimately, in loss of industry competitiveness. A case in point is the US automobile industry, which despite outcry from environmental groups about the negative health and climate impacts of increased fossil fuel use, continued to increase production of sport utility vehicles (SUVs) because of the high profit margin.[78] In the wake of skyrocketing fuel prices, however, sales for SUVs fell 31 per cent in January 2005 and 21 per cent in February 2005, from their 2004 levels.[79] The US auto industry is largely perceived to be in crisis: General Motors plans to close multiple facilities and eliminate 30,000 jobs, Ford will close 14 plants and cut up to 30,000 jobs, and Delphi (a world leader in transportation components and systems technology) is currently in bankruptcy, threatening to close plants, eliminate thousands of jobs, and slash wages and benefits for remaining employees.[80]

Manufacturers outside the US, faced with increasingly stringent emissions standards, are incorporating biofuels and greater fuel economy within their economic strategies and vehicle design (see Chapter 15). Toyota's hybrid technology has already made the company far more competitive than its American counterparts, which have solicited technology transfer agreements to utilize the Japanese automaker's engine designs.[81] Other automakers are investing in next-generation biofuels development. The leading biomass gasification company, Choren, is a joint project of DaimlerChrysler, Volkswagen and oil giant Royal Dutch Shell.[82] Other oil companies are also beginning to invest in biofuels: the leading cellulosic ethanol developer, Iogen, is a joint venture of Royal Dutch Shell and PetroCanada.[83]

As oil prices continue to rise, the higher the energy intensity of a product, the more expensive it will become – a reality that will make energy-efficient production and processes that rely on petroleum alternatives ever more attractive. As such, biofuels may be uniquely positioned to relieve stress on the crop sector and facilitate waste reduction by providing a readily available fuel that is less vulnerable to price shocks, reducing municipal solid wastes and agricultural residues with the development of new technologies, and promoting job creation. Industries developing biofuels and other alternative energy and efficiency technologies will be increasingly competitive and able to profit from global demand for these products (see Chapter 9).

CONCLUSION

Among the main economic and security advantages of biofuels are their potential to reduce costly oil imports, decrease vulnerability to price shocks and disruptions in energy supply, and increase domestic access to energy. These advantages can benefit both industrialized and developing countries. In general, industrialized countries have the option of either producing biofuels domestically or importing them on a large scale as a substitute for greenhouse gas-intensive and increasingly expensive petroleum fuels. These countries can also advance research and development that supports the development of next-generation biofuels and their co-products (as can several biofuels leaders in the developing world, including Brazil, China and India).

Developing countries with large agricultural potentials have the opportunity to substitute for expensive oil imports and, where profitable, develop biofuel exports, as Brazil has begun to do. Export industries could offer substantial rural economic benefits, particularly if the processing facilities are owned and operated by farmers in these countries. However, if developing countries are limited to exporting only the raw agricultural feedstocks for biofuel production, the economic benefits will be reduced. Developing countries that do not have large agricultural potentials might benefit instead from small-scale biofuel production, particularly where limited infrastructure makes petroleum fuels difficult and costly to transport. Development of biofuels for local markets will more effectively displace oil imports and create greater rural employment in poorer developing countries (see Chapter 8).

Recognizing the potential economic advantages of biofuels, some national governments and international actors are already pressing for large-scale biofuel development through blending mandates and renewable energy targets – in effect, guaranteeing a substantial amount of future demand. But biofuel promotion policies will need to consider carefully what type of development is ideal, keeping in mind that once subsidies and incentives are granted, they are often politically difficult to remove. And any country considering increased biofuels development will need to assess the feasibility of adopting different biofuel feedstocks and processing infrastructures based on its unique natural resource and economic context.

The extent to which biofuels development addresses issues of oil dependence largely depends upon the price and quantity of biofuels that can be produced. Historically, biofuels have been more expensive than petroleum fuels, and today nearly all biofuel industries still rely on extensive governmental support (mainly subsidies) to be viable. Moreover, most biofuel crops can displace only a limited amount of oil before rising demand for feedstock puts upward pressure on the prices of agricultural commodities and, therefore, food. These pressures on commodity prices are likely to be beneficial for farmers in the medium and long term despite short-term risks (see Chapter 8).

Thus, while conventional biofuels have great potential to displace oil – particularly in tropical countries – it is only the development of the next generation of feedstocks and conversion technologies that will really determine the full potential of biofuels to diversify the world's liquid fuel supply in an economically viable manner. As these technologies emerge, preference should be given to approaches that use waste streams and perennial energy crops as feedstocks. Use of these feedstocks and technologies can positively affect food commodity prices by taking agricultural land out of production and reducing surplus production, and potentially increasing the price of low-value agricultural products. This would minimize competition with food and animal feed and contribute to healthier social, environmental and economic outlooks (see Chapters 8 and 9).

No matter how successful an emerging biofuels industry is, however, without significant innovation to make the transportation sector more efficient overall, especially in the US and in growing markets such as China and India, demand for liquid transport fuels is likely to produce market conditions that encourage the production of petroleum and biofuels at a level that is not environmentally sustainable. Strong environmental and social protection, rule of law and governance will be key to the overall success of biofuels.

8

Implications for Agriculture and Rural Development

Introduction

Beginning with their earliest advocates, biofuel programmes have aimed to support agricultural economies. Henry Ford and even Rudolph Diesel promoted the use of liquid fuels from plant sources as a way of expanding the market for farm products. Today, most biofuel production efforts are still set up primarily to help domestic agricultural producers and rural economies.

The need to bolster rural areas is critical. Most of the world's hungry people live in farming regions, and smallholder livelihoods are increasingly threatened by the expansion of mechanized industrial agriculture. In the developing world, this mechanization is contributing to a massive migration from rural communities to urban areas, where economic prospects are often no better. Even in the industrialized world, where most people already live in cities, the farming population has declined steadily as larger and more capital-intensive farming operations eliminate agricultural jobs.[1]

This chapter discusses the potential for biofuel development to aid rural areas. It explores how these fuels can increase market demand for agricultural products, as well their potential to boost agricultural employment and substitute for agricultural subsidies. In addition, the chapter briefly examines the merits of the 'food-versus-fuel' debate and highlights some of the risks to rural communities of more concentrated and larger-scale biofuel production.

Expanding markets for agricultural products

Expanded biofuel production can offer particular benefits to people living and working in the world's agriculture regions. Biomass depends upon agriculture and forestry, and it is available in nearly every region of the world. Moreover, as mentioned previously, the economies of scale that dominate the petroleum industry do not apply as readily to the harvesting and processing of biofuels, making the liquid biofuel industry less concentrated and more labour intensive than the fossil fuel industry.

Creating a market for biofuels as a way of increasing the value of the world's farm products is an obvious plus for the agricultural economy as a whole. However, higher crop prices do not automatically translate into better conditions for farmers or rural communities – for instance, they can raise the price of inputs for the meat and agricultural processing industries. They can also fail to trickle down to the poorest participants in the agricultural economy.

Larger markets and higher prices

Historically, biofuel programmes have served the purpose of providing farmers with both a larger market and a price support. During the early 1900s, the French government promoted ethanol production as a way of handling a decline in sugar beet exports. Germany offered a subsidy to keep ethanol prices on par with gasoline, largely to boost demand for domestic grain. And in the US, early fuel ethanol policies were established as a way of dealing with the surplus of grains, potatoes and sugar beets that resulted from agricultural exploitation of virgin western lands.[2]

Today, biofuel production still helps to maintain or increase the price of certain agricultural feedstocks. In the US, rising ethanol production has absorbed a steadily larger share of the country's corn crop, from 12 per cent in 2004 to a predicted 18 per cent for 2005–2006, to a projected 20 per cent plus by 2012.[3, 4] This rising demand for corn feedstock for ethanol is expected to keep crop prices high. According to analysts at the University of Missouri, by 2012 increasing demand could raise the price of corn by an average of €0.11 (US$.013) per bushel and increase net farm income by €246 million (US$298 million) per year.[5,6] Additional demand for corn would also raise the prices of sorghum and wheat by €0.07 and €0.05 per bushel (US$0.09 and $0.06), respectively.[7]

In the European Union (EU), policy-makers have developed the market for biodiesel in large part to support growers of oilseed crops. Limited by the Blair House Agreement, which restricts the amount of acreage that can be planted with oilseeds for food, farmers have instead planted rapeseed and sunflower seed for use in biodiesel fuel. The market has grown so rapidly that more than 20 per cent of EU rapeseed is now sold for fuel.[8] This market expansion has also caused rapeseed oil prices to reach new highs at the Rotterdam market.[9]

The advent of the *Proálcool* programme in Brazil, designed to spur the domestic market for ethanol and keep sugar prices high, led to an expansion of the area planted for sugar cane that still continues today. Since about 50 per cent of the country's sugar is converted into ethanol, the biofuel programme has effectively permitted a doubling of planted acreage – perhaps more since most of the country's mills are integrated facilities that can hedge between sugar and ethanol, and are less risky than sugar-only mills.[10] High gasoline prices and increased demand for ethanol fuel in Brazil over the last few years have been key factors in the rise in the global price of sugar to today's ten-year high.[11]

Other countries are also pursuing biofuel programmes with the aim of expanding the market for common crops. Australia's northern sugar growers have experienced a 20 per cent drop in the price of their sugar, despite high international prices between 1999 and 2004, and have turned to a domestic fuel ethanol programme to provide a more stable market.[12] And French wine growers are hoping that fuel ethanol production will help them to cope with recent overproduction.[13] Likewise, corn producers in South Africa are struggling with a glut of overproduction and are using these surpluses as collateral to finance the construction of eight ethanol facilities, creating a long-term additional market for the crop.[14]

Elsewhere in the developing world, Malaysia and, especially, Indonesia are rapidly expanding their palm oil acreage, hoping that rising European biodiesel demand will help to boost their exports by 30 per cent or more in the coming years.[15] Thailand's ethanol-blending mandate has already increased the price of cassava, and the government is reversing its sugar cane restriction policy to encourage more domestic production.[16] In the Philippines, legislators have been planning to introduce a biodiesel-blending mandate to support the country's nearly 5 million coconut farmers, and an ethanol mandate to help reverse shrinking acreage in the sugar cane industry (see Chapter 17).[17]

While higher crop prices are clearly beneficial to some crop producers, other industries can suffer – in particular, those that purchase agricultural feedstocks. In Europe, the increased demand for biodiesel has caused shortages of rapeseed oil, sending makers of margarine, mayonnaise and salad dressing scrambling for alternative supplies.[18] The Australian beef industry, wary of rising prices for feed grain, has warned than a grain-ethanol programme may lead to greater losses in meat exports than gains from avoided oil imports.[19] And in the US, one study concluded that hog and poultry producers, along with operators of grain processing and exporting facilities, would lose out as more corn is diverted to ethanol.[20]

New feedstocks and new markets

Some of these problems could be avoided as a wider diversity of feedstocks are used and as new technologies for biofuels are developed (see Chapters 4 and 5). This would further diversify the variety of farm products produced, open up new markets for underutilized forms of biomass, and make more land area available for agricultural use. It could also alleviate some of the competition for existing agricultural commodities.

Jatropha, an oil-yielding tree that can grow on degraded soils in arid conditions, is one promising feedstock that would not have to compete for valuable agricultural land. A UK company, D1 Oils Plc, has made arrangements to cultivate it along train tracks in India, on mining-degraded soils in the Philippines and on a wastewater dumping ground in Egypt.[21]

Additionally, a wide range of crop residues, traditionally left to decay in the fields, could be converted to fuel using new cellulosic technologies, offering farmers a second harvest from their crops.[22] In the US, studies suggest that farmers could generate an additional €40 per hectare (US$20 per acre) or more by harvesting corn stalks and leaves for use in biofuels.[23]

Likewise, degraded grazing lands that are not currently arable could become newly productive. Planting hearty perennials such as switchgrass and miscanthus as dedicated energy crops would permit more extensive use of the sun, water and soil of these lands, and also provide new economic opportunities for agricultural workers. In the US, the potential phase-out of the Conservation Reserve Program would make millions of hectares available for reharvesting. While it would be ecologically problematic to replant most of this erosion-prone land with annual crops, perennial grasses or trees could be a harvested on roughly 3.3 million to 9 million hectares (8.2 million to 20 million acres).[24]

In the European Union (EU), non-food crops can be cultivated on set-aside land while receiving the standard agricultural premium. In the 15 EU member states (EU-15) only, dedicated energy crops cultivated on base (not set-aside) land are eligible for a €45 per hectare payment, which makes them more attractive than some food crops. In the future, most crops, including sugar beets, are subject to greater competition as world commodity markets are liberalized (see below and Chapter 9).[25] In 1999, both France and Spain used over 15 per cent of set-aside land to cultivate energy crops, and as reform of the EU Common Agricultural Policy (CAP) moves forward, energy crop cultivation may become an even more important income support for farmers.[26]

The use of cellulosic biomass is likely to rely at least initially on existing crop residues, not newly planted acreage, because of the low cost and accessibility of this feedstock. Brazilian sugar mills, for example, have amassed large piles of bagasse (the residues left over from cane processing), and similar piles of cane leaves and tips could accrue as the practice of burning them is replaced by active harvesting of them.[27] Other likely sources for biomass in the short term are lumber mills and large agricultural processing facilities.[28] New markets for their residues could substantially increase the income of these rural industries.

While cellulosic conversion to biofuel would probably add value to low-quality wastes from forestry and agricultural processing, this could also raise prices for other industries that rely on these wastes as inputs. For example, waste-derived solid fuels, fibre products, mulch, animal bedding, charcoal and composite wood could all become more expensive.[29]

Creating agricultural employment

By generating greater demand for agricultural products, biofuel programmes have the potential to significantly increase employment in rural areas. Already, the US ethanol industry is credited with employing between 147,000 and 200,000 people, in sectors ranging from farming to plant construction and operation.[30] Brazil's ethanol industry employs about half a million workers. And in the EU, although others believe the employment effect will be smaller, a study by the Wuppertal Institute found that when biofuels reach 1 per cent of the fuel supply, the industry is expected to have created 45,000 to 75,000 new jobs, mostly in agriculture.[31]

Research has found that the biofuel industry can generate more jobs per unit of output than the fossil fuel industry, sometimes at lower cost. The World Bank reports that biofuel industries require about 100 times more workers per joule produced than the fossil fuel industry.[32] Even Germany's relatively capital-intensive biodiesel industry generates roughly 50 times more jobs per tonne of raw oil than does diesel production.[33] In terms of job creation costs, a study in Brazil found that a job in the ethanol industry costs 25 times less than one in the petroleum industry.[34] Since the vast majority of employment in biofuel industries is in farming, transportation and processing, most of these jobs will be in rural communities.

In the future, biofuel programmes will contribute to even greater employment in even more countries. For example:

- France expects its proposed biofuel programme to lead to 25,000 additional jobs by 2010.[35]
- In Colombia, government officials hope that a new ethanol-blending mandate will add 170,000 new jobs in the sugar ethanol industry over the next several years, with each farming family increasing its average income two to threefold.[36]
- In Venezuela, an ethanol blend of 10 per cent is expected to provide 1 million jobs in the sugar cane ethanol industry by 2012.[37]
- In China, officials think that as many as 9 million jobs could be created over the long term from the large-scale processing of agricultural and forestry products into liquids fuels.[38]
- In sub-Saharan Africa, the World Bank estimates that a region-wide blend of biofuels – 10 per cent of gasoline and 5 per cent of diesel – could yield between 700,000 and 1.1 million jobs.[39]

Such massive jobs programmes are achievable because biofuels production can be very labour intensive. However, it is not clear that biofuels will produce enough jobs to compensate for the losses being brought about by industrialized agriculture. Much of the early expansion of Brazil's sugar cane area, especially in the northeast, occurred as large plantation owners took over smaller-scale farms. This was a sometimes violent social disruption that led to an increase in unemployment and landlessness in the region.[40] More recently, total employment in Brazil's sugar cane industry declined from 670,000 in 1992 to 450,000 in 2003, largely because of the trend towards mechanical harvesting.[41] In the US, despite an expanding ethanol industry, the farming population of the Midwest has been shrinking for decades and is now one third what it was in 1940.[42]

In developing countries, many agricultural jobs are seasonal, making it harder for workers to maintain steady employment in one area. Sugar cane, in particular, has an ugly history of exploiting temporary workers. To address this problem, some plantation owners in Brazil now provide labourers with off-season work, planting and preparing for the next harvest. This has helped to raise the wages of sugar cane workers above those of other agricultural sectors; but the disparity between wealthy plantation owners and labourers remains striking and subject to continuing criticism.[43]

In industrialized countries, even though full-time farmers are more likely to be tending the land, they are also typically tenants rather than owners. Some 40 per cent of US farmers currently rent their land and facilities. As such, they are unlikely to benefit from higher corn prices – as corn prices increase, so will land-renting rates; nor are they likely to have enough capital to participate in the value-adding process of converting corn into ethanol.[44]

Compared to other biofuel feedstocks, labour-intensive oilseed crops in developing countries may be more amenable to sustainable and equitable job creation. Because tree oilseeds often must be harvested manually, large owners who can purchase advanced harvesting machinery have fewer advantages (see Figure 8.1 for a comparison of the labour intensity of different oilseed crops in Brazil).[45] Moreover, since the process of converting plant oils into biodiesel is fairly straightforward and can happen at relatively low temperatures and pressures, it can often be done on a smaller scale. Thus, feedstock that is labour intensive rather than capital intensive, and the production of biodiesel rather than ethanol, may be the most promising options for supporting poor farmers and providing liquid fuel in remote areas.

Largely for these reasons, the government of India, via its National Biodiesel Programme, is promoting the cultivation of oilseed plants such as pongamia and jatropha. It calculates that a jatropha farm would provide employment for 313 person days per hectare in the first year of plantation and 50 person days per hectare over the next 30–40 years.[46] The government of the state of Uttaranchal has already given more than 5000 impoverished families 2 hectares of land each to develop jatropha.[47]

Brazil's new biodiesel programme, meanwhile, is focusing on castor oil and palm oil. Led by President Lula and launched in 2006, the programme aims to generate employment for family farmers in the north and northeast, with the target of creating 400,000 jobs related to biodiesel production in these areas.[48] To keep the focus on poverty alleviation, rather than agribusiness, the government plans to require biodiesel producers to purchase a minimum percentage of their raw materials from family farmers (with the share varying by region).[49] So far, Lula has said there will be 350,000 people working on the biodiesel 'revolution' by the end of 2006.[50] During the first three years, Brazil's government hopes that

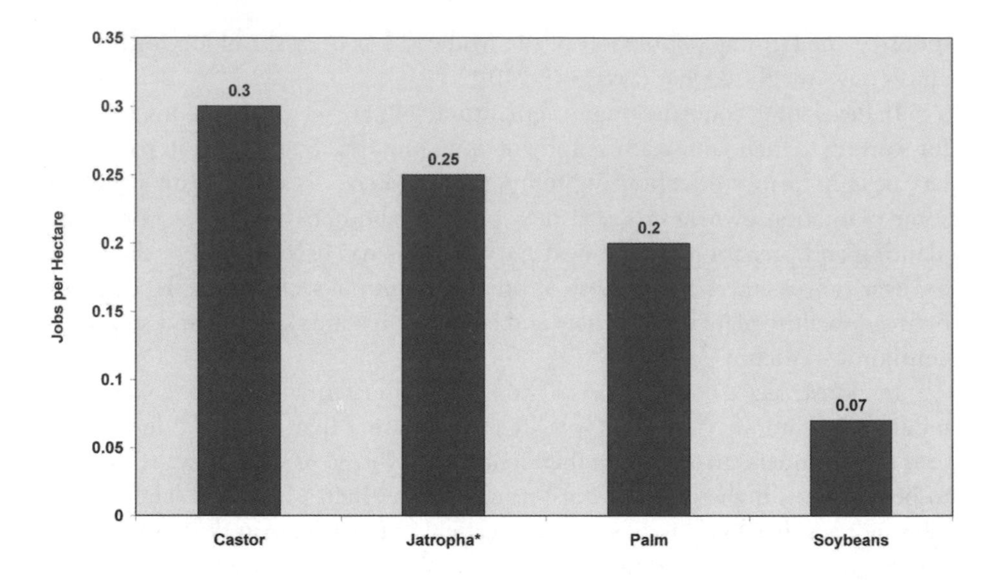

Figure 8.1 *Labour intensity of selected oilseed crops in Brazil*

Note: *Figure is theoretical because jatropha is not yet widely harvested in any country

Source: Kaltner et al (2005) p88; Jones (2006)

200,000 family farms in the northeast, plus 50,000 families from other regions, will start production. It projects earnings of €178 million (US$215 million) for families in the northeast.[51]

In order to succeed, this and similar rural development programmes will probably require extensive and continued government involvement. In Brazil, in particular, large soybean farmers and processors are far better prepared than small family farmers to ramp up production of biodiesel, and soybean oil is currently cheaper than palm oil and, especially, castor oil.[52] Initial results from Brazil are not encouraging. Already, a huge facility that will produce biodiesel from cow tallow, which is 30 per cent cheaper than soybean oil, is being constructed. When it opens, it alone could satisfy nearly 14 per cent of the country's national B2 mandate (for 2 per cent biodiesel blended with 98 per cent diesel).[53] In the northeast, where there is no legal guarantee that smaller-scale castor oil producers will benefit from the profits available downstream in the biodiesel market, small farmers are already selling their oil to middlemen at 25 to 30 per cent below the market price.[54]

Substituting for agricultural subsidies

As described in Chapter 7, a key problem for agricultural regions in the developing world is the surplus production of food crops in industrialized countries. Subsidies and gains in agricultural productivity have helped farmers to boost their crop yields faster than the demand for food has increased.[55] Government supports such as deficiency payments make it economical for industrial farmers to continue overproducing and then to dump their excess product onto the world market at prices well below the cost of production. While this dumping has marginally benefited industrialized agricultural interests, it is stagnating agricultural development in poorer countries, which are far more dependent upon farm incomes. Agriculture provides less than 2 per cent of income and employment in rich countries, while some of the poorest countries rely on agriculture for 35 per cent of their gross domestic product (GDP).[56]

As developing countries put increased pressure on industrialized countries to eliminate or reduce these distorting subsidies, some analysts have offered biofuels as a possible solution. They propose shifting government money out of traditional agricultural subsidies and into programmes designed to increase the domestic supply of biofuels. In other words, rather than sending underpriced crops abroad, industrialized countries could process them into fuels. In doing so, biofuel programmes in industrialized nations could inadvertently do more to boost the agricultural economics of developing countries than biofuel promotion policies in these countries themselves.

In the US and Europe, biofuel support programmes have already acted as a *de facto* substitute for other agricultural subsidies, such as for rapeseed and corn. The EU's growing market for biodiesel has enabled policy-makers to set aside large areas of land from food production, while also reducing international dumping and benefiting domestic farmers. In the US, the ethanol tax credit and various state blending mandates have served to displace some amount of corn deficiency payments.[57] As the demand for ethanol has led to reductions in corn export growth, 'ethanol' has surpassed 'exports' as the bigger user of corn.[58]

Despite the potential for biofuels to mitigate the damage done by industrialized country subsidies, fundamental problems may remain. Even if they grow energy crops for export, farmers in less-developed countries will have a hard time breaking into the markets of industrialized countries due to high tariffs and other trade barriers. Agricultural interests in the EU and US are adamant that their own domestic biofuel industries be protected. In the EU, rapeseed growers are concerned about competition from palm oil growers in Malaysia and Indonesia, who can produce a much cheaper plant oil feedstock. In the US, corn and soybean farmers are concerned about imports of sugar ethanol or plant oils from Brazil or other Latin American countries. And if industrialized countries invest heavily in nurturing a home-grown cellulosic biofuel industry, they will probably protect that industry as well, even if it is not internationally competitive (see Chapter 9).

BIOFUEL PROCESSING: ADDING VALUE TO AGRICULTURAL HARVESTS

While growing and harvesting the feedstock for biofuels can bring income and jobs to rural areas, a large share of the 'added' value of these fuels occurs during the conversion and production stages. As a refinement of agricultural resources, biofuel processing can be a way of bringing more money to agricultural communities, giving farmers the opportunity not only to be suppliers of feedstock, but also to profit from later stages of the value chain.

In the US, for example, the construction of a 150 million litre (40 million gallon) ethanol plant could provide 40 full-time jobs and increase the value of the corn it ferments by 10–25 per cent.[59] Processing can convert €1.65 (US$2) of corn into €4 to €5 (US$5 to $6) worth of products.[60] Biofuel processing facilities not only provide additional jobs in rural areas, they also tend to elevate the prices of nearby crops – by reducing the land area available for these crops and making supplies tighter.[61]

Biofuel processing work tends to pay better than feedstock production, with refiners generally receiving higher wages than typical agricultural labourers. In Brazil, because ethanol production requires more technical skill, these 'value-adding' workers are paid about 30 per cent more than labourers involved in cane harvesting. In the centre-south of Brazil, sugar milling and ethanol fermentation together provide nearly 40 per cent of the jobs in the sugar ethanol industry. Ethanol production provides nearly half of these jobs.[62]

Co-products

Facilities that convert agricultural products into fuels can benefit from the production of 'co-products' as well, further boosting agricultural incomes and the economic viability of biofuel production plants. Some of these co-products are valuable to neighbouring agricultural interests. In the US, for example, dried distillers grain with solubles (DDGS), a high-protein residue from ethanol production, provides a useful feed for cows and pigs. For feedstocks such as jatropha, which is poisonous to eat, the seed cake can make a useful fertilizer. And the utilization of bagasse (sugar cane residues) as energy in Brazil played a key role in keeping the ethanol industry profitable after the government withdrew subsidies during the early 1990s. Since then, large surpluses of bagasse have been a lure for other industries seeking a cheap source of energy.[63]

Other by-products are less useful to the agricultural community, but foreshadow future 'biorefineries' that promise to add even greater value to raw agricultural products (see Chapter 5). Glycerine, a by-product of biodiesel production, can be a useful chemical feedstock for soaps, cosmetics, lubricants and pharmaceutical products. Similarly, ethanol can become ethylene, which is among the most important petrochemical intermediates, contributing to the production of a wide range of chemicals, fabrics and plastics. However, it should be noted that the abrupt expansion of biofuel production can create a market glut of the co-products, as has happened with glycerine in Europe, reducing their price and undercutting their ability to benefit rural economies.[64]

Existing ethanol 'wet mills' in the US are already rudimentary biorefineries, producing not just ethanol fuel but also high-protein feeds, high-fructose corn syrup, sugars for chemical feedstock and a range of other products. Facilities that produce bio-based products may, in the future, resemble petroleum refineries, where fuels comprise the bulk of production, but materials and chemicals provide a disproportionate amount of the profit (see Chapter 5).

Indirect benefits to rural communities

In addition to the direct benefits derived from biofuel harvesting and processing, the creation of new biofuel industries can bring indirect benefits to rural communities. For instance, all of the construction required to build a new biofuel-producing facility – a corn ethanol mill or a biodiesel transesterification plant – can bring a significant one-time boost to the local economy. About as many jobs are produced during the construction phase of an ethanol plant as during its operational phase, and the plant will also require routine maintenance.[65] Additionally, transporting feedstock to the facility and shipping fuels and co-products from the facility can generate extra business for local trucking or rail companies.

Perhaps the biggest multiplier of biofuels, both economically and socially, is the additional money spent by members of the community, who gain new or higher-paying jobs. People buy their basic necessities near where they live; they find local places to buy food, clothes, tools or entertainment; and they pay taxes and contribute to the development of a community. In this realm, biofuel production facilities that are locally owned tend to bring greater local benefits. Farmer-owned enterprises cause more money to circulate in the local economy than traditional cooperatives or corporations owned by city-dwelling stockholders.[66] Workers at ethanol facilities located in smaller communities also seem to support more jobs in their region.[67]

BIOFUELS FOR LOCAL USE

While biofuel production and processing promise many potential benefits, one of the most direct benefits that biofuels can provide rural communities is the fuel itself. Especially in places that are vulnerable to disruptions in the supply of refined petroleum fuels, biofuels can be a more reliable alternative. It can be appropriate both in industrialized countries and in the developing world.

As early as 1906, a senator from the US state of North Dakota dreamed that a new tax reduction on ethanol would mean 'every farmer could have a still' to supply heat, light and power at low prices. More recently, Germany's deputy minister of agriculture has collaborated with John Deere & Co to produce a tractor that farmers can run on their own rapeseed oil.[68] Oil seeds grown in inland parts of Brazil could help farmers to avoid the high cost of diesel, which must often be delivered deep into the country over low-quality roads.[69] In Argentina, some industrial farmers have calculated that home-grown biodiesel can cost half the pump price of fossil diesel.[70] Such home-grown fuel production is reminiscent of the old 'oat model', where farmers grew food for their draught animals.

In more isolated regions, biofuels make particular sense as an alternative to petroleum fuels that must be imported via a long, vulnerable supply line. Many regions have little infrastructure for distributing fuels via train or pipeline. They are particularly dependent upon the liquid fuels that trucks can bring over poorly maintained roads, and the price of these imported fuels can be several times higher than the prices seen at fuel depots in the industrialized world (see Chapter 7).

Remote rural communities typically depend upon imported liquids fuels for applications other than transport, such as heat and power for cooking, lighting and industry. Here, liquid biofuels can play an important role as non-transportation fuels. The World Bank's Millennium Gelfuel Initiative, for example, seeks to promote ethanol stoves as an improvement over stoves that use diesel, kerosene, liquid petroleum gas and conventional fuelwood. In a complementary but so far limited effort, the World Bank has designed a prototype facility that can process sugar cane and sorghum into ethanol, electricity, biogas and cattle feed.[71] In Brazil, organizations including Winrock International are beginning to promote an adapted version of this to help remote areas extract high-quality energy from sugar cane.[72]

Other places are using fuel from oilseed crops. In Mali, the Mali-Folkecentre has facilitated the planting and processing of jatropha trees near villages to spare them the cost of importing expensive fossil diesel (see Box 8.1).[73] Senegal, too, is in the process of pilot testing an innovative rural multi-energy service delivery vehicle based on direct vegetable oil from jatropha.[74] Similar projects are under way in communities around the world. In many places, the oilseeds are ready to be harvested, but people lack crushers that would more efficiently extract the oil.

Since fuelwood collection and cooking are often the responsibilities of women, new 'modern' biomass initiatives can benefit them, in particular, saving many hours of fuel gathering and reducing unhealthy levels of smoke indoors.[75] In Tanzania, community groups distributed jatropha oil lamps and cooking stoves to women in the village of Milmani, who soon raised more than 30,000 jatropha seedlings, which they will soon be able to exploit as a new reserve of fuel.[76]

Box 8.1 Village-scale jatropha oil harvesting for biofuel

In Mali, only 12 per cent of the country's 12 million residents – and just 1 per cent of the rural population – has access to electricity. But biofuels could change that. Equipped with seed crushers, Malian women have increased their social standing by extracting a biofuel that burns cleaner than diesel, arrives more predictably and keeps more money in the local community. Jatropha bushes grow well on marginal lands in arid areas, can be harvested twice annually and remain productive for decades. And the oil from their seeds can be used to fuel generators and vehicles, and provide heat for cooking. Because diesel fuel represents an estimated 50 per cent of the total costs of operating diesel engines, substituting jatropha oil can save communities significant amounts of money and provide energy for many people who otherwise would go without.

Source: see endnote 73 for this chapter

Cellulosic fuels, which can be co-harvested with feedstock for liquid biofuels, offer promise for local energy as well. Sugar cane bagasse is already burned on a large scale for industrial processing of sugar and ethanol in Brazil, Mauritius and Hawaii, among other places. Increasingly, bagasse is seen less as a waste product and more as a good source of electricity, to be sold to the grid or used locally.[77] In China, grain stalks are widely harvested for energy, and in both China and India, wet organic wastes are commonly digested into biogas. Through such efforts, rural communities can begin harnessing a greater amount of local biomass energy now, without waiting for technologies that convert cellulosic matter into liquid form.

Industry concentration and the distribution of profits

At their best, biofuel programmes can enrich farmers by helping to add value to their products. But at their worst, biofuel programmes can expedite the very mechanization that is driving the world's poorest farmers off their land and into deeper poverty. Most likely, the biofuel economy of the future will be characterized by a range of production types – some dominated by large capital-intensive businesses, some characterized by farmer cooperatives that can compete with large companies (and are protected by supportive policies), and some where biofuels are produced on a smaller scale and used within the rural communities themselves. Regardless of the scale of production, however, one thing is clear: the more involved farmers are in the production, processing and use of biofuels, the more likely they are to benefit from them.

Local ownership and small-scale production

Studies show that rural communities benefit considerably more when farmers themselves have a stake in the refining stages of biofuels production. In the US, for example, farmers benefit five to ten times more from the presence of a corn ethanol mill when they are part owners.[78] Yet, historically, farmers have not been the owners of agricultural processing facilities: large processors, with considerable financial capital, have received most of the profits from value-added commodities. In the US, for every dollar spent on food, the share of the final market price going to farmers dropped from nearly US$0.40 in 1910 to US$0.07 in 1997.[79] A typical corn farmer receives less than 10 per cent of the price of corn flakes, while a wheat farmer receives just 6 per cent of the price of bread.[80] Farmers have been squeezed by increasingly concentrated oligopolies at all levels of the industry – seed growers, chemical refiners, machinery manufacturers, food processors and grocery retailers.[81]

As agricultural commodities are refined into higher-value products, international tariffs increase. This hinders market access for value-added exports from developing countries and has prevented farmers in the poorest regions from participating in more profitable agricultural processing.[82]

To help small-scale farmers benefit in the biofuel economy, some governments have enacted policies to ensure that they, too, see profits from fuel production. In Brazil, small growers, who own about 30 per cent of the sugar cane land area, negotiate a revenue-sharing agreement each year with the plantation owners, who own the remaining land area as well as most of the sugar/ethanol mills.[83] Similarly, the US state of Minnesota has sought to favour smaller producers and farmer-owned cooperatives. Starting in the late 1980s, the state provided a special producer payment of €0.65 per litre (US$0.20 per gallon) for the first 57 million litres (15 million gallons) produced by an ethanol mill. As a result, 12 of the state's 14 ethanol mills were formed as farmer cooperatives, and farmer-owned ethanol cooperatives now produce about 40 per cent of the ethanol sold in the US.[84]

Seeking to emulate the success of Minnesota's programme, policy-makers in Saskatchewan, Canada, have approved a biofuel programme that requires distributors to purchase up to 30 per cent of their ethanol from small producers – approximately 25 million litres per year.[85] In the private sector, a Colorado-based company called Blue Sun Biodiesel has worked towards the same end: farmer ownership in biofuel production facilities. By providing farmers a guaranteed market for their crop, as well as additional profit for the sale of biodiesel, Blue Sun persuaded farmers near Denver to invest US$5000 (€4132) each in the enterprise and switch some of their crop to rapeseed.[86] So far, the company is successfully supplying school buses, city buses and delivery trucks with locally produced biodiesel, and the entrepreneurs are now spreading their model to New Mexico.[87] Through such programmes, small growers can collaborate to receive substantially more income for their crop each year.

It is often assumed that stringent environmental or social standards will act as barriers for developing countries to enter industrialized country markets (see Chapters 9 and 18). However, contract farming in Africa has provided a steady source of food products to the European market. In Madagascar, the use of 'micro-contracts' and the provision of support and supervision to small farmers have resulted in higher welfare for farmers, greater income stability and shorter lean periods. Furthermore, improved techniques for resource management and technology transfer have had positive spill-over effects for the production of rice, the staple crop.[88] The possibility exists for agricultural extension services and other policy institutions to use biofuels as a tool for sustainable agricultural development and poverty alleviation in conjunction with the production of biofuels for domestic use or export.

At the same time, rising interest in organic and environmentally sustainable production of biofuel feedstock may also boost rural economic benefits by contributing to increased labour intensity in agricultural production.

Industry consolidation and large-scale production

Despite well-meaning efforts to encourage small-scale biofuel production in many countries, larger-scale owners and corporations will probably still dominate the future biofuel industry. As with many industrial activities, significant economies of scale can be gained from processing and, especially, distributing biofuels on a large scale.

Brazil, for example, is seeing increasing consolidation in its sugar ethanol industry as the largest domestic entities, Cosan and Copersucar, grow and as companies from Europe, Japan and the US invest in Brazilian mills. According to some estimates, in the coming years the industry may be controlled by only six to seven larger groups, compared with about 250 millers today.[89] Even Brazil's embryonic biodiesel economy is dominated by five main producers, one of which, Agropalma, produces more than half of the total output.[90] Dedini, which has built 80 per cent of Brazil's ethanol distilleries, is also the primary builder of the country's biodiesel facilities.[91]

In China, five provinces – Jilin, Heilongjiang, Henan, Anhui and Liaoning – have been using ethanol blends since 2000, primarily to help alleviate the burden of grain surpluses on farmers. However, the country's nascent biofuel industry is comprised by just a few very large ethanol plants, and the extent to which the value-adding benefits have reached rural communities is unclear.[92]

In the US, concentration in ethanol production has fluctuated over the years, although it appears to be again headed in the direction of greater consolidation. During the early 1980s, federal loan guarantees spurred the construction of hundreds of tiny ethanol facilities; but only the largest companies were able to survive competition from low oil prices in the mid 1980s. By the end of the decade, a single company, Archer Daniels Midland (ADM), was producing nearly 80 per cent of the country's ethanol.[93] Since then, cooperative programmes such as Minnesota's have encouraged the construction of jointly owned ethanol mills, such that by 2005, more than half of all ethanol mills were farmer owned and ADM's market share had shrunk to only about 25 per cent.[94] However, about three-quarters of the ethanol plants being constructed in 2005 were not farmer owned, and several large companies, including ADM, have announced plans to increase their capacity dramatically by building larger facilities.[95]

Cargill, an international agricultural processor, is currently building a huge biodiesel plant in the US with a capacity of 189 million litres (50 million gallons), large enough to nearly double the country's 2004 biodiesel production. Simultaneously, Cargill is building large soybean oil production facilities in Brazil and constructing dehydration plants to facilitate imports of Brazilian ethanol to the US via Central America and the Caribbean (regions that enjoy preferential trade status).[96] The largest current US biodiesel distributor, World Energy Alternatives, distributes half or more of the biodiesel in the country and is a subsidiary of Gulf Oil.[97]

Even in Europe, where companies often run under tightly controlled licences, large producers and distributors threaten to dominate the 'value-adding' component of the biofuels industry. ADM is building large biodiesel facilities in Europe, as well, where it is already the second largest biodiesel manufacturer.[98] And in 2005, the EU approved a joint venture between Bunge and Diester, creating the largest biodiesel marketer on the continent.[99]

Concentration in the cellulosic biofuel industry

The next generation of biofuel production facilities will establish a market for far greater amounts of agricultural biomass and promises to create even higher value co-products, further helping rural communities. However, it will also require the development of more capital-intensive production facilities, which could add to the advantages that large companies demonstrate in biofuel production.

The promise of refining previously low-value biomass into boutique fuels and products has already attracted wide-ranging corporate interest. Some companies currently involved in ethanol fermentation hope to become leaders in developing new refining technologies – including Dedini, which is developing a process to hydrolyse and ferment sugar cane bagasse in Brazil, and the enzyme companies Novozymes and Genencor.[100] Other companies are relative newcomers to the field, such as chemical manufacturers Dow and Dupont, both of which have launched efforts to produce a wider variety of biomaterials.[101]

Oil and auto companies are making investments to pioneer the next generation of biofuels, as well. Shell, already involved in biofuel distribution around the world, has invested in the first facilities to demonstrate both enzymatic hydrolysis (with PetroCanada) and gasification of cellulosic matter (in Germany). It has also conducted significant research into a process that can pyrolyse wet biomaterials into a liquid fuel. And auto companies Volkwagen and DaimlerChrysler have partnered to develop a large jatropha plantation in India and are working with Shell on its German demonstration gasification-to-liquid facility (see Chapter 5).

These corporate investments signal that a new 'bio-economy' may, indeed, sprout in the coming decades. It also points to the possibility that still larger companies may enter the rural economy to put the squeeze on farm incomes. If so, the real profits are likely to go not to those who can produce large quantities of biomass feedstock, but to those with the proprietary technology that can ply this feedstock into fuels and products.

FOOD VERSUS FUEL

When considering rapid increases in biofuel production, there is a concern that crops that would otherwise become food might instead become fuel, leaving the world's poorest inhabitants hungry. This concern is important; however, it may be too simplistic. Not only will greater demand for certain crops increase their production, but such demand could bring particular benefits to farmers, who comprise many of the world's poor. While biofuel programmes could raise food prices and contribute to hunger, they could also help to address the root of world hunger: poverty.

So far, biofuel production has, indeed, raised the price of certain foods. For example, biodiesel production in Europe has led to an increase in the price of rapeseed oil, and sugar ethanol production in Brazil has contributed to a rise in the global price of sugar. Such increases in the demand for, and price of, food crops have been a deliberate and fundamental motivation of biofuel programmes as governments aim to protect farmers from excessively low prices (see Chapter 7).

Higher crop prices will not necessarily harm the poorest people. More likely, as with most enterprises, some people will be hurt and some will be helped. While urban slum dwellers are unlikely to benefit from biofuel programmes, many of the world's 800 million undernourished people are farmers or farm labourers, who could benefit.[102] Moreover, if biofuel programmes end up absorbing much of the surplus crop production in industrialized countries, they could spare farmers in the developing world from commodity 'dumping' and artificially low prices.

Poor farmers are more likely to benefit if biofuel production is done in a small-scale, labour-intensive manner – one that keeps them employed and able to afford food. The alternative is large plantations of monocultures controlled by wealthy producers, who could drive farmers from their land without providing new opportunities. In Brazil, where the early years of the *Proálcool* programme did lead to regional food scarcities in the northeast, the government's current embrace of biodiesel is specifically targeted at poverty reduction.[103] By providing families of labourers with a new market for their tree oil crops, the government aims to improve the economic conditions that would otherwise lead to hunger.

In the future, markets for cellulosic biofuel feedstocks offer a promising opportunity to relieve food supplies from direct competition with biofuels. Farmers could preserve the sugary, starchy or oily components of the plant for food and sell the fibrous components as fuels. By adding value to agricultural residues, farmers may even be able to benefit while also selling food at a lower price.

Yet, even cellulosic feedstocks can put pressure on food supplies, particularly if enormous demand for biofuels strains the limits of agricultural potential and productive land. The likelihood of such tension will depend upon a variety of factors, including the ability of agronomists and farmers to further raise agricultural yields, the overall size of the human population, the extent to which calorie-intensive meat and dairy products dominate diets, and the fuel efficiency of people's lifestyles (see Chapter 6).

These factors notwithstanding, the central cause of food scarcity in the world today is, and will likely remain, economic inequality and inadequate food distribution.[104] Since the very poorest people are unable to afford food when prices are set by wealthier consumers, the most immediate question is whether biofuels will help to reduce some of these inequalities.

CONCLUSION

Continued expansion of biofuel production will increase global demand for agricultural products and result in the creation of new jobs at every stage of the production process, from harvesting to processing to distribution. As more countries become producers of biofuels, their rural economies will probably benefit as they harness a greater share of their domestic resources.

But not everyone will benefit equally. Of all the participants in the biofuel economy, agribusinesses are most assured to profit since mechanized harvesting and production chains are the easiest option for rapidly scaling up biofuel production. Large-scale agricultural processors and distributors will be responsible for supplying most of the refined fuels as well. The development of cellulosic conversion technologies will only further exaggerate the advantages of those interests with large pools of financial capital. But the current expansion of biofuel production offers a unique opportunity for policy-makers to avoid some of the pitfalls of existing food industries.

As policy-makers proceed with biofuel programmes, they will need to decide to what extent they want to encourage small farmers or labourers to share in the profits. If this is a priority for governments, then policy options include well-enforced labour standards and profit-sharing agreements, possibly using existing models in the states of São Paulo in Brazil and Minnesota in the US. On the processing side, governments can support smaller-scale producers and farmer cooperatives by requiring fuel blenders to purchase their fuel from them at fair prices.

When considering biofuel programmes for their capacity to promote rural development, decision-makers in industrialized countries must remain mindful of just how important agriculture is to the economies of the developing world. Advocates of rural development at home might consider to what extent they also care about development in other countries that face similar challenges in their agricultural sectors. Restrictive tariffs can benefit rural communities in industrialized countries while harming those in less wealthy countries. At the same time, should industrialized countries begin importing biofuels from developing countries, it may be difficult to enforce international labour standards.

A biofuel industry that is locally oriented – in which farmer owners produce fuel for their own use – is more likely to guarantee benefits to a rural community. In these situations, farmers may risk bad seasons and poor harvest; but, by adding value to their own products and using these goods locally, they are also less vulnerable to external exploitation and disruptive market fluctuations. Although liquid fuels produced at home are often used for cooking or electricity, rather than transportation, it is worth noting that readily available technologies to convert 'modern' biomass into energy promise to be a more directed way of alleviating poverty, especially in more remote oil-dependent regions.

9

International Trade in Biofuels

INTRODUCTION

The current international trade in biofuels is quite small when compared to trade in fossil fuels such as petroleum and natural gas. However, biofuel production is expected to more than double in the coming decade as new government and industry policies promote greater use of renewable energy sources and as the global demand for liquid fuel rises. Growing interest in a wide variety of biomass resources, many of which are underutilized in much of the world, is likely to foster new trading relationships.

The greatest demand for biofuels is concentrated in industrialized regions that consume large amounts of energy, such as the US, the European Union (EU) and Japan, as well as in rapidly industrializing nations such as China and India. The largest potentials for producing these fuels, meanwhile, are found in the tropical countries of South America, sub-Saharan Africa and East Asia, and in Eastern Europe.[1] Trade is a natural outgrowth of such imbalances.

As more countries explore biofuels as an energy source, the costs of meeting energy needs with regionally available feedstocks may guide countries' decisions to trade biofuels internationally. In general, the decision to facilitate trade in these fuels must be balanced with domestic and regional energy needs; large-scale production for export should not pre-empt the development of smaller-scale biofuels production for local use.

This chapter discusses the main risks and opportunities associated with the international trade in biofuels. It describes the current status of this trade and explores the key barriers to its expansion in the future, including tariffs, lack of international fuel quality standards, concerns about environmental and social standards, and a poorly developed market. It also considers biofuel trade in the context of the wider international trading regime.

CURRENT BIOFUEL TRADE

Today, biofuel trade occurs mainly between neighbouring regions or countries, although it is increasingly happening over longer distances. Brazilian ethanol is now exported to Japan, the EU and the US; Malaysia exports palm oil to The Netherlands and Germany; and Canada exports wood pellets to Sweden. This is happening despite the bulky and lower calorific value of most biomass raw material.

Trade in ethanol and related commodities

Only about 10 per cent of the ethanol produced in the world today is traded internationally.[2] Historically, most of this trade has been for non-transportation uses – as a base for alcoholic beverages, as a solvent and for other industrial applications. However, fuel ethanol is becoming an increasingly popular global commodity as oil prices rise and as governments adopt new policies promoting biofuel use.

Ethanol produced from Brazilian sugar cane accounts for the vast majority of liquid renewable fuel traded today. In 2004, Brazil was the world's dominant ethanol exporter, accounting for approximately half of total global trade for all uses.[3] The main recipients of these exports are India, the US, South Korea and Japan (see Table 9.1).[4] Some Brazilian exports also flow into the US indirectly via Central America and the Caribbean, where ethanol is processed and can enter tariff free under the Caribbean Basin Initiative, a regional preferential trading programme.

Several other producer countries, including the Pakistan, the US, South Africa, Ukraine and countries in Central America and the Caribbean, also contribute to ethanol trade, although their relative exports compared to Brazil are quite small. Due to preferential access to the European market, small amounts of ethanol are

Table 9.1 *Brazilian ethanol exports, all grades, 2004*

Importing country	*Exports*[a] *(million litres)*
India	475
US	426
South Korea	239
Japan	209
Sweden	198
Netherlands	156
Jamaica	133
Nigeria	106
Costa Rica	106
Others	361
Total	2447

Note: Figures include fuel, industrial and beverage uses.

Source: see endnote 4 for this chapter

shipped from Africa and Asia to Europe. Pakistan has historically been the largest exporter of ethanol to the EU.[5]

Most of the ethanol traded today is pre-processed ethanol, manufactured in the country where the feedstock is grown, because it has generally not been economical to transport the feedstock long distances for ethanol production. However, some corn from the US is transported to Canada for ethanol production. In the future, the transport of some cereals for ethanol production may take place. Notwithstanding, as sugar is currently the cheapest feedstock, many low-cost producers of sugar cane in Africa, Latin America and Asia plan to increase their share in global ethanol trade (see Chapter 7).

Future ethanol trade will be driven in large part by countries that are not necessarily interested in developing domestic biofuel production, but have a desire to use biofuels to reduce oil dependence and meet carbon emissions targets under the Kyoto Protocol. Likewise, national initiatives to stimulate greater energy independence, boost rural incomes, and support environmental and socially responsible fuel production and use may have marked effects on the international biofuel trade. Japan, for instance, was the fourth largest market for Brazilian ethanol in 2004. In 2005, Brazil's leading oil company, Petrobras, and Japan Alcohol Trading Co launched a joint venture, Nippaku Ethanol KK, to import ethanol into Japan.[6] To accommodate forecasted shipments of 25 million litres (6.6 million gallons) a month, Petrobras will invest €279 million (US$330 million) over the next five years in developing the requisite export infrastructure.[7] Other biofuel-producing nations may develop similar relationships to facilitate trade in ethanol and other fuels.

Trade in biodiesel and related commodities

At present, there is no significant international trade in biodiesel. Germany is the world's largest producer of the fuel (from rapeseed); but this is mainly for use domestically and within the EU. This leaves considerable potential for lower-cost producers to enter the market, including major oilseed producing countries. Of the seven major oilseed crops, just two – soybeans and palm – account for 85 per cent of global oilseed exports.[8] The largest soybean producer and exporter is the US, followed by Brazil, Argentina and China.[9] The largest palm oil producers are Malaysia and Indonesia.

Trade in biodiesel produced from palm oil is projected to increase in the coming years. Both Malaysia and Indonesia have plans to export the fuel to the EU, and Malaysia is also planning exports to Colombia, India, South Korea and Turkey.[10] To satisfy both international and domestic demand – Malaysia aims to substitute all domestic diesel with palm-based biodiesel by 2008 – three new plants are due to begin operation in 2006, producing 100,000 tonnes of biodiesel annually.[11] And the international energy trading company EarthFirst Americas plans to import palm-based biodiesel into the US from Ecuador at a quantity equivalent to half the projected US production of 200 million litres in 2007.[12] This rising trade in palm oil products has raised substantial concerns about forest loss and environmental degradation in producer countries, however (see Chapter 12).[13]

Despite its smaller share of the global market, it appears that the international biodiesel market may also expand rapidly in response to growing global demand. Although Europe currently manufactures more than 90 per cent of the world's biodiesel, both industrialized countries (the US) and developing nations (Ecuador, Indonesia and Malaysia) are building infrastructure to supply regional and international biofuels markets, driven by the need to ensure a secure fuel supply and to reduce dependence upon imported oil (see Chapter 7).

COMPETITIVE ADVANTAGE AND THE BIOFUELS TRADE

As noted earlier, many developing countries have a competitive advantage in biofuels production due to lower land and labour costs, warm tropical climates and a longer growing season. This makes producing biofuels for export a more cost-effective proposition for these countries compared with many industrialized nations. In Brazil, ethanol produced from sugar cane is currently competitive with fossil fuels, at €0.112 (US$0.136) per litre. Several other developing countries have similar costs for ethanol production, including Pakistan, at €0.12 (US$0.145) per litre, and Swaziland and Zimbabwe, at costs close to Brazil's. All of these nations currently export ethanol to the EU.

Some developing nations that currently produce biodiesel for the domestic market have plans to expand their export potential, including Malaysia, Indonesia and the Philippines. Many countries in Central and Eastern Europe also have the potential to produce cheaper biofuel feedstocks (such as wheat or rapeseed), being at a competitive advantage due to low labour costs and high resource availability.[14]

Brazil, Germany, the US and others all have fairly well-developed biofuel industries with advanced technologies for biofuel processing. As biofuel production begins to increase in other countries, these nations will benefit from expanding markets for their technologies (see Chapters 16 and 20).

Biofuels have great promise to provide local energy and substitute for costly oil imports, particularly in those countries that can produce them most cost effectively. But the economic and rural development benefits of producing these fuels in large quantities for export must be weighed against the environmental costs of this production. In general, the fossil energy required to produce ethanol from sugar cane, and biodiesel from palm oil, is lower than that for ethanol or biodiesel produced in Europe, and their corresponding emission reductions are greater. However, this is not the case when virgin forests are razed for biofuel production, as is happening in parts of Southeast Asia. Trade in biofuels is advantageous from a cost and greenhouse-gas minimizing perspective only if it is cultivated on already established agricultural or set-aside lands, or on degraded lands poorly suited for traditional agriculture (see Chapters 11 and 12).[15]

Although developing countries consume much less energy overall than industrialized countries, their demand for liquid fuels is rapidly increasing. This raises the question of whether biomass fuels should primarily be used locally or exported – and whether market forces should have the final say. Brazil, for example, is planning to increase its ethanol production dramatically over the next eight years, and it is beginning to produce biodiesel from soybeans, castor and palm oil. Only a fraction of this will be exported. For most countries, the main drivers determining this choice will be the relative prices of petroleum fuel versus plant oils and ethanol on the international market, as well as incentives offered for the production of renewable fuels. In general, it would be more rational to use biofuels locally and export only the excess – bearing in mind that international competition will force domestic producers to be more competitive.

POLICIES AFFECTING INTERNATIONAL BIOFUEL TRADE

Some countries with large agricultural sectors have shown resistance to opening their markets to foreign biofuel imports. These policies create barriers for many producers in developing countries who wish to significantly expand their production and export of biofuels in the coming years. Policies in the form of tariffs that keep developing country commodities out of industrialized markets or subsidies that support industrialized country biofuel producers may be equally discouraging for market entrants.

However, trade agreements have great potential to increase the global penetration of biofuels into fuel markets. Industrialized countries grant developing countries market access or negotiate trade agreements that lower barriers to trade in biofuels and related commodities, in order to encourage the latter's economic development and to benefit domestic consumers. The future of biofuel trade will be affected by these changing trade policies, especially current negotiations on the production and trade of agricultural products.

Tariff policies affecting biofuel trade

The EU, US and Australia are among the large agricultural exporting economies that have imposed import duties and other restrictions on foreign ethanol and biodiesel and their agricultural inputs (see Figure 9.1).[16] Simultaneously, the EU and US both offer preferential market access to developing countries by way of unilateral tariff reductions that encourage imports of certain agricultural commodities and biofuels.

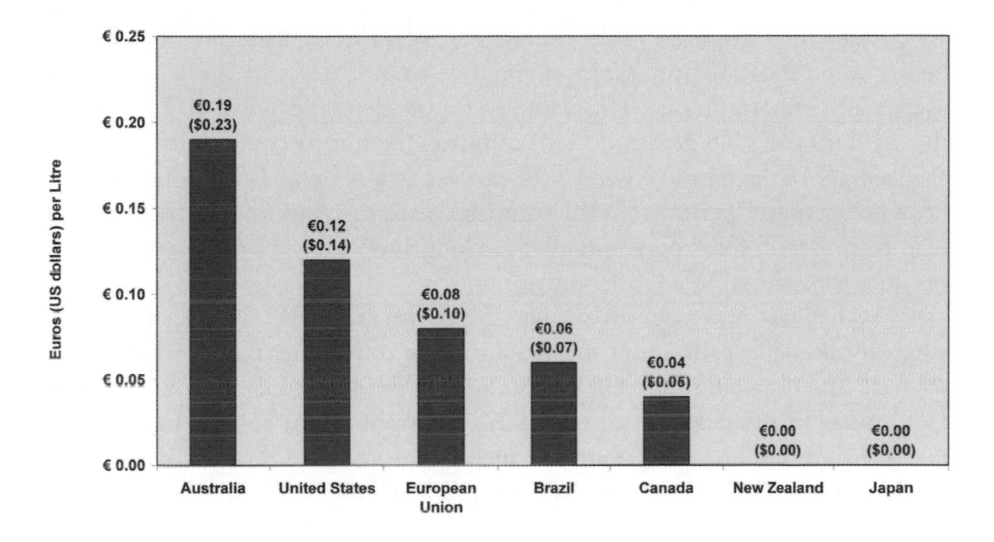

Figure 9.1 *Ethanol import duties in selected countries, 2004*

Source: Fulton et al (2004) p185

Ethanol is taxed at varying rates depending upon its intended use. In the EU, the import duty for undenatured (pure) alcohol is €0.19 per litre, while for denatured alcohol (ethanol with additives) it is €0.10 per litre.[17] Despite the differing tariff rate, both denatured and undenatured alcohol is imported under customs classification 2207 in Europe, making it difficult to identify how much ethanol is used for fuel production. Only fuel ethanol that is pre-blended with gasoline is classified separately under heading 3824 and charged a normal customs duty of around 6 per cent.[18] In the US, ethanol is classified under the agriculture chapter and again under chapter 99 for fuel-grade ethanol. The US taxes ethanol imports at €0.15/US$0.18 per litre (€0.44/US$0.54 per gallon).[19]

Biodiesel imports are also taxed at varying rates, due in part to the different feedstock options. Global trade in whole oilseeds, particularly soybeans, is relatively unrestricted by tariffs and other border measures; however, oilseed meals, and particularly vegetable oils, have higher tariffs. For soybean oil, tariffs average around 20 per cent, while tariff rates for whole soybeans are generally around 10 per cent.[20] In the EU, plant oils for biodiesel face low or no tariffs. For biodiesel in the form of fatty acid methyl ester (FAME) imported from the US, a non-member state duty of 6.5 per cent applies, and there are no quantitative restrictions. In addition, these conditions apply only to the import of the biodiesel (FAME) itself, not to the import of source products such as tallow or used cooking oil. Rules and tariffs governing straight vegetable oils (SVOs) are separate and specific because of the potential for these oils to enter into food production.

Under the Caribbean Basin Initiative (CBI), the US exempts from import tariffs some ethanol from Central American countries and the Caribbean – specifically, imports produced from foreign feedstocks that equal up to 7 per cent of the previous year's US demand.[21] CBI countries have never come close to meeting this ceiling; in the past five years, CBI exports as a share of US production have hovered around 3 per cent.[22] CBI countries also may import feedstocks or fuel (e.g. from Brazil) for export to the US, as long as 35 per cent of the value of the product is produced in a CBI country.[23]

The Central American Free Trade Agreement (CAFTA) will supersede CBI when it takes effect for countries that are party to it, potentially including five Central American countries and the Dominican Republic. Like CBI, it will allow continued tariff-free exports through CAFTA countries for ethanol produced by non-CAFTA and non-CBI countries, such as Brazil, up to the 7 per cent cap of total US production.[24] All other ethanol produced by CAFTA, or CBI country feedstocks, can be imported tariff free. CAFTA was supposed to take effect in January 2006 but has been delayed due to unresolved legal issues, including pending approval by some legislatures in Central America.[25]

In Europe, the EU grants special trading preferences to Africa, the Caribbean and the Pacific (ACP) countries. The EU also has a General System of Preferences (GSP) that encompasses all developing nations. This agreement granted tax preferences for ethanol from 12 countries as a part of an anti-drug regime and had duty-free access for ethanol to the EU market until 2005.[26]

Countries under the so-called Everything but Arms (EBA) initiative are exempted from EU duties on ethanol (and all other exports, except for sugar, rice and bananas that will be covered by exemptions from 2009 onwards). Under this agreement, significant (though erratic) exports come in from the Democratic Republic of Congo, varying from 86,000 litres in 2003 to 19,000 litres in 2004.[27] Altogether, biofuel imports into the EU under preferential trading arrangements nearly doubled between 2002 and 2004 to 3.1 billion litres (see Table 9.2).[28]

As a result of the GSP, Pakistan was the largest supplier of ethanol to the EU for much of the past decade, producing a range of 1.3 million to 2.1 million tonnes of the fuel from sugar cane during the period of 1994 to 2004.[29] In July 2005, however, the World Trade Organization (WTO) ruled that the EU was unevenly granting preferences to the 12 countries included under this policy. As a result, a new GSP Plus system has been designed. Under this regime, 15 countries that ratified certain international agreements on human and workers' rights, as well as on environmental protection, were granted tax-free import preference as of January 2006. Pakistani ethanol is no longer eligible for tax reduction on its exports to the EU, a change that has caused two ethanol plants in Pakistan to close and halted plans for seven new plants. Other countries may step up to fill the gap.[30]

The EU is also in the process of conducting negotiations with Mercado Commun del Sur, or MERCOSUR (the Latin American trade bloc) that would significantly lower or remove trade barriers for these countries; however, negotiations

Table 9.2 *Biofuel imports into the European Union under preferential trading arrangements, 2002–2004*

Trade agreement	*2002*	*2003*	*2004*	*Average, 2002–2004*	*Share of total biofuel trade, 2002–2004*
			(million litres)		*(percentage)*
General System of Preferences (GSP) normal	227	183	288	233	9
GSP Anti-drug preference	553	1569	1413	1178	47.5
Africa, the Caribbean and the Pacific (ACP)	291	269	155	238	9
Everything but Arms (EBA) initiative	30	86	19	45	1.5
Others	107	104	123	111	4
Total preferential	1208	2211	1998	1805	70
Total MFN (most favoured nations)	657	495	1125	759	30
Grand total	1865	2706	3123	2564	100

Source: see endnote 28 for this chapter

have stalled. Conclusion of a MERCOSUR agreement could allow large amounts of Brazilian ethanol to enter the EU.[31] European nations with higher production costs, such as France, Spain and Sweden, have voiced concerns that they could be negatively affected by such a change, so some limits would probably be imposed (e.g. tariff rate quotas).[32]

The relative tariff levels levied on developing country exports can largely determine the degree of success for emergent biofuel industries (over 60 per cent of ethanol imported to the EU was imported tariff free). Similarly, the quantity and placement of agricultural subsidies has a profound effect on the quantity and type of feedstock available for biofuels production.

Agriculture subsidies and the biofuel market

Protection of domestic agricultural interests in many countries is in large part responsible for the hampering of international agricultural trade. The US corn sector – the second largest player in the international ethanol market – is also the largest recipient of the nation's agricultural subsidies, receiving nearly €34.7 billion (US$42 billion) in government payouts over the last ten years.[33] Similarly, the EU's Common Agricultural Policy (CAP) contributes to one third of farmers' incomes, according to the Organisation for Economic Co-operation and Development (OECD) producer support estimate index. Before the recently introduced reform, EU sugar subsidies cost consumers and taxpayers an estimated €3.6 billion

(US$4.4 billion) in 2004, and also compromised the competitiveness of developing countries (see Chapter 7 for more on the role of subsidies).[34] The redirection of export subsidies from classic agricultural commodities toward biofuels can increase the opportunities for developing countries to meet demands for biofuels as their traditional agricultural capacities are affected by changes in the structure of the world market for agricultural commodities.

Removing or reducing these subsidies could change the biofuel trade landscape. The US Department of Agriculture (USDA) estimates that full liberalization of the oilseed trade (meaning both subsidy elimination and tariff reduction) would result in a 5.3 per cent decline in industrialized country outputs, but a 4.8 per cent rise in developing country outputs and an overall increase in world trade volume of 11.4 per cent.[35] Proposed reductions in the WTO's Doha Round of trade negotiations could reduce corn and sugar subsidies and increase opportunities related to these commodity markets.

Although they are not direct subsidies, biofuel targets that ensure demand are projected to have definite effects on global trade flows. Supports via biofuel mandates for the US and EU are projected to increase EU imports of vegetable oil threefold and could cause EU wheat exports to fall by as much as 41 per cent. Canadian wheat and coarse grain exports would be reduced by 13 and 34 per cent, respectively, in 2014.[36] Lower industrialized country exports due to increased demand from government-mandated programmes can thus create increased opportunity for developing country producers.

In preparation for likely future reductions in price supports for traditional agricultural commodities, some governments are considering leveraging their massive funds for agriculture by transforming them into subsidies for the production of renewable fuels (see Chapter 7). The USDA estimates that expansion in US ethanol production has helped to reduce farm programme costs in the US by some €2.65 billion (US$3.2 billion) by reducing food programme supports, a fact that has not been forgotten in anticipation of the next Farm Bill in 2007.[37] But transferring agricultural funds to biofuels may end up subsidizing inefficient production. Under the EU sugar reform plan, sugar producers with costs well above world market prices can receive funds to convert sugar factories into ethanol plants that use either sugar beet or grain as inputs.[38]

If industrialized countries transform traditional agricultural subsidies into supports for domestic energy crop production, emerging developing country producers of energy crops will operate at a significant disadvantage. However, if these subsidies are used to develop more efficient next-generation feedstocks, emerging producers could remain competitive in the near term. Additionally, subsidy schemes can be implemented to periodically re-evaluate supports as products become competitive, and can be used to fund new technologies in early phases of research and demonstration.

Proposed agricultural trade reforms do include some provisions for developing countries. For instance, biofuel exports from ACP countries are planned to receive continued support in the wake of contentious debate over world sugar trade at the WTO. In response to the concerns of countries receiving preferential market access about the reduction in supports to sugar producers in the EU – which have historically kept prices higher in the European market – the EU has promised to provide generous development aid to all 18 ACP countries.[39] The EU sugar reform plan ensures that the production of biofuels 'derived from agricultural crops that can be used to partially replace liquid petroleum products' will not be adversely affected.[40] For energy crop cultivation for biofuels within the 25 EU member states (EU-25), sugar beet will be eligible for EU energy crop aid, worth €45 per hectare. Sugar used for the production of ethanol, as well as by the chemical and pharmaceutical industries, will be excluded from the sugar quota. Since sugar reform is a major challenge not only for EU beet and sugar producers, but also for many ACP suppliers, the European Commission has devised an assistance scheme to respond to the diversity of situations in the different ACP countries covering a broad range of social, economic and environmental actions.[41]

Decisions that industrialized countries make with regard to shifting their own subsidies towards non conventional crops can affect the supply of both conventional and non-conventional crops. What will happen if industrialized countries move towards subsidizing energy crops and away from supporting conventional crops? Will prices for agricultural commodities rise, bringing potential benefits to the rural poor (see Chapter 8)?[42] The answers to these questions will largely affect opportunities for developing country agriculturalists in both conventional agricultural commodities and biofuels export.

Multilateral trade policies

As biofuels markets expand, the international legal framework governing the flows of agricultural commodities, alcohols and alternative energies will be central to the future of biofuels trade. Likely future reductions in agricultural trade barriers have brought increasing pressure on industrialized countries to redesign their national agricultural policies and respond to ongoing trade negotiations. In the WTO's Doha Round of trade talks, ongoing negotiations that will be of particular importance are those covering the liberalization of agricultural trade and environmental goods and services (EGS).

Agricultural versus non-agricultural goods

A key factor in the development of future biofuel policies will be whether a biofuel will be considered an 'agricultural' or 'non-agricultural' good under WTO agreements. If a biofuel (e.g. ethanol) is an agricultural good, its agricultural production may be eligible for so-called 'green box' classification under the WTO's Agriculture Agreement, which covers subsidies for activities that encourage environmental stewardship or land conservation, among other non-trade distorting activities. However, if a biofuel is a non-agricultural good (e.g. rapeseed methyl ester, or RME), then it is subject to the Subsidies and Countervailing Measures (SCM) agreement on non-agricultural subsidies and will not be eligible for this type of environmental exemption. The subsidies could be challenged at the WTO if they are not in line with the SCM.[43] If a biofuel is deemed a non-agricultural good, then the question remains whether it will be treated as an 'environmental' good (discussed below), in which case it could receive significantly deeper tariff reductions than those applied to other non-agricultural goods.[44]

Environmental goods and services (EGS)

Formal talks on environmental goods and services (EGS) began in 2001 as a supplement to the more general discussions on international trade in services under the WTO. Intended to enhance the mutual supportiveness of trade and environment, these negotiations set out to reduce or eliminate tariff and non-tariff barriers to EGS. They are of particular relevance to biofuels – for instance, if materials for the construction of next-generation biorefineries were to be included on the lists of environmental goods and services, construction could be expedited because tariffs on imported technologies and equipment for biofuels and other renewable fuels could be lowered or eliminated.

Negotiations on what constitutes an environmental good are still in flux. The US is proposing a 'list approach' that is based on pre-existing lists of 'environmental goods'. India and other developing countries have actively opposed classifying EGS based on process and production methods (PPMs) (how goods are made). Further discussion of this debate and how it applies to certification is found in Chapter 18; however, it is unlikely that PPMs will be incorporated within the definition of biofuels as environmental goods in 2006 negotiations.[45]

Nevertheless, many European countries would like to specify certain production criteria for biofuels – in particular, sustainable agricultural practices such as cover cropping and practices that help to retain crop residues on the ground for cellulosic feedstocks (see Chapters 12 and 18).[46] Furthermore, classification of certain organic matter streams as biomass fuels may aid WTO classification of biofuels as EGS, although the marginal environmental gains compared to cost must be further explored. It is possible, if not probable, that in future negotiations WTO member countries could mandate certain sustainable practices for generating inputs or other requisites for biofuels to qualify as EGS.[47]

Until such issues are addressed, importing countries may attempt to establish minimum standards for goods entering their countries. Reciprocal agreements between countries with importing and exporting relationships for biofuels are not likely to be challenged at the WTO. For wider global biofuel trade, an important step towards sustainability would be the development of objective environmental performance criteria for biofuels and relevant co-products, set out in credible international, regional or domestic standards.[48] Such an approach is discussed in Chapter 18.

SOCIAL AND ENVIRONMENTAL ISSUES AND THE BIOFUEL TRADE

Less prevalent in trade discussions, although a growing concern among agricultural and environmental groups, is the potential for expanded international biofuels trade to cause social and environmental harm. Like any internationally traded commodity, biofuels will have a greater impact the more widely they are traded around the globe. Here, as with many other trade-related issues, effective national policies will be of critical importance, as international trade often serves to only exacerbate local problems.

The primary social and environmental concerns being raised by environmental and agriculture groups include growing competition for land; the ecological implications of industrial crop monocultures; genetic engineering of biofuel feedstocks; and exploitation of farm workers. There is also concern about lower environmental and labour standards of production in developing versus industrialized countries (so called 'cross-compliance' issues) (see Chapters 12 and 13).

Genetic engineering techniques used to boost the efficiency of crop yields and protect against insect infestations could raise particular concern.[49] In the past, India and the EU have been reluctant to accept genetically modified (GM) agricultural products into their markets (especially for human consumption), and this could influence their willingness to import biofuels from countries that allow GM crops. The Cartegena Protocol on Biosafety under the UN Convention on Biological Diversity (CBD) allows countries to employ the 'precautionary principle' to refuse goods that have used biotechnology anywhere in the production process (see Chapter 12).[50]

RISKS AND OPPORTUNITIES FOR MARKET DEVELOPMENT

Some proponents of biofuels envision a future international biofuel trade that will develop over time into a real 'commodity market' that secures supply and demand in a sustainable way – sustainability being a key factor for long-term security. However, a number of policy and institutional barriers exist that can cause market distortions and harm market entry for biofuels. In addition to the tariffs, subsidies, and social and environmental obstacles already discussed, these potential barriers include increased control of the biofuels market by the oil industry (which could lead to price manipulation) and lack of infrastructure to enable biofuels to be used in vehicles.

Factors leading to unreliable supply and demand also create market uncertainty and could impede biofuel development. Among the economic barriers that these markets currently face are:

- competition with fossil fuels on a direct production cost basis (excluding environmental and social externalities);
- insufficient, unpredictable and/or inconsistent support policies promoting biofuels in many industrialized and some developing countries; and
- relatively immature and unstable markets that are perceived as too risky for long-term or large-volume contracts.

The biofuel market also remains vulnerable to factors outside the control of trade boards and financiers. First-generation biofuels are vulnerable to crop failures and market prices for food. And because they comprise such a tiny share of the global energy trade, they will continue to be price takers in the short and medium term (meaning that prices of biofuels will mirror spikes and dips in oil prices).

In response to these challenges, several mechanisms for reducing risks related to short-term imbalances in biofuel supply and demand are in the early stages of development. In May 2004, the New York Board of Trade took a step towards building institutional support for ethanol in the global market by negotiating an ethanol futures contract; as a result, ethanol is now traded under the symbol 'XA'.[51] This backing from the New York Board – a well-established global futures and options market for internationally traded agricultural commodities – may provide both producers and consumers with a greater level of assurance that their price and quantity needs will be satisfied, attracting more capital to the ethanol industry. However, some have expressed concern that a lack of transparency in commodity trading could hinder the biofuel market.[52]

Harmonized support policies (e.g. on the EU level) and new national incentives for biofuels offer opportunities for formalizing and stabilizing the international biofuel trade by guaranteeing greater overall demand. The EU Strategy for Biofuels, released in February 2006, calls for greater guarantee of supply and demand for biofuels through a framework of incentives for publicly and privately owned vehicle fleets, including city and private bus fleets with dedicated fuel supplies (which can be easily adapted to higher blends of biofuels), farm and heavy goods vehicles (which would receive continued tax exemptions), and fishing fleets and vessels (which offer a potential market for biodiesel).[53] Towards a similar end, the Philippines and Thailand agreed in 2004 to strengthen bilateral and regional cooperation to promote biofuels by moving towards a regional standard for ethanol-blended gasoline, and by pushing the Association of Southeast Asian Nations (ASEAN) to encourage automobile manufacturers to make flexible-fuel vehicles.[54]

Technical and logistical risks of biofuel trade

While countries continue to establish stronger relationships that bolster future trade in biofuels, existing infrastructure in major markets such as the EU and the

US may continue to hamper these international fuel flows. For instance, concerns remain with regard to transit times and costs, as well as the integration of biofuels into existing industrial and consumer transport uses. What follows is a list of some of the major technical and logistical considerations for establishing an international biofuel trade:

- In the longer term, the limited ability to use different fuels may lead to a restricted availability of biomass fuels. Vehicle and industrial installations must be compatible with biofuels. If technology is not available or installed to use biofuels, there may be a limited demand and, thus, less production.
- Local transportation by truck may be very costly (both in biomass exporting and importing countries). For example, in Brazil, new sugar cane plantations are being considered in the centre-west; but the cost of transport and lack of infrastructure may be a serious constraint in the short term. Whether harbours and terminals have the capability to handle large biomass streams may hinder the import and export of biomass to certain regions. End-users located near harbours will be better able to avoid additional transport by trucks. If large-scale biofuels production is developed in new regions, it may take considerable time and capital to build more efficient transport infrastructure.
- Lack of significant volumes of biomass can decrease trade incentives. In order to achieve low costs, large volumes need to be shipped on a more regular basis. Only if this occurs will there be forthcoming investment on the supply side (e.g. the construction of pipelines to transport biofuels), and this will reduce costs significantly.
- Disparities in production, export and import conditions may slow the development of biofuels trade. The most efficient producers of biomass (e.g. certain developing countries) may not have the highly developed infrastructure necessary for trade and may be geographically further from regions with highest demand. Investment in countries with low production costs to facilitate the export of first-generation biofuels may be unattractive to countries producing these fuels and transporting them using existing infrastructure.

In response to these problems, many governments are offering incentives to those entrepreneurs willing to develop next-generation biofuels (see Chapters 4 and 17). Foreign direct investment may also encourage the development of more stable infrastructure for biofuels transport. Already, China has invested more than €49.6 billion (US$60 billion) in Brazil, Argentina and Angola over the last five years to meet demand for agricultural products, €24.8 billion (US$30 billion) of which was directed to Brazil largely for infrastructure improvements to facilitate agricultural exports.[55]

CONSISTENT FUEL STANDARDS FOR THE INTERNATIONAL BIOFUEL TRADE

Fuel quality standards are an essential building block of successful international trade in biofuels for transport. Any fuel designed for use in modern vehicle engines, which have tight tolerances and advanced control systems, must meet stringent quality-control standards. Because biofuels are a relatively new fuel alternative, there is greater uncertainty regarding the performance of these fuels than for conventional gasoline or diesel fuels. Without fuel quality standards, there is a higher potential that bad batches of fuel will enter fuel distribution channels and that resulting negative experiences might damage consumer confidence in biofuels – as occurred in Australia and set back the industry for a substantial period of time.

Biodiesel, in particular, can show significant differences in basic feedstock material. The saturated fat content of different vegetable oils varies considerably, which can affect key physical characteristics of the biodiesel produced, such as viscosity and 'cloud point' (where wax crystals begin to form as fuel temperatures drop). The most commonly used oil for biodiesel in Europe, rapeseed, has a low saturated fatty acid content, which makes it much less prone to fuel-gelling problems in cold winter operating conditions. However, biodiesel fuels made from other vegetable oils that are more highly saturated can raise issues of clouding or gelling that must be addressed in order to avoid problems in vehicle start-up and operation (see Chapter 15).

For this reason and others, the European standard for biodiesel, though not raw-material specific, has technical parameters that limit the non-RME content to a maximum of around 25 per cent. Expecting that EU demand for biofuels will exceed economical levels of production and encourage greater imports, the EU Strategy for Biofuels, released in February 2006, suggests amending the biodiesel quality standard EN 14214 to facilitate a wider range of vegetable oils, as long as there are no significant negative effects on fuel performance.[56] From 2007 onwards, German oil refineries will be required to blend 2 per cent biofuel content in petrol until 2009, and 4.4 per cent biodiesel content in conventional diesel.[57]

Already, other biofuel standards are in place, although they are not applied globally. The American Society for Testing and Materials (ASTM) initiated the development of biodiesel standards in 1994.[58] Recognizing the need for biodiesel quality control, many countries use the ASTM standards to assure the quality of biofuels (see Table 9.3).[59] For example, the US has adopted ASTM standard D 6751 for pure biodiesel fuel used in blends of up to 20 per cent with diesel fuel.[60] Biodiesel fuel that meets this standard and is legally registered with the US Environmental Protection Agency is considered safe for sale in the US. Similarly, Germany established a stringent biodiesel pre-standard specifically for fuel made from rapeseed oil (rapeseed methyl ester) in 1997. European standards for biofuels are set by the European Committee for Standardization (CEN) and implemented by national standardization bodies – for example, the Deutsches Institut für Normung eV (DIN, the German Standards Institute). For B100, the quality control standard is DIN EN14214; the technical fuel specification for diesel, DIN EN 590, allows blends of up to 5 per cent biodiesel, fulfilling EN 14214. As standards are set by the market stakeholders, policy has only an indirect influence.

An example of a system of fuel quality standards that addresses biofuels directly is the BQ-9000 Quality Management Program, established in the US in 2005 by the National Biodiesel Accreditation Commission. The intended benefit of this voluntary system is to provide biodiesel users, as well as engine and vehicle companies, with a feeling of confidence about the fuel. Key objectives of BQ-9000 are to:

- promote the commercial success and public acceptance of biodiesel;
- help ensure that biodiesel fuel is produced to and maintained at the industry standard, ASTM D 6751; and
- avoid redundant testing throughout the production and distribution system.

Table 9.3 *Standards and specifications for biodiesel and ethanol in selected regions or countries*

Country/region	*Fuel type*	*Standards and specifications (year adopted)*
Australia	Biodiesel	Adapted from the European Committee for Standardization (CEN) and the American Society for Testing and Materials (ASTM)
Austria	Biodiesel	ONORM C1191 ON (1997)
Brazil	Biodiesel	ANP 255 (based on ASTM D6751 and CEN EN 14212)
Czech Republic	Biodiesel	CSN 656509 (5% rapeseed methyl ester, or RME); CSN 656508 (30% RME) (1998)
European Union	Biodiesel	EN 14214 and Deutsches Institut für Normung (DIN) EN 590 (diesel standard allowing 5% biodiesel)
	Ethanol	DIN EN 228 (gasoline standard allowing 5% ethanol or 15 ethers)
France	Biodiesel	Gazette Officielle (1993)
Germany	Biodiesel	DIN E 51606 (1997); replaced by EN 14214
Italy	Biodiesel	UNI 10946
Japan	Biodiesel	Not yet implemented, but under consideration
Philippines	Ethanol	Regional standard (2004)
South Africa	Biodiesel	Standard based on EN 14214
South Korea	Biodiesel	Modified ASTM 121-99 (plan to develop own standards by 2006)
Sweden	Biodiesel	SS 155436
Thailand	Ethanol	Regional standard (2004)
US	Biodiesel	ASTM D 6751 (100%)

Source: see endnote 59 for this chapter

This system accredits companies (both producers and marketers) rather than fuel, and helps to ensure that all biodiesel produced and sold in the US will meet the same standard, D 6751. The programme is a unique combination of the ASTM standard for biodiesel, ASTM D 6751, and a quality systems programme that includes storage, sampling, testing, blending, shipping, distribution and fuel management practices.[61]

Quality-control standards have been adopted for ethanol, as well. For example, in Europe, ethanol blends of up to 5 per cent in gasoline are subject to the technical fuel standard DIN EN 228. In 2004, the Philippines and Thailand established a regional standard for ethanol-blended gasoline in order to strengthen bilateral and regional cooperation in promoting consumer confidence and overall use of biofuels.[62] Several other countries have adopted standards for biofuels as well, while others are in the process of doing so.[63]

The International Energy Agency's Task 40 may contribute to biofuels standardization by collecting information on technical specifications required by consumers and conveying them to potential suppliers.

Conclusion

In the future, international trade policies and changes in the world market for agricultural goods are expected to positively affect growth in the international biofuels trade. As the demand for non-fossil liquid fuels grows, countries will increasingly adopt and refine standards for biofuel quality and will support advancements in compatible transportation infrastructure. However, careful policy planning will be needed to ensure opportunities for sustainable trading relationships that support socio-economic development in the world's rural and agricultural regions.

In the near term, it is likely that a few large-scale producers, such as Brazil and Malaysia, will be the major exporters of biofuels. Importing nations may have the opportunity to set requirements for biofuel production and harvesting in these countries through bilateral agreements. Alternatively, these countries may prefer to keep and use their own biofuels to substitute for costly oil imports and conserve foreign exchange. Biofuels trade could also be hampered by countries that wish to protect domestic agricultural interests by imposing tariffs and other trade barriers that restrict foreign imports.

Trade policies can play an important role in advancing human and economic development in the world's rural areas. Preferential trade agreements and other policy instruments can be used to alleviate poverty in developing countries by helping these countries to generate revenue through increased biofuel exports, or by allowing freer movement of technologies related to biofuels and other renewable energy technologies. Industrialized countries can also help poorer nations by transitioning distorting agricultural subsidies towards domestic energy crop production. These subsidies should be used to develop more efficient next-generation feedstocks that do not compete with traditional export commodities and hurt developing country producers.

Choices made in the trading arena will influence the overall sustainability of biofuel markets as well. There is a danger that large multinational cartels – whether oil companies or agribusiness companies – will simply replace unsustainable oil production with unsustainable biofuels production, maximizing profit but yielding undesirable social and environmental outcomes.[64] Ramping up biomass production for biofuel export could exacerbate many of the same problems caused by the export of traditional cash crops and energy resources, including ecological damage and fragmentation of agricultural communities.

One of the most touted energy security benefits of biofuels vis-à-vis oil is the creation of a more diversified and dispersed liquid fuel supply. While some suggest that ramping up large-scale production in tropical countries would achieve this end most efficiently, others suggest that biofuel production that meets local demand, rather than duplicating the centralized market control that oil companies display, is preferable. As production expands, the issue of whether biofuels should best be used locally or exported must be considered carefully, particularly in light of the various rural development and environmental implications. In general, trading relationships that benefit all contributors to the biofuels production chain – including small-scale farmers – will yield the greatest social, economic and environmental benefits.

In the next five to ten years, a technological transition is expected to take place that will replace first-generation biofuels with a second generation of fuels with substantially improved economic and environmental performance. The prices of these fuels relative to fossil fuels will be crucial in determining the rate of development of the world market for biofuels.

As biofuel production expands, additional trade-related questions will arise, including how to develop standards for fuel feedstocks that are also food commodities, and whether biofuel standards should be voluntary or mandatory. There are also questions regarding how to develop these policies in accordance with WTO regulations, while also meeting demands for biofuel production that is both environmentally and socially sustainable. These issues are addressed in greater detail in Chapter 18.

Part IV

Key Environmental Issues

10

Energy Balances of Current and Future Biofuels

INTRODUCTION

One of the primary incentives for expanding the production and use of biofuels worldwide is the potential environmental benefit that can be obtained from replacing petroleum fuels with fuels derived from more-renewable biomass resources. From an energy perspective, however, not all biomass is created equal, nor are all biofuel production processes equally efficient. When considering a biofuel promotion strategy, it is therefore useful to know which biofuels require more or less energy.

While biofuels themselves consist solely of energy photosynthesized with sunlight, producing them requires human effort and outside energy resources. Farmers capture the 'free' energy of the sun by seeding, watering and fertilizing plants. The biomass that grows must then be harvested, transported and refined into a liquid fuel. All of these activities use energy; but people can choose between more and less efficient biofuel production pathways. Some feedstocks are more efficient and easier to process than others, and some farming and refining methods are more energetically frugal than others.

This chapter examines the relative energy requirements of both current and future biofuel production pathways. It discusses ways of measuring the energy performance of biofuels and compares different fuel options based on these criteria. It also explores potential pathways for improving both the energy efficiency and fossil energy balance of biofuels, including options for achieving energy savings from these fuels in the near term.

MEASURING ENERGY PERFORMANCE

There are two primary measures for evaluating the energy performance of biofuel production pathways:

1. *Energy balance* – the ratio of energy contained in the final biofuel to the energy used by human efforts to produce it. Typically, only fossil fuel inputs are counted in this equation, while biomass inputs, including the biomass feedstock itself, are not counted. A more accurate term for this concept is fossil energy balance, and it is one measure of a biofuel's ability to slow the pace of climate change.
2. *Energy efficiency* – the ratio of energy in the biofuel to the amount of energy input, counting all fossil and biomass inputs, as well as other renewable energy inputs. This ratio adds an indication of how much biomass energy is lost in the process of converting it to a liquid fuel, and helps to measure more and less efficient conversions of biomass to biofuel.

To illustrate the difference between these concepts, consider the example of wheat ethanol. The energy balance represents the number of joules contained in the ethanol divided by the number of joules used by people to plant, nurture, harvest and refine the wheat grain. The energy efficiency, as defined here, represents the number of joules in the ethanol divided by the number of joules used by people to plant, nurture, harvest and refine the wheat grain, as well as the number of joules contained in the grain that is harvested. While the energy balance can exceed 1, the energy efficiency can never exceed 1 because some of the energy contained in the feedstock is lost during processing.

Table 10.1 *Definitions of energy balance and energy efficiency*

Energy balance	*Energy efficiency*
Joules contained in the biofuel. Joules used by people to plant seeds, to produce and deposit agricultural chemicals, and to harvest, transport and refine the feedstock.	Joules contained in the biofuel. Joules used by people to plant seeds, to produce and deposit agricultural chemicals, and to harvest, transport and refine the feedstock (including joules contained in the feedstock).

Currently, tropical plants have more-favourable energy ratios because they grow in more ideal conditions for using sunlight and water and because they are often cultivated manually, with fewer fossil energy requirements and fewer inputs of fertilizer and pesticides. Temperate biofuel production pathways are usually less efficient, although they have become significantly more efficient in recent decades as agricultural practices have improved and fuel production mills have streamlined their operations. In the future, the energy cost of refining biofuels from lignocellulosic biomass will probably continue to exceed that of producing biofuels with conventional starch, sugar and oil; but these lignocellulosic biofuels will bring with them greater quantities of residue bioenergy to use as processing energy.

Fossil energy balance

For biofuel promotion efforts aimed at reducing the use of fossil fuels, the fossil energy balance is a useful metric, although it is often called simply 'energy balance' because fossil fuels are the predominant energy inputs in the US and the European Union (EU). Because they release sequestered greenhouse gases, sulphur, particulates, volatile hydrocarbons and metals, fossil fuels generally do more harm to the environment than renewable fuels; thus, the more fossil fuel inputs a certain biofuel requires, the less energetically desirable it is.

Table 10.2 compares the approximate fossil energy values for various biofuels.[1] Some production pathways are much more favourable than others, depending upon the productivity of the crop, its responsiveness to fertilizer and irrigation inputs, the need for chemical pesticides, and the difficulty of harvesting and refining it into a fuel. Most importantly, a biofuel's fossil energy balance depends upon how much of these energy needs are provided by fossil fuel energy. Although both Brazilian sugar cane and cellulosic feedstocks such as switchgrass are productive plants, their especially favourable fossil energy balances are largely due to the fact that they are processed using the energy of biomass residues available at the mill.[2]

Ethanol feedstocks such as sugar beets, wheat and corn have been criticized because their fossil energy balance is close to 1, a threshold many consider the line between an energy sink and an energy source. But this view fails to account for two important nuances. First, ethanol is a liquid fuel that has qualities that make it useful in the existing transportation infrastructure. Since the natural gas and coal used to produce ethanol do not have this quality, it can be practical to lose energy in the process of converting these fuels into ethanol. Second, even crude petroleum must be refined into usable liquids. Diesel and gasoline have fossil energy balances between about 0.8 and 0.9, numbers that are more relevant for comparison than 1.

Energy efficiency

While the fossil energy balance accounts for just fossil energy inputs, measures of energy efficiency also include the biomass energy used to produce the fuel. Because this ratio includes the energy contained in the feedstock, energy efficiency must always be less than 1. Since petroleum feedstocks are counted in fossil energy balance as well, the energy efficiencies of producing gasoline and diesel are equivalent to their fossil energy balances.

While solid biomass energy may be more climatically benign than fossil fuels, it is not exactly 'free'. It can be used for other purposes, and it can also be used inefficiently. As discussed in Chapter 6, there are competing uses for biomass beyond the provision of transportation fuels, which could be more efficient. As cellulosic conversion processes develop, the efficiency of using them for liquid fuels should continue to be compared to the efficiency of using them for electricity, heat or material production.

Table 10.2 *Fossil energy balances of selected fuel types*

Fuel (feedstock)	*Fossil energy balance (approximately)*	*Data (bracketed) and source of information*
Cellulosic ethanol	2–36	(2.62) Lorenz and Morris (1995)
		(5+) DOE (2006)
		(10.31) Wang (2005)
		(35.7) Elsayed et al (2003)
Biodiesel (palm oil)	~9	(8.66) Azevedo (2005)
		(~9) Kaltner, cited in Azevedo (2005)
		(9.66) Azevedo (2005)
Ethanol (sugar cane)	~8	(2.09) Gehua et al (2005)
		(8.3) Macedo et al (2005)
Biodiesel (waste vegetable oil)	5–6	(4.85–5.88) Elsayed et al (2003)
Biodiesel (soybeans)	~3	(1.43–3.4) Azevedo (2005)
		(3.2) Sheehan et al (1998)
Biodiesel (rapeseed, EU)	~2.5	(1.2–1.9) Azevedo (2005)
		(2.16–2.41) Elsayed et al (2003)
		(2–3) Azevedo (2005)
		(2.5–2.9) BABFO (1994)
		(1.82–3.71) depending upon use of straw for energy and cake for fertilizer; Richards (2000)
		(2.7) NTB Liquid Biofuels Network (undated)
		(2.99) ADEME/DIREM (2002)
Biodiesel (sunflower)	3	(3.16) ADEME/DIREM (2002)
Biodiesel (castor)	~2.5	(1.5) Kaltner, cited in Azevedo (2005)
		(2.1–2.9) Azevedo (2005)
Ethanol (wheat)	~2	(1.2) Richards (2000)
		(2.05)ADEME/DIREM (2002)
		(2.02–2.31) Elsayad et al (2003)
		(2.81–4.25) Gehua et al (2005)
Ethanol (sugar beets)	~2	(1.18) NTB Liquid Biofuels Network (undated)
		(1.85–2.21) Elsayad et al (2003)
		(2.05) ADEME/DIREM (2002)
Ethanol (corn)	~1.5	(1.34) Shapouri et al (1995)
		(1.38) Wang (2005)
		(1.38) Lorenz and Morris (1995)
		(1.3–1.8) Richards (2000)
Ethanol (sweet sorghum)	~1	(0.91–1.09) dos Santos (undated)
Diesel (crude oil)	0.8–0.9	(0.83) Sheehan et al (1998)
		(0.83–0.85) Azevedo (2005)
		(0.88) ADEME/DIREM (2002)
		(0.92) ADEME/DIREM (2002)
Gasoline (crude oil)	0.80	(0.84) Elsayed et al (2003)
		(0.8) Andress (2002)
		(0.81) Wang (2005)
Ultra low sulphur diesel	0.79	Elsayed et al (2003)
Gasoline (tar sands)	~0.75	Larsen et al (2004)

Notes: These ratios do not count biomass inputs. Here, petroleum fuels cannot have a balance greater than 1 because crude oil is counted as an energy input, while biofuels processed entirely with non-fossil fuels could have a balance of infinity. The ratios for cellulosic biofuels are theoretical.

Source: see endnote 1 for this chapter

The energy efficiency of carbohydrate-to-ethanol pathways is relatively low, while that for oilseed-to-biodiesel pathways can be quite high. Converting plant oils into biodiesel is a simple process that yields an amount of fuel similar to the amount of plant oil put into the process. In contrast, fermenting sugars to ethanol alone entails a loss of about half the feedstock's mass and energy, which is released as carbon dioxide (CO_2) at the mill. Fuels derived from cellulosic fibres are even less energy efficient because the fibres are more difficult to catalyse into sugars.[3]

These ratios would be different if the energy efficiency counted all the biomass contained in the crop as a biomass input. Here, sugar cane's efficiency ratio would diminish – thanks to the burning of cane leaves in the field – while the efficiency of fibrous biofuels would increase since they would utilize a greater portion of the plants' total biomass.[4]

In general, petroleum fuels have a higher energy efficiency than biofuels. This is because petroleum has been collected and refined by geological forces over millions of years. Oil fields contain vast reservoirs of relatively pure hydrocarbons, and oil can be transported more efficiently than solid biomass feedstocks (through pipelines more often than trucks); oil refineries are also many times larger than even big ethanol mills.[5] In contrast, obtaining fuels from biomass can require cultivating, collecting and refining a more diffuse form of energy, preventing comparable economies of scale.[6]

In the future, the energy efficiency of biofuels is likely to compare increasingly favourably to that of petroleum fuels. While biofuel feedstock production and conversion processes are becoming more efficient over time, petroleum fuels are likely to become less efficient. Crude oil is becoming more difficult to extract, and stricter fuel emissions standards are requiring more intensive refining. The energy efficiency of even the world's largest oil fields has declined significantly as field operators now push the deepest oil out the wellhead by pumping vast quantities of steam or methane below. The energy cost of producing fuels from tar sands, a thick feedstock often hailed as the most promising source of 'unconventional' oil left to be tapped, is much higher than conventional oils (see also Chapters 11 and 12).[7]

Meanwhile, biofuel production processes have become much more energy efficient in recent decades. Better plant breeds, more parsimonious farming methods and larger processing facilities have all yielded increasing quantities of fuel since the 1970s. As the biofuel industry expands, fibrous residues such as sugar cane bagasse, dried distillers grain (DDG), wheat stalks and lignin extracted during cellulosic conversion will probably become more important energy supplies. Although unlikely, it is possible that a biofuel's fossil energy balance could approach infinity – if it is produced using entirely renewable energy.

ANALYSIS OF ENERGY INPUTS

The energy inputs for biofuels fall into three main categories: the agricultural energy required to cultivate and/or harvest the feedstock; the processing energy required to convert the feedstock into fuels; and the transportation energy required to deliver the feedstock to the refinery and deliver fuels to commercial depots.

Agricultural energy includes the natural gas used to manufacture fertilizers and pesticides, the fuels used to pump irrigation water to the fields, and the petroleum used to propel the ploughs, seeders/planters and harvesters. In most cases, the production of nitrogen fertilizers is the predominant requirement of agricultural energy: in the US, for example, nitrogen fertilizers consume 70 per cent of the energy that farmers use to grow corn.[8] Processing energy includes the energy used in crushing and grinding the feedstock, producing reactive agents and purifying the fuel.

Agricultural and processing energy are the two dominant energy uses in biofuel production, and their relative proportions vary depending upon the type of fuel being produced. Producing ethanol from sugar beets or wheat typically requires around 20 per cent agricultural energy and 80 per cent processing energy. For biodiesel, the breakdown is 40 per cent agricultural energy and 60 per cent refining energy.[9] Transportation is also a small contributor to biofuels' overall energy requirements.

In addition to these three areas of energy use, some analyses of energy balance also include the energy required to construct buildings and machinery and, in some cases, the energy used in farmers' manual labour. Adding these inputs increases the completeness of the energy picture; but rarely do these activities comprise more than a small share of total energy use.[10]

Agricultural energy

The qualities and amounts of agricultural energy input vary significantly depending upon the feedstock used. Wastes collected at forestry mills or agricultural processing centres (including sugar mills) may not technically require any additional energy input at all since they are already part of existing processes.

The agricultural energy requirements of dedicated plant feedstocks vary by species. Both corn and rapeseed can require large quantities of fertilizer, but soybeans, as nitrogen-fixing plants, need little to no nitrogen fertilizer. Corn grown in the western states of the US corn belt (e.g. in Nebraska) uses a great deal of irrigated water, as does sugar cane grown in the north-eastern region of Brazil.[11] But other corn belt states (e.g. Iowa) and the centre-south of Brazil use rainwater, reducing their energy requirements. A plant such as jatropha can be very energy efficient, producing a high yield of oil in arid conditions with very little fertilizer.[12]

On the harvesting side, some biofuel feedstock is more difficult to harvest than others. Harvesting large monocultures of grain can be done with significant economies of scale, using industrial-size machines. By comparison, collecting oilseeds manually from trees is more arduous. However, combine harvesters utilize fossil fuels, while human and animal labourers often expend a form of renewable energy.[13]

Processing energy

The inputs for processing energy vary by biofuel production pathway. Transesterifying plant oils into biodiesel is much easier than hydrolysing starches and fermenting sugars into ethanol. This is because the ethanol process entails the breakdown of molecules, as well as the distillation of ethanol from water (starches must be hydrolysed into sugars, the sugars fermented into ethanol, then the ethanol distilled from water several times). In contrast, vegetable oils can be mixed with methanol and a catalyst to produce biodiesel. This is a simpler process, and it is worth noting that most of the energy required is used to refine methanol from natural gas. In Europe, rapeseed methyl ester (RME) biodiesel uses only half as much overall energy, and one third as much processing energy, as wheat ethanol.[14]

Transportation energy

Transportation energy makes up only a small proportion of the energy used to produce biofuels. One reason for this is that biofuels are not transported very far – processing facilities are usually located within 200km of feedstock fields, and so far the fuels are consumed largely within the producing country. However, the relative share of transportation energy could increase as international trade in biofuels expands (see Chapter 9).

Energy tabulations have found that transportation accounts for 2–5 per cent of biofuels' energy inputs.[15] In contrast, as much as 10 per cent of the inputs for petroleum fuels can go to transportation. This is because crude oil is a more concentrated resource and is often refined and used far from the wellhead. By transporting crude oil over longer distances, oil companies can benefit from greater economies of scale at the refining stage; however, as oil reserves become more concentrated in regions such as the Middle East, the transportation costs for petroleum fuels will probably increase further.[16]

Fossil versus biomass energy

In the US and Europe, most of the energy used to process biofuels comes from fossil fuels. Natural gas provides the bulk of the energy for producing fertilizers and pesticides, and diesel fuel powers most of the tractors and trucks involved in production and transport. In general, larger wet-milling facilities are powered mainly by coal, while dry mills are powered largely by natural gas.[17]

In contrast, virtually all of the processing energy used in Brazil comes from renewable sources. Bagasse, the fibrous residue that remains after sucrose is extracted from sugar cane stalks, contains a vast amount of energy. It can provide all of the energy used to process ethanol, and, in more efficient operations, a substantial quantity is left for other purposes, such as electricity generation for export to the grid. Wastes from Brazil's distilleries are also used to fertilize a portion of the sugar cane fields, reducing the energy needed to manufacture fertilizers. The result is that Brazilian sugar cane ethanol has a much better fossil energy balance than other biofuel production pathways.

Co-products

An accurate accounting of a biofuel's energy balance also considers the energy required to manufacture the co-products that are often produced alongside these fuels. In a typical dry mill for ethanol production in the US, about one third of the energy used to process corn kernels goes to the production of dried distillers grain, an animal feed.[18] Similarly, biodiesel facilities extract large quantities of glycerine from the plant oils (for use in soap, etc.), offsetting the energy use of the biodiesel fuel by about 20 per cent.[19] These offsets are now included in most calculations of the fossil energy balance of biofuels, with some studies also accounting for the energy gains obtained from not producing an animal feed or chemical feedstock via another pathway.

IMPROVEMENTS IN EFFICIENCY AND FOSSIL ENERGY BALANCE

The energy efficiency of biofuel production has been improving for several decades and is likely to continue to improve as producers seek to reduce their energy costs. Simultaneously, the use of fossil fuel energy can be replaced by energy from biomass co-mingled in biofuel production systems (as with bagasse in Brazil). Both of these lead to improvements in the energy balance of biofuels.

The farming of biofuel feedstock has become more efficient as farmers have improved crop yields while minimizing energy inputs. In the US, corn yields have increased at an average annual rate of 1.2 per cent over the last 40 years, and increased eightfold over the last 100 years. The amount of corn grown per kilogram of nitrogen fertilizer has increased by 70 per cent in the last 35 years, and production of the fertilizer itself has become more efficient, as well. Today, 1kg of nitrogen fertilizer can be produced for half the energy required in the 1970s.[20]

Converting crops into fuels has likewise become more efficient. Ethanol yields per kilogram have increased by 22 per cent in the US since the 1970s (for corn) and by about 20 per cent in Brazil (for sugar cane).[21] This is because of better crushing methods and, in the US, the use of enzymes that hydrolyse starches into sugars more effectively.[22] Larger production plants also bring greater economies of scale: they can house their own cogeneration plants to produce both electricity and steam, which not only use fossil fuels more efficiently, but also provide the option of co-firing with biomass residues.[23]

These improvements appear likely to continue. Agronomists expect crop yields to continue increasing, while better soil management practices, such as no-till farming, can save energy by reducing the need to plough.[24] In the future, to reduce the use of fossil fuels, biomass could replace natural gas as a feedstock for nitrogen fertilizers.[25]

Biorefining will also continue to become more efficient. In the US, advanced enzymes are able to convert starch into sugars at much lower temperatures than before, while newer strains of yeasts are able to tolerate higher concentrations of ethanol, reducing the energetic requirements for distilling the ethanol from water. In Brazil, fuel production procedures have been intentionally inefficient since they were a way of burning off huge piles of 'waste' bagasse. But as the value of bagasse as an energy source is better appreciated, it is being used more frugally.[26] Likewise, utilizing other co-products will bring further improvements in energy use. For example, cattle raised near ethanol plants can eat wet distillers grain, skipping the energy-intensive process of dehydrating this into a more transportable dry form.[27]

As new fuel conversion technologies develop, refiners will be able to exploit large amounts of 'raw' biomass. Several mill owners in the US state of Iowa already have plans to burn dried distillers grain, wood wastes such as sawdust, and corn stover and wheat stalks (normally left to deteriorate on fields) for processing power.[28] In Brazil, sugar cane growers could salvage cane leaves and tips, which contain as much energy as bagasse but are traditionally burned off before harvest. In processes that use new enzymes to hydrolyse fibres into sugars, the impregnable lignin residue (about one third of the plant's biomass) would be most effective as a source of processing heat.[29] In new gasification processes, refiners could easily burn a portion of the biomass feedstock for process energy (see Chapter 5 for more detail on these technologies).[30]

Agricultural and forestry residues, as well as other 'wastes', would require no additional cultivation energy. In some cases, as with bagasse or pulp mill wastes, the feedstock has already been collected by procedures intended to yield other products. In other cases, farmers could collect residues at the same time that they harvest crops, using new 'single-pass' harvesters being developed for this purpose. The efficiency advantage of wastes will be reflected in more favourable energy balances.[31]

Energy crops, for their part, require less maintenance. Because they are usually perennial crops, they do not need to be replanted, and they require less pesticides and fertilizers. Cultivating them thus entails fewer tractor trips. Agronomists also expect that they can increase yields of these crops significantly – by perhaps more than double – since breeders will be able to focus on simpler characteristics (quick growth of the entire plant) and since plants such as switchgrass and miscanthus have not yet benefited from intensive breeding.[32]

CONCLUSION

Clearly, some biofuel production pathways are more efficient than others, with geography being the principle determinant of efficiency. Since transportation energy accounts for only a small share of a biofuel's overall energy use, this suggests that it would be more energetically efficient for countries with temperate climates to import biofuels (e.g. made from sugar cane or palm oil) than to produce them at home. It would be more efficient to transport the final fuels, rather than just the feedstock, because the fuels are more energetically dense.

It is generally acknowledged that biofuels produced from temperate oilseeds, sugar beets, wheat and corn have limited ability to displace other fuels because of either their low yields or high input requirements. However, this feedstock is still more energetically efficient than cellulosic biofuels when considering all of the energy inputs, including the biomass used to provide the energy needed for the conversion process. While cellulosic conversion technologies will improve over time, in the near term, cellulosic biomass has the greatest potential as a fuel to provide process energy for conventional (first-generation) biofuels, providing a means to significantly improve the overall fossil energy balance of these fuels. As cellulosic conversion becomes more viable, analysts should continue to evaluate the most efficient uses of cellulosic biomass, raising the importance of 'energy efficiency' metrics.

When considering strategies for slowing the pace of climate change, the fossil energy balance of different biofuel production pathways can be a useful measure of their relative effectiveness. It is worth emphasizing that the fossil energy balance of biofuels could theoretically approach infinity, but only if renewable energy alone is used to cultivate, harvest, refine and deliver biofuels. However, fossil energy balance does not take into account other ways in which biofuel production contributes to climate change, such as changes in land use.

11

Effects on Greenhouse Gas Emissions and Climate Stability

INTRODUCTION

One of the major drivers of biofuel developments worldwide is concern about global climate change, caused primarily by the burning of fossil fuels. There is substantial scientific evidence not only that the Earth is warming, but that this warming is happening at an accelerating rate as emissions of greenhouse gases (GHGs) continue to rise.[1]

Transportation, including emissions from the production of transport fuels, is responsible for about one quarter of global energy-related GHG emissions, and that share is rising.[2] Transport accounts for 27 per cent of total emissions in the US (including 42 per cent of carbon dioxide emissions) and 28 per cent of total emissions in the European Union (EU).[3] According to the United Nations, the EU's GHG emissions declined overall between 1990 and 2003; but the share of emissions from the transport sector increased by 24 per cent.[4] It is estimated that transport-related GHG emissions in the 15 EU member states (EU-15) will increase 34 per cent above 1990 levels by 2010 if no further measures are taken to slow their growth.[5]

In rapidly industrializing developing nations such as China and India, which now lead global growth of vehicle sales, emissions from the transport sector will probably rise far faster over the coming years. For the near term, at least, unless human behaviour patterns change significantly, biofuels and improvements in energy efficiency offer the only options for dramatically reducing demand for oil and transport-related GHG emissions.

This chapter discusses the current and potential impacts of biofuels on the global climate, both through the different stages of biofuels production and use and over the entire life cycle. It provides estimates of potential GHG emissions reductions associated with biofuels for transport relative to emissions from petroleum fuels.

BIOFUELS AND THE GLOBAL CLIMATE

In the case of petroleum products, such as gasoline and diesel, a life-cycle analysis of the climate impact includes all GHG emissions associated with the following life-cycle stages: the exploration and production of oil; the transport and refining of oil for use; the storage, distribution and retail of oil; the fuelling of a vehicle; and the evaporative and exhaust (tailpipe) emissions associated with using the oil in a vehicle. For biofuels, the stages to be considered include the planting and harvesting of crops (including impacts on soil carbon storage, emissions associated with energy required for irrigation, and the production and use of fertilizers and pesticides); processing the feedstock into biofuel (including co-products); transporting the feedstock and the final fuel; storing, distributing and retailing biofuel; and, finally, the impacts of fuelling a vehicle and the evaporative and exhaust emissions resulting from combustion.[6,7]

The climate impact of biofuels depends greatly upon their fossil energy balance – that is, how much energy is contained in the biofuels themselves versus how much fossil fuel energy was required to produce them (see Chapter 10).[8] This, in turn, depends upon the energy intensity of feedstock production (including the type of farming system and inputs used), processing, and transporting the feedstock and final product.

Unlike fossil fuels, which contain carbon stored for millennia beneath the Earth's surface, biofuels have the potential to be 'carbon neutral' over their life cycles, emitting only as much as the feedstock absorbs. This is because biofuels are produced from biomass, and exactly the same amount of carbon dioxide (CO_2) that is absorbed from the atmosphere by the plants through photosynthesis is set free through combustion.[9] This accounts for an almost closed CO_2 cycle (see Appendix 7 for a flow chart of bioenergy versus fossil energy).[10]

Outside of combustion, the primary sources of GHGs during the life cycle of biofuels occur during production of these fuels. CO_2 and other greenhouse gases are emitted from the cultivation of crops, the manufacture of nitrogen fertilizers, and the consumption of fossil fuels in the machines used for growing the feedstock and refining it into biofuel. On the other hand, biofuel production results in the generation of co-products, which can substitute for products manufactured conventionally and the non-renewable primary energy used in their production.[11]

With the exception of a few studies that report associated increases in GHG emissions, most studies find a significant net reduction in global warming emissions from both ethanol and biodiesel relative to conventional transport fuels.[12] There exists broad agreement that the use of biofuels, made with today's technologies, can result in significant net reductions in carbon emissions, and that reductions with next-generation feedstocks and technologies will be even larger.

However, figures vary widely due to differing assumptions about factors such as management practices, conversion and valuation of co-products. Estimates differ depending upon assumptions about the feedstock used; land-use changes (where and how much additional land will be required, and what the bioenergy crops are replacing); crop management (including use of fertilizer and tilling of soil); crop yields; processes and their efficiencies (including fossil inputs for refining); the relative efficiencies of gasoline and ethanol (diesel and biodiesel, including blends, have about the same vehicle efficiency); credits attributed to co-products; and the methodologies used to calculate total life-cycle emissions.[13]

Estimates also vary depending upon the GHGs that are considered and their relative impacts. Most studies consider emissions of CO_2, nitrous oxide (N_2O) and methane, but many omit ozone. Ozone affects climate directly; but it is not emitted during the fuel cycle. Instead, it is formed by photochemical reactions with other gases that are emitted, including nitrogen oxides (NO_x), carbon monoxide (CO) and non-methane organic compounds. In addition, most studies do not look at other gases that might also be GHGs – including hydrogen (via its effect on ozone) and particulate matter – which could alter the calculated life-cycle GHG balance of these fuels.[14] It is important to keep in mind that the reference system (petroleum-based fuels) is rarely ever evaluated with full GHG accounting, such as methane emissions associated with production.[15]

The equation could be even more favourable for biofuels if waste streams and/or agricultural and forestry residues are used as feedstock (see Chapter 4). However, while the conversion technologies necessary to convert cellulosic residues and other wastes to biofuels currently exist, they have not yet been commercialized.[16] Most estimates for these technologies come from engineering studies; but it is assumed that net GHG emissions will be dramatically lower with these technologies.

Feedstock production and harvest

The production of feedstocks, particularly the change in land use and the use of fertilizer, is generally the most GHG-intensive stage in the life cycle of biofuels. Gases released as a result of feedstock production include CO_2, N_2O (from nitrogen fertilizer application and decomposition of leaf litter) and methane.[17]

Different crops have different GHG emission or carbon sequestration characteristics (the ability to capture and store carbon), depending upon factors such as fertilizer requirements and root systems. Associated emissions also vary depending upon where the feedstock is grown because climate, solar resources and soil productivity all affect crop yields and fertilizer application rates.

The following sections discuss the climate-related impacts of biofuels feedstock production and harvest, looking more closely at the affects of land-use change, crop management and selection, and harvesting.

Land-use change

On a global basis, organic matter in soils contains more than twice the carbon in atmospheric CO_2, and additional carbon is stored in biomass. Because these pools of carbon are so large, even relatively small increases or decreases in their size can be of global significance.[18] The amount of carbon stored (or sequestered) in plants, debris and soils changes as land use is altered, including when biomass is grown and harvested. Because increases in the land area used to produce feedstocks can result in large releases of carbon from soil and existing biomass, they can negate any benefits of biofuels for decades.[19] And because changes could extend over long periods of time and then reach a new equilibrium, it is important that any analysis be time dependent.[20]

To highlight the importance of dramatic one-time changes in land use, it is relevant to note that the burning associated with forest clearing in Indonesia and Malaysia was one of the largest contributors to global GHG emissions in 1997.[21] As discussed in Chapter 12, much of this clearing is to open the land for new palm oil plantations. In addition to the potential impacts for the climate on a global scale, devastation of vast forestlands could cause significant climatic change on a regional basis. Studies reveal that wide-scale destruction of forests can affect the hydrological cycle and regional climate, reducing precipitation and increasing temperatures.[22] For example, destruction of the Amazon could lead to serious disruptions in hydrological cycles, threatening to reduce rainfall in inland areas such as Brazil's *cerrado*, a vast expanse across the high plains that is home to some 935 species of birds and nearly 300 mammal species, including many that are threatened or endangered.[23]

In general, converting land from natural cover to intensive agriculture with annual crops reduces plant biomass above ground and, over time, emits carbon from the soil (the reverse of a GHG benefit) as inputs of debris decline and as increased soil temperature and aeration result in further losses. Even converting to 'sustainable' energy crops can reduce soil carbon content – for instance, if wild forests are levelled to produce biofuel feedstocks.[24]

If, on the other hand, land is converted from existing annual crops to perennial herbaceous species, such as native grasses, organic matter in the soil progressively increases; with woody crops, organic matter increases yearly over the term of the rotation.[25] This is because perennial crops tend to deposit more carbon in the soil as roots, and the absence of tillage slows decomposition of soil matter. Thus, it is likely that the cultivation of perennial biomass crops in areas previously used for annual crops will increase the organic carbon content of soil.[26]

In fact, soils have a far greater capacity to store carbon than the temporary CO_2 bond of biomass crops. But any changes will take time.[27] In addition, the potential to sequester additional carbon is very site specific and depends upon former and current land uses, agricultural practices, climate and soil characteristics.[28] And the storage reservoir in soil exists only when potential reservoirs are not filled; in other words, 'all else being equal, there is a maximum amount of carbon that can be stored in vegetation and soils in any climate'.[29] Furthermore, carbon sequestration is reversible. Any carbon that does accumulate in soil or biomass could be released if the use or management of land is later converted back to previous uses, such as annual crops.[30] Thus, carbon storage in soils can be only a short-term option for reducing GHG emissions.[31]

Crop management

Crop management – including the level of fertilizer and pesticide use, the fuels used to drive farm machinery, means of irrigation and treatment of the soil – also plays an important role in determining the climate impact of biofuels. The burning of diesel fuels to drive tractors and other farm machinery releases CO_2, in addition to NO_x and hydrocarbons, which help to create ozone. And irrigating crops using fossil energy also releases CO_2. The western corn/soybean belt of the US, in particular, requires high levels of irrigation, as do large areas of rapeseed cultivation in Europe and the smaller sugar production areas in north-eastern Brazil. Seed cultivation, meanwhile, requires a share of all the energy inputs during a previous crop cycle.

The most significant factor in terms of climate impact, however, is chemical fertilizers, which require large amounts of fossil energy input. Typically, fertilizers and pesticides are manufactured using natural gas as an input, and nitrogen (N) fertilizer, in particular, can require vast amounts of natural gas to produce. Pesticides are generally fossil fuel based, increasing energy inputs and, therefore, associated GHG emissions (see Chapter 12 for more on the environmental impacts of chemical fertilizers and pesticides).

Some of the nitrogen fertilizer used on fields is eventually emitted as nitrous oxide (N_2O), which is released directly from the soil or through run-off water. N_2O is a potent greenhouse gas that accounts for 6 per cent of anthropogenic GHG emissions, and inorganic N fertilizer accounts for about 60 per cent of this total; leguminous crops such as soybeans account for another 25 per cent. Atmospheric concentrations of N_2O are increasing at a rate of 0.2–0.3 per cent annually.[32] Emissions rates depend upon soil type, climate, crop, tillage method and application rates.

Primary energy demand for producing N fertilizers varies from place to place, as does fertilizer use. Use also differs by crop type.[33] For example, large amounts of fertilizers are used for corn farming, and at high enough quantities that fertilizer production and distribution alone can account for 70 per cent of all agricultural energy inputs and an even greater proportion of the GHG emissions due to its degradation into N_2O.[34] In contrast, soybeans, which are leguminous nitrogen fixers, require substantially less N fertilizer. Research is needed to determine whether intercropping of legumes, such as soybeans, with perennial biomass crops can reduce the need for N fertilizer.[35]

Any efforts to reduce the use of fertilizers and pesticides, or to replace fossil fuel use in powering farm equipment and tilling the soil, could significantly lower associated emissions. For example, biofuels and renewable power can be used to fuel and power equipment instead, reducing the climate impact. Rather than tilling the soil, farmers could plant perennial crops and/or adopt conservation tillage practices involving direct seeding, which actually increases soil carbon storage by leaving organic matter relatively undisturbed.[36]

Feedstock selection

The feedstock selected for ethanol and biodiesel production is critical. It determines everything from the energy yield per unit of land, to the use of fertilizer, to the amount of carbon that can be sequestered in the soil.

For ethanol, grains such as corn, barley and wheat generally produce high concentrations of starch; but they are less efficient in their use of land and fertilizer than is sugar cane.[37, 38] European wheat, for example, yields about 2500 litres of ethanol per hectare; European rapeseed, used to make biodiesel, yields about 1200 litres of fuel per hectare.[39, 40]

In contrast, palm oil trees can produce 5 times as much oil per hectare as rapeseed, and more than 13 times as much as soybeans. While less productive than palm trees, jatropha bushes are still significantly more productive per hectare than rapeseed.[41] And test plots of switchgrass in the US have yielded enough for about 10,900 litres of ethanol per hectare each year.[42] Miscanthus and hybrid poplars also have the potential for high yields.[43]

In addition to their high yield potential per unit of land, such energy crops require less fertilizer and pesticide input, do not require tilling and can sequester significant amounts of carbon in the soil.[44] Because perennial crops also have the capacity to restore soil carbon contents over time, they could help to improve the quality of degraded lands. Switchgrass, for example, actually expands its deep root system when harvested, thereby increasing organic matter in soils and increasing carbon sequestration.[45]

But the life-cycle GHG impact of energy crops ultimately depends upon what these crops are replacing. If they replace natural grasslands or forests, GHG emissions will probably increase; if, on the other hand, energy crops are planted on unproductive or arid land where conventional crops cannot grow, or in place of annual crops (e.g. in place of corn grown for ethanol, or rapeseed for biodiesel), they have the potential to significantly reduce associated emissions. In addition, many plants, such as jatropha and pongamia, can thrive on unproductive or arid lands where conventional crops cannot grow.

Finally, crop and forest residues, animal wastes and the organic part of municipal waste could all be used to produce fuel with next-generation technologies (see Chapters 4 and 5). These are feedstocks that might otherwise have no other uses, do not in themselves require land, chemical inputs or irrigation, and can provide useful energy.

Harvesting

The way in which crops are harvested, including the procedure, timing and machinery used, all affect the level of GHG emissions associated with the process.

For example, on some soils the removal of crop residue can result in the release of soil carbon (the opposite of sequestration), in addition to having detrimental impacts on other factors that are important for healthy soil function. There is evidence that residue removal changes the rate of physical, chemical and biological processes in the soil by causing more fluctuations in soil temperatures and increased water evaporation. One study estimates that residue harvesting can reduce corn-derived soil organic carbon by 35 per cent relative to when it is retained; this is averaged over all tillage systems.[46] Other studies have found that soil organic carbon with corn crops is greatest under no-till systems in which residues are allowed to remain on the ground, while it is lowest in no-till when stover is removed.[47]

While surface matter is important, roots are the largest contributor to soil organic matter and, thus, the most significant part of the plant for carbon accrual.[48] In fact, one of the benefits of the perennial plant switchgrass is that its root system actually grows and sequesters more carbon in the soil after the crop is harvested each year, as noted above.[49]

The timing of harvest can also be important – for example, trees should be harvested in winter so that leaves are not removed from the area. In addition, tillage may play an even greater role than residue removal in the release of carbon from soils.[50] Leaving some residue on the ground, and planting cover crops which prevent soil loss after harvest, can help to limit this problem.

The use of machinery, which generally runs on diesel, also results in emissions of GHGs. Pre-harvest burning of sugar cane, a common practice in Brazil, releases as much as one third of the crop's biomass into the air as CO_2, with some amount of methane and NO_x also emitted.[51] This represents a lost opportunity because the dry matter could be used as process energy to offset the use of fossil fuels.

Refining feedstocks into biofuels

In refining, the most significant factors with regard to climate impact are the conversion efficiency of the refining process or facility, energy inputs and outputs, source of process energy (e.g. fossil versus renewable power), and the emissions attributed to co-products. The efficiency of conversion from feedstock to biofuel is important because it drives the amount of feedstock required for a given volume of biofuel, which in turn affects the amount of land and fossil fuel input needed to grow, transport and process the crops.[52]

Significant amounts of energy – in the forms of process heat, mechanical energy and electricity – are needed for the refining process. In North America and many other regions, most of this energy is derived from natural gas and coal, used both directly and for electricity generation.[53] However, residual biomass could be burned, instead, to generate electricity or steam, reducing or even eliminating the need for external energy inputs. In Brazil, bagasse (a by-product of sugar cane crushing) is used for energy production, enabling mills and distilleries to be almost entirely energy self-sufficient; some even sell their surplus electricity to the power grid.[54] Sugar factories in the US state of Hawaii also burn bagasse to provide steam and electricity for sugar processing, and sell their excess power to local utility companies.[55] Transitioning to renewable energy for the refining process in this way could significantly reduce life-cycle GHG emissions associated with biofuels production, particularly if it replaces coal.[56]

When biofuel plants produce co-products as well – such as animal feed (from ethanol production) or glycerine and fatty acids for soaps (from biodiesel production) – the GHG emissions released during the refining process are 'shared', meaning that the amount attributable to biofuels is lower than would otherwise be the case. Co-products can replace competing products that require energy to produce; thus, they can offset energy needed to make these products another way. For example, animal feed made while refining corn into ethanol or soybeans into biodiesel can reduce the need to grow corn or soy specifically for animal feed production. The energy saved can partly offset the energy needed to produce biofuels.[57] Co-products, such as straw, could also be converted into more fuel. Current processes make little use of this resource; but the potential for saving conventional energy is significant.[58]

Sensitivity analyses reveal that net energy calculations (and, thus, associated GHGs) are most sensitive to assumptions about the allocation of co-products.[59] Therefore, whether or not they are included, and how, can make a significant difference in estimations of the emissions associated with biofuels production. According to Fulton et al (2004), most studies that look at the life-cycle GHG emissions of ethanol from corn, for example, assume that various co-products reduce the net emissions of ethanol by 5–15 per cent.[60] Armstrong et al (2002) mention co-product credits for rapeseed methyl ether (RME), including animal feed and glycerine, ranging from 5 to 14 per cent.[61]

Transport of feedstocks and fuel

Biomass feedstocks are generally transported from fields to biorefineries by truck, travelling a few dozen to a few hundred kilometres.[62] Each truckload is much smaller than an oil tanker's load and thus does not benefit from the same economies of scale. And trucks must carry a great deal of excess water, fibre and protein contained in the plant feedstock, which increases transport energy requirements. Transport by train or pipeline, where feasible, could significantly reduce associated emissions. But today the fuel requirements – and, thus, associated emissions – are minimal for distribution of biofuels to the refuelling station.[63] A study by the US

National Renewable Energy Laboratory (NREL) comparing soybean diesel to petroleum diesel concluded that transporting crude oil consumes about five times more energy (per unit of final fuel) due to longer transport distances, meaning that resulting GHG emissions for transportation are far lower for soybean diesel than conventional diesel.[64]

While there is relatively little trade of biofuels on a global scale today, this situation is changing very rapidly. Trade will increase significantly as consumption rises and as domestic demand for biofuels in some countries exceeds their potential to produce them (something already occurring in Europe) (see Chapter 9). Then, long-distance transport will become a more considerable factor for biofuels. At the same time, the distances that feedstocks and, particularly, fuels are transported have only a small impact on life-cycle CO_2-equivalent emissions because the net energy requirements of long-distance transport (generally via ship) are relatively small per volume of fuel shipped.[65] Hamelinck et al (2005) conclude that shipping of refined solid biomass and biofuels is possible at relatively low costs and modest energy losses.[66] Most important will be minimizing the transport of wet untreated biomass.[67]

Combustion

The combustion of biofuels results in the release of carbon dioxide into the atmosphere; but because the emissions are already part of the fixed carbon cycle – absorbed by plants as they were growing – they do not contribute to new emissions of CO_2. NREL estimates that biodiesel from soybeans emits almost 10 per cent more CO_2 than does petroleum diesel due to more complete combustion and 'the concomitant reductions in other carbon-containing tailpipe emissions'; however, most of this is renewable or recycled in growing soybean plants.[68]

LIFE-CYCLE IMPACTS OF CURRENT-GENERATION BIOFUELS

Ultimately, it is the net emissions over the full life cycle of biofuels – from changes in land use to combustion of fuels – that determine their impact on the climate. Research on net emissions is far from conclusive, and estimates vary widely. Calculations of net GHG emissions are highly sensitive to assumptions about system boundaries and key parameter values – for example, land-use changes and their impacts; which inputs are included, such as energy embedded in agricultural machinery or the energy needs of farm labourers (generally not included); and how various factors are weighted.

According to Quirin et al (2004), who reviewed more than 800 studies and analysed 69 of them in detail, the primary reasons for differing results are different assumptions made about cultivation, and conversion or valuation of co-products.[69]

Larson (2005), who also reviewed multiple studies, found that the greatest variations in results arose from the allocation method chosen for co-products, and assumptions about N_2O emissions and soil carbon dynamics.[70] In addition, GHG savings will vary from place to place – according, for example, to existing incentives for GHG reductions.[71] And the advantages of a few biofuels (e.g. sugar cane ethanol in Brazil) are location specific.[72] As a result, it is difficult to compare across studies; however, despite these challenges, some of the more important studies point to several useful conclusions.

The majority of life-cycle analyses carried out thus far look at grains and oilseed crops in North America and the EU. The exceptions are a study on sugar cane ethanol in Brazil, one on sugar cane ethanol in India and one on biodiesel from coconut.[73] Furthermore, most studies have looked at ethanol, biodiesel and ethyl tertiary butyl ether (ETBE).[74] A limited number have considered vegetable oil and biogas, dimethyl ether (DME) and biomass-to-liquid (BTL) fuels. But there have been no studies to date on biodiesel from palm oil, cassava or oilseed plants such as jatropha and pongamia, or on pyrolysis oil diesel or hydrothermal upgrading (HTU) diesel.[75] In addition, no locally relevant lignocellulosic energy crop studies have been done in the developing country context.[76]

It should be noted that a few studies stand out from the rest in that they have reported increased emissions from biofuels relative to conventional petroleum fuels. For example, Pimentel (1991, 2001) has estimated that ethanol derived from corn results in a 30 per cent increase in life-cycle GHG emissions over gasoline.[77] Other studies reporting an increase are by Pimentel and Patzek.[78] They stand apart from the rest because they incorrectly assume that ethanol co-products should not be credited with any of the energy input (and, thus, associated emissions) in feedstock growing and fuel processing. They also include data that are outdated and do not represent the current agricultural and refining processes, and/or are poorly documented and thus cannot be fully evaluated.[79]

The other notable exception is a series of studies by Delucchi, who also finds that biofuels from many of the current feedstocks have higher life-cycle emissions than petroleum fuels.[80] Delucchi (2005) includes co-products in his analysis and assumes that production processes will continue to become more efficient and will switch to low-emitting process fuels (such as renewable power), and he is continuously updating his model and data.[81] His work differs from other studies primarily in that he includes a detailed accounting of the entire nitrogen cycle, uses comprehensive CO_2-equivalency factors (accounting for many GHGs that most other studies do not incorporate), and has a comprehensive and detailed accounting of land-use changes and resulting impacts on the climate.[82]

This analysis notwithstanding, the vast majority of studies have found that, even when all fossil fuel inputs throughout the life cycle are accounted for, producing and using biofuels made from current feedstocks result in substantial reductions in GHG emissions relative to petroleum fuels.[83] The following subsections consider ethanol and biodiesel separately.

Ethanol

As mentioned earlier, there are significant variations in findings for life-cycle GHG reductions associated with ethanol (see Table 11.1).[84] However, most studies show that as ethanol blend levels rise, any emissions benefits associated with ethanol increase as well. According to Wang et al (1999), this occurs because more ethanol (and, thus, less petroleum fuel) is used and because vehicle fuel economy improves as the share of ethanol increases.[85]

Farrell et al (2006) looked at six representative studies of fuel ethanol from corn, adjusting them for commensurate system boundaries. They found that, depending upon the study input parameters (such as energy embodied in farming equipment), switching from gasoline to corn ethanol yielded anywhere from a 20 per cent increase in emissions to a 32 per cent decrease. Their best estimate, with today's yields and technology, is that life-cycle emissions decline by 13 per cent.[86]

Larson (2005), who reviewed more than 30 life-cycle assessment studies for various biofuels, found that ethanol from wheat ranged from a 38 per cent benefit to a 10 per cent penalty.[87] Delucchi (2005) estimates that emissions from corn ethanol can range from a 30 per cent reduction to a 30 per cent increase relative to those from petroleum fuels.[88]

In general, of all potential feedstock options, producing ethanol from corn results in the smallest decrease in overall emissions. The greatest benefit, meanwhile, comes from ethanol produced from sugar cane grown in Brazil (or from using cellulose or wood waste as feedstocks, as discussed later in this chapter).[89] Several studies have assessed the net emissions reductions resulting from sugar cane ethanol in Brazil, and all have concluded that the benefits far exceed those from grain-based ethanol produced in Europe and the US. Kaltner et al (2005) estimate that the total life-cycle GHG emissions reductions associated with Brazil's ethanol industry are equivalent to 46.6 million tonnes annually (12.75 million tonnes of carbon per year), or approximately 20 per cent of Brazil's annual fossil fuel emissions.[90]

Fulton et al (2004) attribute the lower life-cycle climate impacts of Brazilian sugar cane ethanol to two major factors. First, cane yields are high and require relatively low inputs of fertilizer since Brazil has better solar resources and high soil productivity. Second, almost all conversion plants use bagasse for energy, and many recent plants use cogeneration (heat and electricity), enabling them to feed electricity into the grid. As such, net fossil energy requirements are near zero and, in some cases, could be below zero (in addition, less energy is required for processing because there is no need for the extra step to break down starch into simple sugars; because most process energy in Brazil is already renewable, this does not really play a role).[91]

Table 11.1 *Estimated change in life-cycle greenhouse gas emissions per kilometre travelled by replacing gasoline with various ethanol blends*[a]

Feedstock and blend (country or other specifics, where available)	*Emissions change (per cent)*	*Source*
Corn		
E10 (US)	–1	Wang et al (1999)
E10 (China)	–3.9	Wang et al (2005)
E85 (US)	–14 to –19	Wang et al (1999)
E85 (China)	–25	Wang et al (2005)
E90 (US, 2010)	+3.3	Delucchi (2005)
E95 (US, 1999)	–19 to –25	Wang et al (1999)
E100	–13	Farrell et al (2006)
E100	–21	Marland et al (1991)
E100 (wet milled)	–25	Wang (2001)
E100	–30 to –33	Levy (1993)
E100 (dry milled)	–32	Wang (2001)
E100	–38	Levelton Engineering Ltd (2000)
Sugar beet		
E100	–35 to –56	Levy (1993)
E100 (Northern France)	–35[b] to –56[c]	Armstrong et al (2002)
E100	–41	GM et al (2002)
E100	–50	EC (1994)
E100	–56	Wuppertal Institute (2005)
Molasses		
E10 (Australia)	–1 to –3[d]	Beer et al (2001)
E85 (Australia)	–24 to –51[d]	Beer et al (2001)
Sugar cane		
E100 (Hydrous; Brazil)	–87 to –95	Macedo et al (2004a)
E100 (Anhydrous; Brazil)	–91 to –96	Macedo et al (2004a)
Wheat		
E100	–19	EC (1994)
E100	–32 to –35	Levy (1993)
E100	–45	Wuppertal Institute (2005)
E100	–47	Gover et al (1996)
E100 (UK)	–47	Armstrong et al (2002)

Notes: [a] assumes use of current-generation ethanol fuels in conventional spark-ignition vehicles; [b] average case; [c] best case; [d] range depends upon credits for co-products.

Source: see endnote 84 for this chapter

It important to note that using ethanol to make ETBE results in even greater GHG savings than blending ethanol directly with gasoline, according to Edwards (2005) and Quirin et al (2004). This is because ETBE replaces methyl tertiary-butyl ether (MTBE), which has a relatively high energy demand, whereas ethanol often replaces gasoline, which requires less energy for production than MTBE.[92]

Biodiesel

The range of estimates for GHG emissions reductions from biodiesel is also large. Most studies show a net reduction in emissions, with waste cooking oil providing the greatest savings (see Table 11.2).[93] The exception is Delucchi (2003), who estimates that biodiesel from soybeans will lead to significant emissions increases by 2015.[94] Depending upon assumptions (including land-use change), he believes that soy biodiesel could result in net emissions ranging from zero (relative to fossil fuels) to an increase of more than 100 per cent.[95]

Other studies show major reductions in emissions from soybean diesel. Larson (2005) found that estimates for emissions reductions from soybean methyl ester (SME) are similar to those for rapeseed methyl ester (RME), which provides a

Table 11.2 *Estimated change in life-cycle greenhouse gas emissions per kilometre travelled by replacing diesel with 100 per cent biodiesel*[a]

Feedstock and country (other specifics, where available)	*Emissions change (percentage)*	*Source*
Rapeseed		
Germany	–21	Armstrong et al (2002)
Netherlands	–38[b]	NOVEM (2003)
	–40	Larivé (2005)
	–44 to –48	Levy (1993)
	–49	GM et al (2002)
	–51	Scharmer and Gosse (1996)
Australia	–54	Beer et al (2001)
	–56	ETSU (1996)
	–56 to –66	Scharmer and Gosse (1996)
	–58	Richards (2000)
	–68	Wuppertal Institute(2005)
Pure plant oil		
Unspecified oil source	–42	Wuppertal Institute (2005)
Soybeans		
US (2015)	+107	Delucchi (2003)
Netherlands	–53[b]	NOVEM (2003)
	–63	Levelton Engineering Ltd (1999)
Australia	–65	Beer et al (2001)
US	–78[b]	Sheehan et al (1998)
Tallow		
Australia	–55	Beer et al (2001)
Waste cooking oil		
Australia	–92	Beer et al (2001)

Notes: [a] assumes current-generation biodiesel used in conventional compression-ignition vehicles; [b] only CO_2 emissions are considered.

Source: see endnote 93 for this chapter

15–65 per cent reduction per vehicle kilometre travelled.[96] Again, varying results are due to different assumptions, as described above.

Summary of life-cycle impacts

The production and use of biofuels, *per se*, does not necessarily result in a net reduction in GHG emissions relative to petroleum fuels. However, most studies have found that the majority of biofuel feedstock-to-fuel pathways, with existing commercial technologies, have a solidly positive GHG balance. The higher the blend of biofuels to conventional fuels, the greater any savings in GHG emissions will be.

According to Larson (2005), conventional grain- and oilseed-based biofuels can offer only modest reductions in GHG emissions. The primary reason for this is that they represent only a small portion of the above-ground biomass. Larson estimates that, very broadly, biofuels from grains or seeds have the potential for a 20–30 per cent reduction in GHG emissions per vehicle kilometre, sugar beets can achieve reductions of 40–50 per cent, and sugar cane (average in southeast Brazil) can achieve a reduction of 90 per cent.[97]

Quirin et al (2004), meanwhile, considered both current and future vehicle technologies, and used 2010 as their time reference. They looked only at studies that included methane, N_2O and CO_2, analysed impacts of all relevant agricultural sources (fertilizer production and emissions from field), and accounted for co-products. Their conclusion is that the GHG emissions balances of all biofuels considered are favourable compared to fossil fuel counterparts.[98] More specifically, they found that ETBE has advantages over all other biofuels; whether ethanol is better than biodiesel depends upon the feedstock used; and biodiesel from rapeseed is preferable to pure rapeseed oil because the glycerine co-product can be substituted for technically produced glycerine.[99]

In general, the viability of biofuels as low-carbon replacements for oil depends less upon the amount of energy required in production than upon the type of energy used – assuming the same system boundaries (e.g. no land-use changes and the same level of final output). Corn-derived ethanol, for example, may indirectly emit as much fossil carbon into the atmosphere as gasoline if the corn is grown with nitrogen fertilizers derived from petroleum sources; irrigated, harvested and delivered with vehicles run on conventional fuel; and processed using energy generated from coal. If, however, the corn is grown with manure or other natural fertilizers, harvested and delivered with biofuels, and distilled with renewable power, the associated life-cycle emissions could drop to near zero.[100] This highlights the importance of choice of feedstock, selection of refining processes, and the careful planning and designing of the entire biofuel pathway, integrating it within the context of the biomass energy system.[101]

Reducing the climate impact

In the future, there is the potential to further reduce GHG emissions associated with biofuels through a variety of means. These include improved yields with existing feedstocks, improved process efficiencies, new energy crops, new technologies and an increase in co-product development.

Improved yields with existing feedstocks

Over the past several decades, significant yield improvements have been achieved with a variety of crops – from sugar cane and corn, to soybeans, oil palm and willow – and advances are expected to continue.[102] Yield increases are due to several factors, including breeding (particularly hybridization), genetic improvement, better farming practices and farming conservation measures (see Chapter 6).[103] Yields for miscanthus, switchgrass and other energy grasses are expected to increase significantly as well. So far, energy crops such as switchgrass and poplar trees have not been bred intensively, and some experts believe that breeding could result in a doubling of their productivity.[104]

As crop yields improve, the amount of land and other inputs required to produce a given amount of biofuel decline, generally reducing the climate impact (see Chapter 12 for other potential environmental implications).

Improved process efficiency

Advances in technology and process efficiencies offer the potential for additional reductions in associated carbon emissions. Improvements to date have been significant, as seen in both the US and Brazil.

Over the past 30 years, the US ethanol yield per bushel of corn has increased steadily, from less than 9 litres (2.4 gallons) per bushel in the 1970s to between 9.8 and 10.6 litres (2.6 and 2.8 gallons) by the mid 2000s. This represents an efficiency increase of 8–16 per cent; where one falls in this range depends upon the starch content of the corn and process efficiency.[105]

In Brazil, the improvements have been even more significant. Fulton et al (2004) note that the ethanol yield from 1 tonne of sugar cane increased 23 per cent between 1975 and 2002, from 73 litres per tonne in 1975 to 85 litres in 1995 and 90 litres in 2002. The best values are 10–20 per cent higher than average, and it is expected that these will become the average over the next several years.[106] According to other sources, the increase in yield, due to technological innovations and efficiency improvements, has been far greater. Nastari (2005) estimates a near tripling over the past 30 years, from about 2000 litres of ethanol per hectare of sugar cane in 1975 to 5000 litres in 1999 and 5900 litres in 2004, for an average annual increase of 3.8 per cent.[107] Some put the current yield as high as 7000 litres per hectare under good conditions.[108]

A study by the Dutch Energy Agency (NOVEM) and Arthur D. Little (ADL) estimated that life-cycle CO_2-equivalent emissions would decline significantly for many processes by the 2010–2015 period, with most pathways leading to high reductions relative to gasoline or diesel. In many cases, GHG emissions reductions would exceed 100 per cent, due mainly to the use of biomass for process energy. They projected that the greatest reductions would result from production of cellulosic ethanol using enzymatic hydrolysis in biorefineries, but that biomass gasification and conversion to final fuels such as diesel and DME would provide similar reductions.[109]

While Quirin et al (2004) project that the GHG benefits of biofuels will increase with time, they also note that higher conversion efficiencies will mean fewer co-products, reducing advantages somewhat.[110] In addition, it will be important to ensure over time that the benefits from technological progress are not outweighed by the rising costs of obtaining an ever growing supply of feedstock through unsustainable means – such as replacing tropical forests with palm oil plantations.

New energy feedstocks

Improvements in technologies and process efficiencies could bring about significant further reductions in emissions; but they will not be enough to change relative benefits of given types of biomass and land-use changes.[111] Such improvements might also have difficulty counterbalancing the negative impacts of expanding feedstock supply and associated land use if not sited, selected, planted and managed in a sustainable manner.

Thus, it is important to focus on new energy crops, such as short-rotation forests and perennial grasses, which offer significant potential for further reducing the life-cycle emissions of biofuels. Such crops, if planted in place of annual crops or on degraded lands or unimproved pasture, can increase standing biomass growing above ground and the amount of biomass under ground and, hence, carbon sequestration.[112] Despite their current lower ethanol yields, at least one study has determined that making ethanol from hay and switchgrass results in some of the lowest life-cycle GHG emissions because these crops sequester carbon in the ground – assuming they are grown on unimproved pasture or in place of annual crops.[113] Because these perennial energy crops also generally require less fertilizer and less irrigation than other feedstock crops do, they effectively reduce CO_2 emissions associated with the final product (biofuel).

The use of short-rotation forestry and logging residues on forest land (the tops and branches that remain after trunks are harvested at felling sites) can reduce GHG emissions relative to other feedstock crops because the trees increase carbon sequestration in the soil and their biomass, and N_2O emissions drop because of the reduced use of fertilizers. However, Börjesson and Berndes (undated) caution that the recovery of logging residues can result in higher emissions than if the branches and other residues are left to decay on site.[114] More research may be needed to determine the life-cycle impacts of such practices (see Chapter 12 for more on the impacts of residue removal).

Technological advances that enable the use of cellulosic and other feedstocks – including animal manure, the organic portion of municipal wastes and waste restaurant grease – could dramatically reduce the life-cycle GHG emissions associated with some biofuels, while providing the added benefit of reducing the amount of land and other resources required to store or dispose of these products. Quirin et al (2004) argue, however, that alternative uses of waste materials must be considered when determining the life-cycle impacts of using them as biofuel feedstock (e.g. for BTL). Studies, to date, have ignored this issue.[115]

Advanced technologies

New technologies under development offer the potential to dramatically increase yields per unit of land and fossil input, and further reduce life-cycle emissions. Cellulosic conversion processes for ethanol offer the greatest potential for reductions because the feedstock can come from the waste of other products or from energy crops, and the remaining parts of the plant can be used for process energy (see Table 11.3).[116]

Larson (2005) projects that future advanced cellulosic processes – to ethanol, Fischer-Tropsch process (F-T) diesel or DME – from perennial crops could bring reductions of 80–90 per cent and higher.[117] According to Fulton et al (2004), net GHG emissions reductions can even exceed 100 per cent if the feedstock takes up more CO_2 while it is growing than the CO_2-equivalent emissions released during its full life cycle (e.g. if some of it is used as process energy to offset coal-fired power).[118] Delucchi, too, believes that next-generation feedstocks (such as switchgrass and poplar) and processes can result in substantial reductions compared with petroleum fuels, assuming that all major production processes and the use of fertilizer inputs become more efficient, and that biomass is used as process energy.[119]

Typical estimates for reductions from cellulosic ethanol (most of which come from engineering studies since few large-scale production facilities exist, to date) range from 70 to 90 per cent relative to conventional gasoline, according to Fulton et al (2004), although the full range of estimates is far broader.[120] Where exactly cellulosic ethanol falls in such a range depends upon the feedstock used to produce it, assumptions regarding fertilizer input, end-use efficiency of vehicles, and co-products.[121] Wang et al (1999) estimate that cellulosic ethanol made from woody biomass (such as poplar trees) can achieve reductions larger than those from herbaceous biomass.[122]

Table 11.3 *Estimated change in life-cycle greenhouse gas emissions per kilometre travelled by replacing gasoline with ethanol from cellulose*[a]

Feedstock (country or other specifics, where available)	*Emissions change (percentage)*	*Source*
Wheat residue (straw)	–57	Levelton Engineering Ltd (2000)
Corn residue (stover)	–61	Levelton Engineering Ltd (2000)
Hay	–68	Levelton Engineering Ltd (2000)
Grass	–37.2	Delucchi (2005)
	–66 to –71	GM/ANL (2001)
	–71	Levelton Engineering Ltd (2000)
	–73	Wang (2001)
Waste wood (E85; Australia)	–81	Beer et al (2001)
Crop residue (straw)	–82	GM et al (2002)
Wood	–51	GM et al (2002)
Poplar tree	–107	Wang (2001)

Note: [a] in conventional spark-ignition vehicles.

Source: see endnote 116 for this chapter

A number of other advanced technologies are also being developed to convert biomass into gaseous and liquid fuels for vehicle use. As mentioned earlier, a NOVEM–ADL study estimated life-cycle CO_2-equivalent emissions that might be typical for a variety of processes by the period 2010–2015. They projected that cellulosic ethanol made using enzymatic hydrolysis in biorefineries would provide the greatest reductions, followed by biomass gasification (see Table 11.4).[123]

The company Genencor has developed a new kind of alpha amylase enzyme that dramatically reduces the temperatures required for processing corn starches – from 105–150°C to 32°C.[124] This 'no-cook' hydrolysation process could significantly reduce energy input requirements, thereby reducing emissions directly or increasing the amount of excess renewable energy that can be fed into the local electric grid.

High-oil algae colonies offer a potentially large source of feedstock for biodiesel that could result in significant emissions reductions. While precise numbers are disputed, it is believed that algae can produce far more oil per hectare than any other biodiesel feedstock, and can do so even in deserts. Significant emissions reductions could be possible, particularly if agricultural waste streams or power plant emissions are used to feed the algae (see Chapter 4).[125]

Hydrous anaerobic pyrolysis, which mimics the geological conditions that created crude oil, can use 'wastes' that other industries must dispose of (often at

Table 11.4 *Estimated change in life-cycle greenhouse gas emissions per kilometre travelled by replacing gasoline or diesel with advanced biofuels (2010–2015)*

Fuel	*Feedstock/location*	*Process*	*Emissions change (percentage)*
Diesel			
Biodiesel	Rapeseed (local)	Oil to fatty acid methyl ester (FAME) (transesterification)	–38
Biodiesel	Soybeans (local)	Oil to FAME	–53
Diesel	Biomass – eucalyptus (Baltic)	Hydrothermal upgrading (HTU) biocrude	–60
Diesel	Biomass – eucalyptus (Baltic)	Gasification/ Fischer-Tropsch process (F-T)	–108
Diesel	Biomass – eucalyptus (Baltic)	Pyrolysis	–64
Dimethyl ether (DME)	Biomass – eucalyptus (Baltic)	Gasification/DME conversion	–89
Gasoline			
Gasoline	Biomass – eucalyptus (Baltic)	Gasification/F-T	–104
Ethanol	Biomass – poplar (Baltic)	Enzymatic hydrolysis	–112
Ethanol	Biomass – poplar (Brazil)	Enzymatic hydrolysis	–112
Ethanol	Biomass – poplar (local with feedstock from Brazil)	Enzymatic hydrolysis	–101
Ethanol	Corn (local)	Fermentation	–72

Source: see endnote 123 for this chapter

cost). The process releases few GHGs, largely because the conversion process does not oxidize organic matter.[126] Changing World Technologies, based in the US, claims to use the same feedstocks for process energy and fuel.[127] And gasification followed by Fischer-Tropsch (F-T) synthesis (also called biomass to liquid, or BTL), which uses carbon-fixing perennial plants, can reduce GHG emissions by more than 100 per cent if the process is powered by the biomass feedstock, and if the feedstock are perennial plants that sequester carbon or wastes. It can also convert lignin into liquid fuel – something that cellulosic processes cannot do (see Chapter 5).[128]

Figure 11.1 shows the range of estimated possible reductions in emissions from wastes and other next-generation feedstocks relative to those from current-generation feedstocks and technologies.[129]

Yet another possible means for improving the GHG benefits of biofuels is carbon capture combined with storage. According to Faaij (2005), during the fermentation process, about half the biomass in sugar- and starch-rich sources is converted into ethanol and the remainder is converted into CO_2. In addition, with regard to F-T production, about half the carbon in the original feedstock can be captured before conversion of syngas to F-T fuels. CO_2 capture and storage during

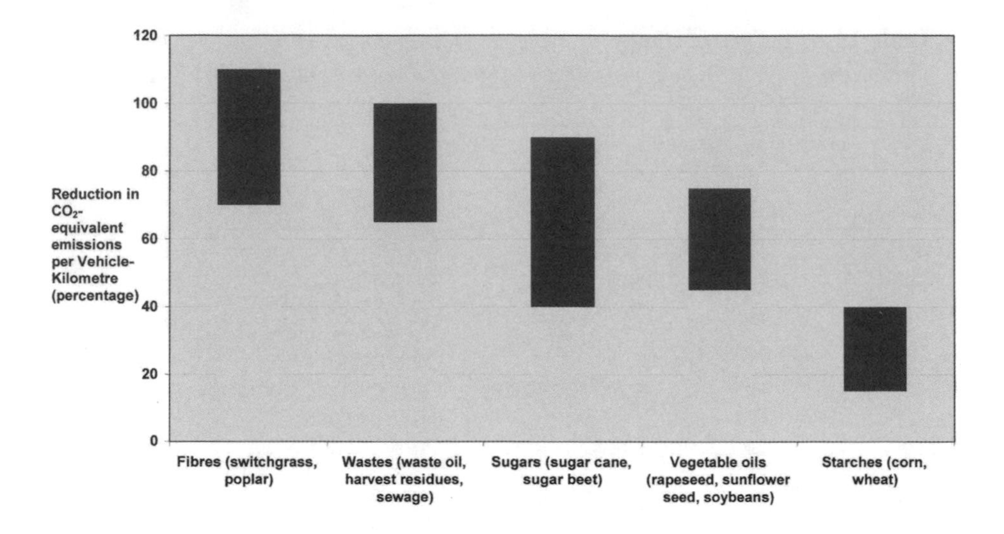

Figure 11.1 *Reductions in greenhouse gas emissions per vehicle kilometre by feedstock and associated refining technology*

Source: see endnote 129 for this chapter

these processes could allow for negative emissions per unit of energy produced on a life-cycle basis.[130] Larson (2005) also projects that this option would enable reductions to exceed 100 per cent.[131]

Co-products

As discussed above, the production of additional co-products can reduce GHG emissions, as well. In particular, renewable lignin from energy crops can reduce or eliminate the need for coal or gas required for processing, directly reducing GHG emissions. If excess electricity is available to feed into the local utility grid, offsetting fossil-generated power, resultant emissions reductions could be even greater.

Trade-offs

Despite the potential climate-related benefits associated with biofuels, and the major role that the transport sector plays in the production of global GHGs, many experts have questioned the wisdom of converting biomass to transport fuels if the primary aim is to reduce global warming gases. They cite the high cost of emissions reductions relative to other options, and note that land and biomass resources would be more efficiently used for other purposes.

Economic costs

Biofuels are currently a relatively expensive means of reducing GHG emissions compared to other mitigation measures, according to analyses from many countries, with the cost of CO_2-equivalent emissions reductions exceeding €135 per tonne (US$163 per tonne), according to estimates analysed by Fulton et al (2004) (see Figure 11.2).[132] The one exception is Brazil, where pure ethanol sold for nearly 40 per cent less than the gasoline–ethanol blend in late 2005 (even accounting for the lower energy content in ethanol).[133]

Even within the transport sector, there are several more cost-effective options for reducing carbon emissions, including investments in and promotion of public transportation, increased use of bicycles and other non-motorized vehicles, improvements in vehicle fuel efficiency, and changes in urban planning and land use. Not only could such measures provide larger reductions at lower costs, but they are also of strategic importance for many developing nations, where demand for transport is increasing rapidly.[134]

Several studies in the EU have concluded that GHG emissions savings from fuel ethanol, based on domestic production using wheat and sugar beets, have cost a minimum of €200 (US$242) per tonne of CO_2 – about ten times the marginal abatement cost in the EU emissions trading scheme of €20 per tonne.[135] Larivé

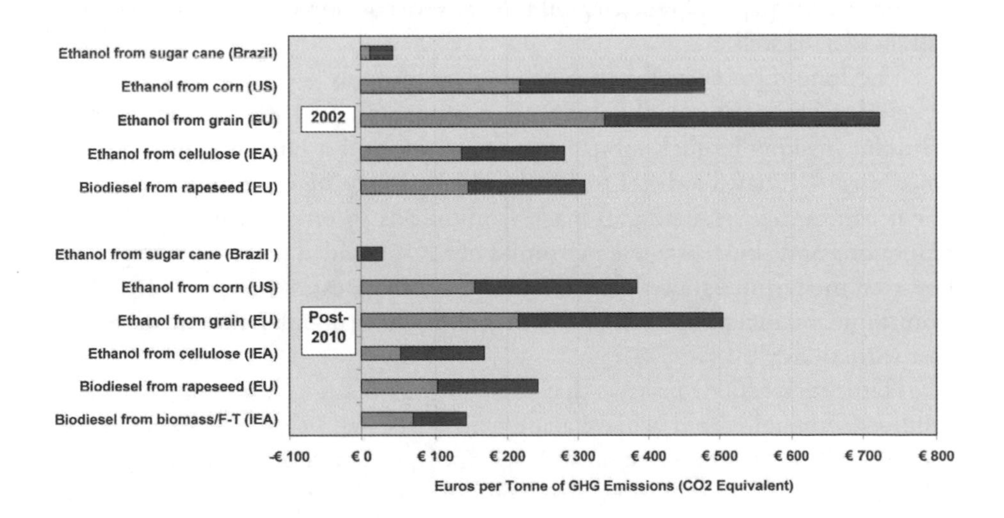

Figure 11.2 *Biofuel cost per tonne of greenhouse gas reduction*

Note: Ranges were developed using highest cost/lowest GHG reduction estimate, and lowest cost/ highest GHG reduction estimate for each option, then taking the 25 and 75 percentile of this range to represent the low and high estimates in this figure. In some cases, ranges were developed around point estimates to reflect uncertainty.

Source: Fulton et al (2004)

(2005) notes that, even with oil at €50 (US$60.5) per barrel, there are no biofuel options that cost less than €100 per tonne of CO_2 avoided.[136] Compare this to the Kyoto Protocol's Clean Development Mechanism, which, according to a World Bank report, provides carbon market prices of less than €8 (US$10) per tonne of CO_2 and an expected maximum price of €12.4 to €16.5 (US$15 to $20) per tonne of CO_2 over the coming decade.[137]

Other studies have noted that using biomass to produce combined heat and power is currently a far more cost-effective means for reducing GHG emissions than converting it to biofuel. And a 2003 study prepared by Mortimer et al for the UK Department of Environment, Food and Rural Affairs (DEFRA) concluded that biofuel is one of the most expensive options for reducing CO_2 emissions from biomass. Its authors estimated that the greatest savings, of all options considered in the report, comes from the use of glass-fibre loft insulation which, for the same amount of money, could reduce CO_2 emissions by 92 to 141 times more than was possible using biodiesel from oilseed rape, and by 24 times more than was possible heating with wood chips from short-rotation coppice.[138, 139]

At the same time, the CO_2 avoided by using biofuels is only a part – albeit a significant part – of the societal benefit derived from these fuels. It is important to note that many options exist to substitute for coal in the generation of heat and electricity; but biofuels offer the only realistic near-term option for displacing and supplementing liquid transport fuels. Assuming that oil prices remain high, the economic advantages of replacing oil for transport as opposed to fossil electricity will increase as well.

The long-term costs for next-generation biofuels – such as F-T biodiesel or cellulosic ethanol – could also be significantly lower than production costs are today, making biofuels more cost competitive as a means of reducing GHG emissions.[140] Today, biodiesel from oilseed crops is quite expensive to produce; but it can outperform ethanol made from grains in terms of potential GHG reductions and thus costs less per tonne of CO_2 avoided. Over the long term, however, fuels from lignocellulosic biomass have the greatest potential (of all fuels from biomass, including methanol, ethanol, hydrogen and synthetic diesel) for cost reductions.[141]

Hamelinck (2004) projects that these next-generation biofuels, particularly cellulosic ethanol, could achieve abatement costs well below €41 (US$50) per tonne of CO_2; with oil prices at €41 (US$50) per barrel or higher, biofuels are fully competitive and mitigation costs could actually be negative.[142] The potential for such great cost reductions is the result of several factors: the feedstock can be grown on less valuable land, has lower energy requirements for cultivation and harvest, and has the potential for much higher energy yields per hectare.[143] It is also possible to move greater distances per hectare on cellulosic fuels than it is on fuels from rapeseed or sugar beet, a difference that will become greater with time.[144] Fulton et al (2004) note that because it could bring down per tonne costs of GHG reductions so dramatically compared to today's technologies, 'supporting the use of a relatively expensive fuel that provides large emissions reductions, like cellulosic ethanol, may be worthwhile'.[145]

Moreover, as global oil prices rise, biofuels will become more cost competitive and could ultimately be less expensive than conventional transport fuels. As mentioned earlier, this has already been the case in Brazil. Where and when this occurs, the use of biofuels could actually save consumers money, meaning that the costs of reducing GHG emissions through the use of biofuels would be negative.

Finally, it is important to keep in mind the environmental and associated economic costs of non-biofuel alternatives to conventional petroleum fuels. Estimated reserves of unconventional oil from tar sands and oil shale in Canada and the US, respectively, are far greater than combined proven reserves and estimated possible (undiscovered) reserves of conventional oil (see Chapter 12).[146] By some estimates, more than six times as much CO_2 is released per barrel of oil produced from oil sands than per barrel of conventional oil.[147]

Biomass and land resources

Several studies have concluded that, even aside from the more favourable economics of combined heat and power (CHP), CHP is the best option for reducing GHG emissions with biomass.[148] The biggest impact can come from using biomass resources to replace coal, rather than petroleum fuels. For example, Armstrong et al (2002) estimate that using biomass as fuel to produce steam and electricity, or CHP, provides up to 200 gigajoules (GJ) of energy per hectare of land, compared to a maximum potential of 30–60GJ per hectare from ethanol or biodiesel.[149] According to Graham et al (1992), using cellulosic crops to produce ethanol provides about half the CO_2 reductions that could be achieved by using them to displace coal.[150]

Over the medium term (2020 and beyond), however, Faaij (2005) argues that using biomass to produce transport fuels may be a more effective means for reducing GHG emissions than using biomass to generate power:

> *This can be explained by the partly observed and partly expected reduction in carbon intensity of power generation due to large-scale penetration of wind energy, increased use of highly efficient natural gas-fired combined cycles and deployment of CO_2 capture and storage (in particular, at coal fired power stations).*[151]

Others recommend that, unless the goal is short-term reductions in CO_2 emissions, afforestation and reforestation measures should be considered before using land for the production and use of biofuels or even electricity. Tampier et al (2004) estimate that, over a 50-year period, 'trees can sequester more carbon dioxide out of the air than can be displaced through transportation fuels or electricity from biomass'.[152]

On the other hand, an International Energy Agency (IEA) study notes that afforestation and forest protection are only conditional mitigation options that are 'subject to future management regimes'.[153] Afforestation is a temporary measure, whereas bioenergy can provide an irreversible mitigation impact. Schlamadinger and Marland argue that, over the long term (40 years or more), there is a relative advantage to using surplus agricultural land for biofuel production (and substitution of fossil fuels) rather than afforestation. They note that:

> *The net carbon advantage depends on the growth rate of the site and on the efficiency with which fossil fuel carbon emissions are reduced through the use of biofuels. Biofuel production is the better choice, especially with efficient use of biomass and for high growth rates.* [154]

Also in the future, cascading biomass over time – recycling biomass materials for various uses – can help to optimize the CO_2-mitigation effects of biomass resources. It is possible to displace more fossil fuel feedstock and, thus, to derive a far greater carbon benefit by using biomass for material production, such as plastics, and subsequently using that material (when it has reached the end of its useful life) for energy production. Dornburg and Faaij (2004) have studied in detail the climate and economic impacts of cascading biomass and conclude that this practice could provide CO_2 benefits up to a factor of five compared to biomass used for energy alone.[155]

Conclusion

A dramatic increase in the production and use of biofuels has the potential to significantly reduce overall GHG emissions associated with the global transport sector; alternatively, it could intensify the threat of a warming world. The overall climate impacts of biofuels will depend upon several factors, the most important of which are associated land-use changes, choice of feedstock and how it is managed (including its energy yield per hectare and level of fossil inputs), and the refining process (including co-product output and source of process energy). If, for example, perennial crops replace annual crops – such as corn now grown to produce ethanol – and are processed with biomass energy that offsets coal-fired power, the resulting biofuel can significantly reduce GHG emissions compared to petroleum fuels.

Alternatively, if prairie grassland is converted to corn or soy, treated with chemical fertilizers and pesticides, and refined with coal and natural gas, the resulting fuel could have a greater impact on the climate over its life cycle than do petroleum fuels. However, even 'sustainable' energy crops can have a negative impact if they replace tropical forests, resulting in large releases of carbon from soil and existing biomass that negate any benefits of biofuels for decades. In general, crops that require significant energy inputs (such as fertilizer) and valuable farmland, and have relatively low energy yields per hectare, should be avoided.[156]

The greatest potential for reducing GHG emissions and associated costs lies in the development of next-generation feedstocks and biofuels. In the future, next-generation technologies – particularly advanced cellulosic technologies – offer the potential to reduce transport-related GHG emissions significantly. Assuming that oil prices remain high, it will be possible to achieve negative CO_2-abatement costs in the process, while providing a host of other environmental and social benefits as well. Government policies should focus on commercializing these advanced technologies and driving down their costs as rapidly as possible.

Governments must also do whatever possible to protect virgin lands such as grasslands and forests, and to encourage the use of sustainable feedstock and management practices in order to minimize associated GHG emissions – indeed, such policies should extend beyond biofuel production to the agricultural sector, in general. In addition, a certification scheme needs to be developed that includes GHG verification for the entire life cycle of biofuel products; it will be a challenge to find ways of implementing such a scheme in a generally accepted manner; but this should not deter governments from making the effort. This is not the case in most countries today, and biomass is often automatically considered 'carbon neutral' without accounting for upstream emissions. The UK is now contemplating a scheme for imported biofuels that includes the entire supply chain in emissions accounting, and Belgium has already put such a scheme into legislation (see Chapter 18).[157]

Further research is needed to fill gaps in the existing body of life-cycle studies, including more analyses that are locally relevant to developing countries, and studies that cover the range of the biofuel feedstocks and pathways (including biodiesel from palm oil or jatropha). It is also essential to have a better understanding of the scale of N_2O emissions from feedstock production and their potential impact on the global climate in order to attain a more accurate picture of the full impacts of feedstock production on emissions. And more studies are needed that consider the relative efficiency of land use, for 'While GHG mitigation per vkm [vehicle kilometre] is an important measure, land-use efficiency in achieving GHG reductions may be the most important consideration'.[158]

Finally, biofuels must be used to replace petroleum fuels rather than to supplement them. If biofuels are able to substitute for petroleum, they can provide a far greater benefit to the global climate than if they are produced and burned simply to meet a share of the world's rapidly increasing demand for transport fuel. In general, any plan to promote the production and use of biofuels on a large scale must be part of a broader strategy to reduce total energy use in the transport sector. In addition to ending subsidies for conventional fuels (and for unconventional petroleum fuels), governments must encourage the development of lighter, more fuel-efficient vehicles, and promote and support smarter urban design and mass transit.

12

Environmental Impacts of Feedstock Production

Introduction

One of the greatest benefits of using biofuels is the potential to significantly reduce greenhouse gas (GHG) emissions associated with the transport sector. One of the greatest risks, however, is the impact on land used for feedstock production, particularly virgin land, and the associated effects on habitat, biodiversity, and water, air and soil quality. These concerns are particularly valid with current first-generation feedstocks. On the other hand, bioenergy production offers the potential to reduce the environmental load relative to conventional industrialized agriculture if farming practices are adjusted to maximize total energy yield rather than the oil, starch or sugar contents of their crops, diversifying plant varieties and reducing chemical inputs.

In contrast to the environmental costs of fossil fuels on land and wildlife, the impact of biofuels on the landscape has been relatively small. But as production levels rise and as more nations increase their use of these fuels, the environmental trade-offs seen thus far will probably be experienced on a far larger scale. This creates an urgent need to develop biofuels in a more environmentally sustainable manner.

This chapter discusses the main environmental issues associated with the production of biofuel feedstocks. For comparison, it first briefly discusses some of the environmental costs associated with obtaining petroleum-based fuels. It then examines the costs and benefits – to the land, water and air – of generating biomass feedstocks for biofuels (the chapter does not address climate change issues, which are discussed in Chapter 11).[1]

Environmental costs of oil exploration and extraction

Crude oil represents a fraction of 1 per cent of ancient biomass that has been stored in vast quantities over tens of millions of years.[2] Locating it and then removing it from the ground can entail high costs to the landscape. The first step in extracting oil from the ground is finding deposits that are large and concentrated enough to remove profitably. This exploration not only requires the construction and use of large machines, but it involves drilling in sites that are sometimes in ecologically sensitive areas. Most of the holes drilled come up dry: historically, oil exploration companies drilled five holes for every 'wet' one, while today the ratio is about one in three with the advent of three-dimensional (3-D) seismic imaging.[3]

The impact of the machinery used to extract petroleum itself is relatively small – the energy embodied in this machinery and the energy used for pumping, compressors, etc. consume only about 1 per cent of the total energy it can process during its lifetime.[4] However, the processes required to extract 'enhanced' oil from the ground with secondary and tertiary methods are more energy intensive and can use more than 15 per cent of the energy in crude oil. This includes the energy needed to drill deep, often dry, holes (usually diesel fuel); to construct concrete pipes to hold the hole open (coal and electricity); and, occasionally, to heat the underground oil deposits into a manageable liquid, and to inject chemicals, carbon dioxide (CO_2) or even steam (natural gas).[5]

On land, oil extraction can destabilize terrain and disrupt underground aquifers by removing large volumes of oil and methane from the ground. Land subsidence is common where oil has been extracted from soft sandy formations. Oil drilling near the ocean, meanwhile, can suck salty seawater into freshwater aquifers: in the Gulf of Mexico, for example, coastal drilling has transformed uplands into wetlands, and wetlands into open water.[6] In addition, much of the water used to push oil out of reservoirs, once polluted, remains underground, where it is either treated onsite or simply cleansed by evaporation. But oil fields leak waterborne ions and chemicals into the surrounding ecosystems, particularly sodium, chloride, boron, benzene and arsenic.[7]

Oil deposits often overlap with natural methane gas deposits. If nothing is done to prevent the methane's escape, it rises into the atmosphere, where it acts as a powerful greenhouse gas. Oil companies can burn it, releasing CO_2, or, if the proper infrastructure is available at the wellhead, they can contain it to use in the oil extraction process or to sell.[8]

Drilling muds, formed from drill cuttings and containing toxic aromatic compounds, accumulate around the base of marine oil platforms. Where tidal currents are weak, cuttings remain intact enough to form large 'cutting piles'. In the deeper waters of the northern North Sea, these piles can contain as much as 40,000 tonnes of contaminated sediment, with devastating consequences for marine habitat in the surrounding areas. Adverse affects of the early muds, based on diesel oil, were found to extend out as far as 5km from the point of discharge. Less toxic alternatives are now used where possible, and cuttings are generally re-injected down wells or removed for treatment onshore.[9]

Once crude oil reaches the Earth's surface, it causes other environmental problems at the drilling site. Earlier production techniques did not contain oil once it was struck, and often 'gushers' would last for hours or days, occasionally catching fire and spreading clouds of smoke and pollutants. Today, oil is usually – but not always – contained in the drill hole.[10] But smaller overflows continue during the lifetime of an oilfield. In jungle regions, extraction operations can be especially toxic for ecosystems and native lands.[11] Drilling offshore can result in the dredging of sensitive marine ecosystems, including coral reefs, and oil leaked directly into the ocean is far more difficult to contain.[12]

As oil prices rise, it is becoming more economical to extract oil from non-conventional resources, such as tar sands and oil shale. Estimated reserves of these unconventional petroleum fuels are enormous – Canada's oil reserves are considered second only to Saudi Arabia's; but the environmental costs of extracting and using them are enormous, as well.[13, 14] With Canadian tar sands, for example, about 20 per cent of the resource is close enough to the surface to be strip mined. The largest such pit in Alberta is a 50 square mile 'moonscape' of slag heaps and tailing ponds. Once the pits are depleted, extraction involves reaching deeper deposits by pumping in steam to dissolve the thick oil so that it can be brought to the surface. This requires large amounts of both water and energy, and results in significant quantities of wastewater. The oily slurry that separates from grains of sand must then be refined to make usable oil, as discussed in Chapter 13. Two tonnes of this sand are required to produce a single barrel of oil.[15] Impacts on air quality are great as well, with both direct emissions from the mining process and indirect impacts associated with electricity generation.

Whether conventional or unconventional, oil exploration and extraction have negative impacts on wildlife and plant populations due to local pollution, as well as the loss or fragmentation of habitats, and obstruction or elimination of migration routes that result from infrastructure (from roads to facilities to pipelines) required to access these fuels.[16] In the future, a greater share of the world's oil will also come from more remote areas – such as offshore, and the Arctic – which are more sensitive to disruption and pollution, and are often the habitats of rare species.

BIOFUEL FEEDSTOCK PRODUCTION AND LAND-USE CHANGES

The environmental problems associated with obtaining the feedstock for biofuels can also be serious – indeed, this is probably the most environmentally disruptive stage of biofuel production. However, the net environmental impact of land use for feedstock production on habitat, biodiversity, and soil, water and air quality depends upon a variety of factors, including the choice of feedstock, what the feedstock replaces and how it is managed.

Threatening wild habitat

The impacts of agricultural land-use practices for conventional crops can be dramatic, affecting everything from biodiversity to the global climate. The biggest threat posed by expanding the amount of land under cultivation for energy or any other use is the irreversible conversion of virgin ecosystems. Deforestation, for example, causes the annihilation of species and their habitats, and the loss of ecosystem functions. Studies reveal that wide-scale destruction of forests can affect the hydrological cycle and the climate, reducing regional precipitation and increasing temperatures.[17]

Rising demand for biofuels is motivating the expansion of agriculture onto previously unexploited lands in some regions and could cause intensification of land use in others. In the near term, the ecosystems at greatest risk include the rainforests of Malaysia, Indonesia and Brazil, and the savannahs of southern Brazil. These regions are of great value due to their species richness, containing a large share of the world's diversity of flora and fauna. Also under threat are natural forests, grasslands and wetlands, from rural England to Tanzania, which are homes to large numbers of mammals, songbirds and wild plants.

Expanding into Brazil's cerrado

Sugar cane, in particular, has a poor environmental track record. Over the past several hundred years, it has had as large an impact as perhaps any other agricultural commodity, with the greatest consequence being the loss of biodiversity.[18] This is particularly true in tropical regions such as the Caribbean, where diverse ecosystems were destroyed largely by sugar planted during early European colonization.

In Brazil, the cultivation of sugar cane for ethanol, and increasingly of oilseed crops for biodiesel, is occurring as agricultural pressure is also increasing to meet rising demand for sugar and soy in food and feed markets. The expansion of sugar cane production via large monocultures has replaced pasturelands and small farms of varied crops.[19] Plantations for sugar and ethanol production have expanded predominantly into areas once used for cattle grazing, as cattle move on to new pastureland (often cleared rainforests).[20] In the future, the *cerrado*, a largely wild central savannah that covers more than one quarter of Brazil's land area, is considered the natural expansion area for sugar cane production.[21] The *cerrado* is home to half of Brazil's endemic species (found nowhere else on Earth) and one quarter of its threatened species. Expansion of agricultural production into the region's complex ecosystems could result in irreversible ecological damage.[22]

A similar trend is occurring with soybeans, which are replacing vast stretches of both wild *cerrado* and rainforest. Over the decade between 1994/1995 and 2004/2005, Brazilian soybean production nearly doubled, from 11.7 million to 22.3 million hectares.[23] Many of the country's large soybean plantations have been developed by acquiring smaller plots of land with more varied crops, while about half of the nation's soybean crop has moved into *cerrado* in the middle-west region, where climate and soil are considered suitable for cultivation.[24]

Expanding into rainforests

The *cerrado* is not the only Brazilian ecosystem at risk. In the state of Mato Grosso do Sul, in the country's southwest, the state assembly is debating the construction of ethanol plants along the upper Paraguay River. The river runs through the Pantanal, the world's largest wetland area, and there is concern that plant construction will threaten the environmental balance of this delicate ecosystem.[25] As transportation infrastructure advances towards Brazil's Amazon region to facilitate the flow of soybeans, sugar cane and other products to processing facilities and ports in the east, it is more likely that these crops will expand into sensitive areas where they are not currently economically viable.[26]

Already, rainforests from Brazil to Southeast Asia are being cleared to grow soybeans and palm oil.[27] While these crops are now produced primarily for food rather than fuel, they are being used increasingly to produce biofuels for transportation and electricity.[28] Their cultivation adds to the predominant forces of deforestation in the Brazilian Amazon, which are cattle ranching and illegal timber cutting.[29] According to Schneider et al (2000), increased soybean production is triggering the expansion of pastureland for cattle into new forest areas, which is considered 'one of the main causes of tropical deforestation in the Brazilian Amazon'.[30] To date, nearly 20 per cent of the Amazon, home to an estimated 30 per cent of the world's species of plants and animals, has been burned or otherwise destroyed, much of it due to large-scale agriculture.[31]

Palm oil expansion, meanwhile, is one of the 'leading causes' of rainforest destruction in Southeast Asia and 'one of the most environmentally damaging commodities on the planet', according to Simon Counsell, director of the UK-based Rainforest Foundation. Palm plantations are expanding rapidly in eastern Malaysia as well as in Indonesia where, despite laws prohibiting clearing for palm oil plantations, natural forests are being felled at a rapid pace.[32] Rather than planting on abandoned agricultural land, palm oil producers are instead expanding into forestland, a more attractive prospect since recently cleared forests need less fertilizer and the timber can be sold for capital.[33]

This expansion poses a tremendous threat to the region's biodiversity, including endangered mega-fauna such as tigers, Asian elephants and the Sumatran rhinoceros.[34] According to the environmental organization Friends of the Earth, palm oil plantations in Indonesia and Malaysia are the primary cause of the decline of the world's orang-utans and could drive the species to extinction within the next ten years.[35] Environmentalists and other groups in Europe, particularly in the UK and The Netherlands, have expressed concern that rising EU demand for biofuels could accelerate this deforestation, undermining the environmental benefits sought through policies to promote these fuels.[36]

The palm oil industry has noted that there is far more biodiversity in palm oil plantations than in fields of annual grains, vegetables and other short-term crops because palm oil trees are perennial crops cultivated in tropical areas.[37] This is relevant, however, only if palm oil plantations replace these crops, not wild forests (for information regarding the Roundtable on Sustainable Palm Oil, formed to address concerns related to palm oil production, see Chapter 18).

Expanding into cropland, wild lands and conservation reserve land

While expansion of biofuel crops into tropical forests might be the greatest land-use concern due to potential effects on biodiversity, biofuel expansion could cause problems elsewhere, as well. In the UK, for example, feedstock production could threaten the diversity in farmland uses and might have a detrimental impact on the nation's bird population.[38] In Tanzania, as in many other countries, there is concern that biomass crops could replace existing forest or grasslands, and that small farms of varied crops could be replaced by large tracts of monoculture plantations.[39] In the US, rising demand for soybeans and corn could increase the 'duo-culture' cropping style that already dominates much of the Midwest.

Expansion of the corn-ethanol market in the US also threatens to put millions of hectares of conservation reserve land back into production, increasing erosion and eliminating many of the wildlife benefits realized since the reserve programme began. Much of this land was put into reserve during the mid 1980s because its rough, sloping ground makes it difficult to plant.[40] In 2004, the European Commission estimated that 5 per cent of European Union (EU) transport fuel needs could be met by growing energy crops on currently unproductive agricultural lands, while forests, grassland and the use of wastes could provide yet more.[41] The EU is now considering sustainability impacts of these plans, and even the possibility of expanding nature conservation areas; thus, the potential biofuels resource base will probably be reduced.[42]

Given the possible threats to both land and wildlife, it will be necessary to carefully manage the risks associated with the use of set-aside lands. The trade-off is not necessarily easy: as demand for biofuels increases, if feedstock crops are not produced in the US and Europe, then they will simply be cultivated elsewhere, leading to potential landscape and ecosystems effects in other regions.

Fortunately, as crop yields continue to increase, land requirements per litre of biofuels (and possibly water and other requirements, as well) will decline, along with related impacts on habitat and wildlife. To date, many first-generation energy crops have been bred for their starch or oil content, not for their leaves, stems or other resources; there has been little or no effort to optimize plants for mass, a trait that is far easier to manipulate. Aggressive breeding programmes could significantly increase yields of cellulosic feedstocks, such as corn stover or switchgrass for next-generation biofuels, without the use of genetic modification.[43] Furthermore, crops that are bred specifically for energy content may require less fertilizers and pesticides.[44] Another important option for next-generation biofuels will be to grow short-rotation woody crops and other perennial plants instead of annual crops, thereby increasing yields while reducing chemical inputs and overall environmental impacts.

But some efforts to improve crop yields could have a cost. If use of fertilizers and other inputs are increased, for example, GHG emissions will rise and other negative impacts result. Plant breeding that selects only the most productive crops can be done at the cost of genetic diversity (although cross-breeding can increase diversity), bringing a greater susceptibility to disease.[45] And genetic engineering, while offering the potential to increase yields, remains controversial, with concerns about possible cross-pollination and other long-term effects (see Box 12.1).[46]

Box 12.1 Benefits and costs of genetically modified crops

Genetically modified (GM) crops are being developed and used increasingly in the US and elsewhere because of the desire to increase yields and reduce the need for water and pesticides. As of 2003, about 80 per cent of the US soybean crop and 40 per cent of its corn crop were genetically modified, and GM rapeseeds and sugar beets are now available as well. Use of GM crops is also increasing elsewhere, although there is considerable opposition to them in much of the world, including in several European countries and much of Asia.

Among possible benefits of GM crops, studies cite the potential for higher yields (even under difficult growing conditions), qualitative improvements in crops and diversification of plant uses. According to some reports, the use of RoundUp Ready soybeans (engineered to tolerate potent herbicides that previously would have destroyed crops along with weeds) has reduced the need for herbicide and increased yields by as much as 10 per cent, while reducing soil erosion by up to 50 per cent because the lack of weeds makes it easier to employ no-till farming methods.

Others contend that RoundUp Ready soybeans have hurt crop yields and that the appearance of more-resistant strains of weeds has increased per-hectare rates of glyphosate (herbicide) use. According to Benbrook (2004), chemical use has increased with the introduction of GM organisms. While Bt crops (resistant to specific insect pests) may have reduced insecticide use somewhat, herbicide-tolerant crops have actually increased herbicide use much more substantially (after an initial reduction). Scientists at the Minnesota-based Institute for Agriculture and Trade Policy note that use of GM crops only helps to prevent yield decreases in years with very specific infestations. They also note that no-till adoption has been flat for several years in the US, and there is no evidence that RoundUp Ready crops have increased its use. Additionally, there is concern that the use of GM crops accelerates negative effects on plant biodiversity that already existed due to monoculture farming and pest and weed resistance.

Any benefits of GM crops must be weighed carefully against the risks they pose to wildlife, wild and organic plants, and human health. One three-year study in the UK, completed in 2003 (the largest then undertaken), demonstrated that GM crops have severe implications for wild birds. The government panel involved in the study concluded that, while there was not enough evidence to predict long-term impacts, the potent herbicides used on GM crops posed an unquantified risk to wildlife. Scientists linked the potential risk to the way in which GM crops were sprayed: spraying GM sugar beets caused more damage to the environment than spraying conventional sugar beet varieties. Further, GM herbicide-resistant rapeseed had the same number of weeds overall as conventional rapeseed, but more grass weeds and fewer broadleaved weeds, whose seeds are an important food source for wildlife.

The same study concluded that GM and other crops might not be able to coexist because of potential contamination, and that future GM generations could pose new risks. Ecologists have expressed concern that GM plants might breed with wild relatives, altering their identity or even creating fast-growing 'super weeds', with increased resistance to some pesticides. Studies in the US and Europe have shown that the potential exists for genes to flow from GM crops (including oilseed crops) into wild populations because of historical hybridizations and the movement of pollen. Two separate UK studies found that bees carrying GM pollen from rapeseed had contaminated conventional plants more than 26km (16 miles) away. And a 2002 US survey by the Organic Farming Research Foundation found that up to 80 per cent of organic farmers in the Midwest (where millions of hectares of GM crops are grown) reported direct costs or damages associated with GM contamination, which disqualified their products as organic.

Others contend that GM crops could accelerate the pesticide treadmill – whereby pervasive cultivation and spraying of herbicide-tolerant crops could build resistance in weeds, requiring ever-increasing applications of herbicides and introducing hardier weed varieties into the environment. Other impacts can include the killing of important soil organisms and dangers to human health due to increased pesticide use. A recent report from the Africa Centre for Biosafety and Friends of the Earth–Nigeria linked intensive cultivation of GM soybeans in Latin America with decreased soil fertility and increased soil erosion and deforestation. In Argentina, according to a local non-governmental organization (NGO), the consequences of growing GM soybeans include 'a massive exodus [of people] from the countryside and ecological devastation'. Air spraying of pesticides (particularly glyphosate) on RoundUp Ready soybeans has destroyed other crops, killed farm animals and caused human illnesses ranging from allergies to vomiting and diarrhoea.

Genetically modified industrial micro-organisms (GMIOs), created to flourish under manufactured conditions (such as high temperatures) rather than in nature, will probably play a significant role in future biofuel development – particularly in the fermentation of carbohydrates to manufacture cellulosic ethanol. But because GMIOs would probably have a competitive disadvantage in the wild, there is less opposition to them than to GM crops. Even the German Green party supports the use of GMIOs, while continuing to oppose GM crops. However, there remains the potential that these organisms might pose a yet undetermined threat to the environment, making testing and the development of appropriate safeguards essential.

Source: see endnote 46 for this chapter

Minimizing land-use and wildlife impacts

Despite the multiple risks posed to land and habitat through production of biofuel feedstocks, many options exist for minimizing these costs or even improving biodiversity. One possibility is to expand onto degraded lands that cannot currently support agriculture; another is to use agricultural and forest residues and wastes as feedstocks, a practice that is possible today only on a small scale, but that offers significant potential when processing and biofuel conversion technologies are further developed.

Expanding onto marginal lands

Over the short term, India plans to cover 400,000 hectares across several hundred districts – approximately 0.13 per cent of the nation's land area – with jatropha plantations.[47] This is not expected to affect plant or animal biodiversity, as the bushes will be planted on current wastelands. However, no studies have been done, to date, to determine any potential biodiversity impacts, and The Energy Resources Institute (TERI) (2005) notes that there is concern that once India moves beyond the initial demonstration phase and increases the scale of jatropha and pongamia plantations, this could interfere with the nation's natural forest ecosystems.[48]

Jatropha is also being used in African countries such as Mali to reverse desertification while providing income and a new source of fuel for both electricity and transport.[49] The benefits of jatropha and pongamia are discussed in greater detail elsewhere in this chapter and in this book.

Greater land efficiency of 'next-generation' feedstocks

Oilseed and starch crops grown in temperate areas use land inefficiently and can require large amounts of fertilizers and pesticides. But tropical oil and sugar plants, which are more land efficient and need fewer chemical inputs, can also raise environmental concerns, primarily because they can expand into wild habitats. And all biomass crops are of concern if they are grown in monoculture habitats and/or with genetically modified organisms. However, there exists the potential to dramatically increase the land efficiency of biofuel feedstocks.

The best option for reducing land use and related impacts is commercializing the production of cellulosic ethanol, as well as gasification technologies to produce biofuels from woody biomass and waste products, using Fischer-Tropsch synthesis (see Chapters 4 and 5). Cellulosic feedstocks could represent a significant improvement over conventional feedstocks. The 'wastes' are essentially free, residues can be extracted from land that is already under cultivation, and perennial energy crops can provide a more diverse habitat for wildlife. They provide better ground cover, improving the soil and water retention. And they can yield a greater quantity of fuel per hectare than sugars, starches or vegetable oils, meaning that they use the land far more efficiently.[50]

Cellulosic technologies will enable the use of agricultural and forest residues not just as process energy (as bagasse is now used in Brazil), but also as feedstocks for biofuels. However, it is important to note that, unlike other residues, residues from forests do not create 'waste' problems if they are not removed and used because they are recycled naturally. However, more research is needed to determine how much can safely be 'harvested' without affecting soil quality.

Additionally, many cellulosic energy crops, such as perennial grasses or woody biomass, offer greater diversity and variability in biofuel products and can be grown economically on less valuable land and with fewer inputs.[51] These crops could potentially be grown in place of food crops (assuming they are no longer needed for food) that are now used to produce biofuels, such as the increasing share of US cropland that is used to grow corn for ethanol, or the rapeseed, sugar beets, wheat and sunflower that are grown in Europe for the production of biodiesel. It will be critical to focus such efforts on native species of grasses that are appropriate for the soils and climate where they are planted, and to avoid the potential for invasive species, such as kudzu, to take over crops and natural areas.

Suitable crops vary from one climate or region to another. One challenge will be to find more biomass crops that are appropriate for use in more arid regions, such as the Mediterranean region. At this point it appears that most perennial crops are not suited for production in very dry summers without irrigation, or that they increase fire risk where this is already a serious threat – though this might be mitigated through appropriate timing of wood harvesting.[52]

Advanced technologies offer other options as well. Some experts believe that oil-rich micro-algae can be grown on a massive scale atop buildings and even in deserts and used to produce biodiesel, with the remainder after pressing used as animal feed and ethanol feedstock.[53] Between 1978 and 1996, the US National Renewable Energy Laboratory sponsored an Aquatic Species Program that focused on the potential to produce biodiesel from high-lipid algae. The programme's final report estimated that 28.4 billion litres (7.5 billion gallons) of biodiesel could be produced on 200,000 hectares of desert land; producing that amount of biodiesel from rapeseed would require nearly 117 times that much land and, of course, rapeseed cannot be grown in the desert.[54]

In addition, waste products such as waste oils, restaurant grease and the organic part of municipal solid wastes can be converted into fuel, saving valuable land area that might otherwise be used for landfills. To give a sense of the potential scale, New York City alone produces enough garbage to fill 1 hectares of landfill in fewer than ten days.[55] Converting this to biofuels could potentially result in a significant amount of fuel while saving land and money. It is estimated that just converting US agricultural waste into oil and gas would result in the equivalent of 4 billion barrels of oil each year, or nearly 13.6 per cent of global oil consumption in 2004.[56] And as discussed in Chapter 5, there already exist technologies to process animal wastes and carcasses into fuel.

Benefits of next-generation feedstocks for wildlife

Despite the enormous ecological costs that could potentially come from increasing production of biofuels feedstocks, some crops under some circumstances can actually encourage increased wildlife populations and diversity – particularly perennial crops that provide a stable environment, are well managed and replace existing annual crops (such as corn or rapeseed) or are planted on marginal lands.[57] In fact, the limited studies done to date suggest that the more stable environment provided by perennial crops can increase population sizes and diversity of birds, small mammals and soil fauna.[58]

A study by the Wisconsin Department of Natural Resources in the US found that switchgrass (a prairie grass native to North America) has the potential to provide high-quality nesting cover for many grassland bird species currently in decline.[59] When the grass was harvested for bioenergy production, suitable habitat remained for a number of birds, particularly species that prefer short- to medium-height vegetation. Maintaining some unharvested areas within the field would permit habitat for birds that prefer tall vegetation.

But even more-sustainable energy crops such as woody crops and switchgrass cannot substitute for natural forests or prairies, and they cannot support the same mixture of bird and small mammal species that would be found there.[60] Furthermore, perennial crops such as short-rotation tree plantations might be no better for biodiversity than annual crops if large tracts of monocultures replace numerous small fields of a variety of annual crops.[61] However, they can serve some of the functions of these natural systems and can enhance regional biodiversity if planted in landscapes dominated by annual crops, as long as they do not displace an important food source provided by the original land use.[62]

In general, as one study has noted, 'the effect of biomass plantations on biodiversity may depend as much on how they fit into the landscape as on the particular species and management systems selected'.[63] For example, biomass crops can benefit wildlife if they are used as buffers along waterways and between forests or natural grasslands and annual crops, or as protective corridors that allow plants and wildlife to move from one natural area to the next.[64] Tree plantations sited alongside natural forests can expand the habitat for forest bird species, even in regions dominated by agriculture.[65] Intercropping with grasses and trees can help to maintain the diversity of plants and animals (or at least reduce loss) on energy plantations, while also reducing soil erosion.[66] Patches of vegetation left because of less-frequent mowing or rotational harvesting also result in greater diversity and abundance of small mammals.[67] And the creation of forests and grasslands that are structurally and species diverse can help to reduce the negative impacts of weeds, insects and diseases.[68]

EFFECTS ON SOIL QUALITY

In general, when land is converted from natural cover to intensive annual crop production, the organic matter content of the soil decreases over time. The use of chemical fertilizers to add nutrients back into the soil and pesticides to deal with weeds, insects and blights reduces soil biodiversity. Use of nitrogen fertilizers also causes acidification of soils and surface waters. According to the United Nations

Environment Programme (UNEP), these problems are increasingly caused by nitrogen emissions in industrialized countries as sulphur emissions are brought under control.[69] Intensive farming also causes soil erosion, which is especially a problem in areas with prolonged dry periods followed by heavy rains, and with steep slopes and unstable soil, such as the Mediterranean.[70] Erosion causes a loss of organic soil substances, and the resulting nutrient losses can cause eutrophication in nearby surface waters, affecting other plants and wildlife.[71]

Soil impacts during feedstock harvesting depend upon both the intensity of the activity and the length of crop rotation periods. Intensive harvesting methods can compact the soil (which affects soil structure and biodiversity and can cause waterlogging), deplete soil nutrients and organic matter, and affect the soil's capacity to hold moisture; short crop rotation periods, meanwhile reduce soil fertility.[72] Soil exposed during and after harvesting is vulnerable to erosion, so frequency of harvesting and replanting are significant in determining environmental impact.

On some soils, the removal of crop residue can reduce soil quality, increase associated GHG emissions through the loss of soil carbon and promote erosion; thus, the amount that can be removed sustainably varies by crop type. There is evidence that residue removal changes the rate of physical, chemical and biological processes in the soil by causing more fluctuations in soil temperatures and increased water evaporation. These changes, in turn, affect crop growth. One study estimates that corn-derived soil organic carbon is reduced by 35 per cent when residue is harvested versus when it is retained; this is averaged over all tillage systems.[73]

Soil quality benefits of perennial energy crops

Where perennial energy crops such as some trees or native grasses replace annual crops, they can improve soil quality in a variety of ways, including increasing soil cover (and thus reducing erosion), reducing soil disturbance, improving organic matter and carbon levels in soil, and increasing soil biodiversity. This is particularly true if application of inputs such as fertilizers and pesticides are reduced, and if crops provide year-round soil cover.[74]

Tree plantations can provide many benefits. For instance, there is evidence that plantations of some tree species can reduce evaporative water losses and improve soil moisture conditions to allow for cropping on previously degraded lands. Tree species that fix nitrogen, such as leucaena and acacias, can reduce the need for nitrogen fertilizer while improving soil quality and producing food for farm animals.[75]

Oilseed trees, such as jatropha or pongamia, require little input or rainfall and can thrive in infertile soil. High-yield wild jatropha varieties have been found in Mexico and Mali.[76] Pongamia trees flourish in dry areas with poor or saline soils, including much of Asia, the Middle East and many islands in the Pacific and Indian Ocean, and produce a high oil yield while improving soil quality.[77]

Tree leaves can add organic matter and improve soil fertility and physical properties, and tree cover can protect land from water and wind erosion.[78] Short-rotation coppice (SRC) could reduce erosion even more if cover crops are used during the first two growing seasons to stabilize soils.[79] The positive effects that trees have on the soil's ability to retain water improve with time; therefore, longer rotations are preferable from this standpoint.[80] Palm trees can also protect soil from erosion and can be grown in poor soils, establishing tree cover fairly rapidly and mimicking tropical rainforest, according to one report.[81]

A 1987 study by Pimentel and Krummel estimated that erosion from SRC is one order of magnitude less than that from row crops, and erosion from hayland (or switchgrass) is one order of magnitude less than for SRC.[82] Perennial grasses, which have extensive root systems and do not require tilling, and plantations with some species of trees (including jatropha and pongamia) cannot only reduce soil erosion, but also help to increase soil productivity. They can also reduce the need for chemical inputs and water, compared with more intensive crops such as corn, wheat and soybeans. Such crops can often grow on marginal or erosion-prone lands, reduce chemical run-off and provide good habitat for wildlife. This is especially true if they are allowed to grow for several years before being harvested.[83]

Sustainable management and harvesting practices

Achieving a sustainable biofuels industry – including sustainable cellulosic ethanol based on crop residue – will require management practices that maintain long-term soil productivity and reduce chemical inputs. Conservation tillage and no-till planting – which involve the use of machinery that injects seeds directly into the soil, thus avoiding the ploughing of fields – have become more common for some crops in the US, Brazil and Europe. Their use can reduce soil erosion and the leaching of fertilizer, while saving the energy required for ploughing fields.[84] Using a cover crop between rows of trees further reduces the potential for erosion. Switchgrass also provides erosion control, but is most beneficial after it has become established.[85]

Germany has seen successful results with a new farming practice called 'double-cropping'. The fundamentals of this system include at least two crops and two harvests annually on the same field, no ploughing, year-round ground cover, little to no chemical pest management (weeds can also be used as feedstock), and the use of fermentation residues for closed cycling of nutrients.[86] Mixing of crops (varieties, heights, etc.) can be beneficial by helping to reduce the need for nutrient inputs, to enhance diversity of landscapes and crops, to reduce the use of heavy machinery and water, and to create year-round coverage, minimizing negative impacts on soils.[87] Importantly, crop mixing also preserves and enhances the diversity of crop species and can provide shelter and food for a larger variety of wildlife species than does the use of single crops.[88] As mentioned earlier, diversity also reduces susceptibility to pests (thereby reducing need for pesticides) and disease.[89]

Furthermore, the use of intercropping, crop rotation and bio-fertilizers can help recirculate nutrients in the soil that are lost through the growing and harvesting of crops such as sugar cane and sweet sorghum.[90] In Brazil, it was expected that long-term cultivation of sugar cane would reduce soil productivity; however, the opposite has proven to be true, most likely due to good soil preparation, superior varieties of sugar cane and recycling of nutrients through the application of vinasse (the nutrient-rich water waste left over from sugar milling and ethanol distillation).[91] In addition to vinasse, nutrients can be returned to the fields via ash produced during processing (assuming it does not include toxins that are absorbed by some plants while growing and that could become concentrated in soils).[92]

When crops are harvested, organic matter and nutrients can be maintained in the soil if sufficient biomass is left to conserve nutrients and organic matter (e.g. if only a small portion of branches and treetops are removed in short-rotation forests, and a portion of crop residue is allowed to remain in agricultural fields).[93] It should be noted that while some agricultural residue is safe to remove for biofuel feedstock use, it is important to leave some on the ground to minimize soil loss and run-off. Numerous field studies analysed by Benoit and Lindstrom (1987) suggest that a 30 per cent removal rate would not significantly increase soil loss under a no-till system. However, they found that no-till without residue cover could allow more soil erosion than conventional tillage, while no-till with residue cover usually results in less soil erosion.[94] In general, sustainable crop residue removal rates depend upon a variety of factors, including yield, management practices and soil type.[95]

Similarly, in forests, logging residues reduce exposure of the soil to sun, wind or rain, lowering the risk of erosion. Dead wood and residues also help to regulate water flow through forests and are increasingly seen as important for the protection of biodiversity. Thus, it is essential to leave tree roots in the ground and some of the branches as mats to protect the soil. Residues and dead wood are also important sources of nutrients. Generally, the lowest concentration of nutrients is in wood and the highest is in tree foliage, so the rate of extraction and degree to which foliage remains on site play a major role in determining forest health.[96]

Other factors that can affect soil quality include the frequency of residue removal and the degree of tillage.[97] Because soil is more likely to be disturbed or compressed by heavy harvesting machinery when it is wet, farmers can also reduce damage by minimizing the use of heavy machinery, and harvesting when the soil is dry or frozen.[98] In addition, winter harvesting of trees reduces soil nutrient loss because the leaves are not removed.[99]

WATER USE AND POLLUTION

Worldwide, the agricultural sector accounts for an estimated 70 per cent of global freshwater use and as much as 90 per cent of water resources in some developing countries, much of it for highly inefficient irrigation.[100] As with agricultural production for food purposes, the growing of most feedstocks for today's biofuels affects water supplies in two main ways: first, large amounts of water are required for feedstock production, potentially depleting valuable freshwater resources; and, second, the run-off of agro-chemicals and other waste products can pollute nearby waterways, threatening wildlife and speeding eutrophication.

Water use for irrigation

Cultivation of sugar cane, in particular, is highly water intensive. However, a recent World Bank report notes that water use in Brazil's sugar cane industry is declining, in part due to the creation of a legal framework to establish charges for water use. This is due primarily to reduced consumption in São Paulo, where most of the country's cane is grown; in fact, water use for sugar cane production is gradually increasing in other regions of Brazil.[101]

Heavy water use during dry spells, which occur often in Brazil and many other countries, intensifies water shortages and damages river ecosystems.[102] Irrigation also results in soil loss and leaching of nutrients and agro-chemical residues from the soil.[103] As demand for biofuels increases, water consumption will rise where feedstock crops are grown.

Water quality issues

Feedstock production also affects water quality, primarily through run-off of chemical inputs. Corn requires more pesticides than other food crops, and corn hybrids need more nitrogen fertilizer than any other crop. Run-off from these chemicals can find its way into the groundwater, causing contamination and affecting water quality.

Typically, less than half the nitrogen applied to crops in the form of fertilizer is actually taken up by the plants; the remainder is dissolved in surface waters, absorbed into groundwater or lost to the air.[104] Eutrophication (rapid plant growth in water that results in oxygen deprivation) of surface waters from excess nitrogen run-off is a major concern, as is pollution from chemical pesticides. In the US Midwest, the nation's corn belt, chemical run-off enters the Mississippi River and is carried to the Gulf of Mexico, where it has 'already killed off marine life in a 12,000 square mile area', according to writer Michael Pollan[105] (for more on the impacts of fertilizer and pesticide use, see Box 12.2).[106]

Reducing impacts on water

Careful crop selection can affect both water use and quality by reducing the need to irrigate, and lowering if not eliminating the need for chemical fertilizers and pesticides – meaning that the water draining from their soils will have lower

BOX 12.2 IMPACTS OF FERTILIZER AND PESTICIDE USE

The environmental impacts of fertilizer and pesticide use are many and varied, affecting everything from air quality and climate, to human and animal health, to the health of waterways and soils. Despite these problems, use of these chemical inputs continues to rise worldwide.

Nitrogen (N) fertilizer is generally made from natural gas, and pesticides are made from oil. Not only is fossil energy required to produce these chemical inputs, but it is also needed for their transportation and application, resulting in emissions of CO_2 and a host of other pollutants. Inorganic N fertilizer accounts for about 60 per cent of total anthropogenic nitrogen, which contributes to both climate change and ozone depletion. According to UNEP: 'There is a growing consensus among researchers that the scale of disruption of the nitrogen cycle may have global implications comparable to those caused by disruption to the carbon cycle.'

Fertilizer application rates vary depending upon crop, soil type, temperature and other factors. For instance, palm oil requires less fertilizer per unit of output than other oilseed crops. According to the US Department of Agriculture, fertilizer makes up about 45 per cent of the energy required to grow corn, despite the fact that the use of fertilizer for grain production has declined since the early 1980s. Fertilizer inputs are highest for oilseed rape, sugar beet, wheat, corn and potatoes, and generally lowest for non-wheat cereals, soy, linseed and sunflower.

The use of nitrogen fertilizer results in emissions of nitrogen oxides, which lead to acidification and eutrophication. As production of first-generation feedstock crops increases, and drier, less fertile areas are increasingly used, the demand for irrigation and chemical fertilizers will rise. While some crops can expect improved yields through the development of hybrids, they will probably require more nitrogen fertilizer and pesticides, as well.

In addition to polluting groundwater and water bodies, pesticides can kill beneficial soil and wildlife species and cause damage to neurological, reproductive and endocrine systems in humans and wildlife. According to the World Health Organization (WHO), as many as 220,000 people die each year from pesticide poisoning, and millions more suffer from mild to severe affects of poisoning annually.

By 2000, corn growers in the US were using 20 times more pesticide than they did in 1950, while also losing twice as much of their crop to pests. Bt (*Bacillus thuringiensis*) corn, a genetically modified plant, was introduced in 1996 to address this problem; but it will probably only continue the pesticide treadmill (in which pests become resistant, requiring the use of more pesticides). By 1997, eight insect pests were already resistant to Bt corn.

Sugar cane also requires a significant amount of pesticide (in the form of herbicide) – about the same amount as soybeans – and even more herbicide per hectare than corn. Almost all cane farmers in Brazil have switched to no-till practices, which improve soil quality and reduce erosion, but also increase weed pressure over time. As a result, the use of chemical herbicides has risen, along with the number of herbicide-resistant weeds, forcing farmers to rotate crops and to control weeds through mechanical means, but also to increase their toolbox of herbicide varieties.

Chemical inputs for feedstocks can be reduced in a variety of ways. In Brazil, the use of chemical fertilizers has declined significantly due to regulations that control their use and the recycling of vinasse and filter cake into fertilizer. Fertilizer use can also be reduced through a careful matching of application with soil type and yields, and through careful timing and placement. Although most dedicated energy crops also require pesticide (particularly herbicides in the early years in order to control weeds) and fertilizer inputs, research thus far suggests that needs are low relative to annual crops. Some crops do better with organic manure than with petroleum fertilizer. Jatropha can be fertilized with the de-oiled cake that remains after oil is expelled from the seeds.

Insecticide use can be significantly lowered for some crops through a variety of strategies, including crop rotation, manual operations and optimal selection of species. For example, in the US, 97 per cent of corn receives herbicide treatments; but application rates decline to less than 25 per cent with crop rotation and to under 5 per cent when corn is grown in rotation with small grains. Pesticide use has also reportedly declined in the Brazilian sugar cane industry due to genetic developments of sugar cane species. And in India, although studies have not been done to determine the amount of pesticides that might be required for jatropha and pongamia plantations, jatropha oil itself has been used as a biodegradable pesticide in some parts of the country.

Where pesticides and fertilizers are used, trees and perennial grasses can act as filters of agricultural chemicals when they are planted between annual crops and waterways. Another option altogether is to shift to production of cellulosic ethanol, the feedstock for which – ranging from corn stover and rice bagasse to forest residue and municipal waste – requires few if any additional chemical inputs.

Source: see endnote 106 for this chapter

concentrations of chemicals.[107] Furthermore, some crops can filter the nutrients that leach off adjacent farmland, ensuring that they do not reach nearby water bodies. Thus, the use of perennial crops and no-till buffer zones near waterways can reduce the biological and chemical oxygen demands generally placed on watercourses in agricultural areas.[108]

In addition, crops that are well managed can regulate water flows and reduce the risk of floods and droughts. This is particularly true for woody crops that are harvested over long periods of rotation.[109] The large leaf area of most woody crops, which is maintained for more of the year than leaves of annual crops, combined with their deeper root systems, can increase evapotranspiration and reduce the potential for run-off and leaching.[110] Another study suggests that growing such crops over large areas could improve water storage in dry regions.[111]

On the other hand, there is also the potential for negative impacts on local and regional hydrology from the introduction of energy crops. A 2001 study by Lyons et al found that short-rotation coppice in southeast England actually increased the interception and use of rainfall, reducing rainfall infiltration and causing negative impacts on regional aquifers.[112] This is an issue that requires further study.

As mentioned earlier, plants such as jatropha and pongamia can be grown in arid and semi-arid areas on marginal lands. Yet, even these crops require irrigation in some regions, and irrigation could increase yields in most locations. The need for irrigation can place additional pressure on already scarce water resources. In India, it is expected that jatropha plantations will need to be irrigated once monthly during the summer for the first two to three years after planting.[113] However, if such plants replace more water-intensive feedstock crops, there is the potential that water demands per unit of fuel could decline significantly. An added benefit of some oilseed crops such as jatropha and other perennial energy crops is that they tend to require fewer chemical inputs than annual feedstock crops. As a result, they have far less impact on water quality than do annual grain and oilseed crops such as rapeseed, soybeans and corn.[114]

Air Quality and Atmosphere

The most significant sources of local and regional air pollution associated with biofuels occur during combustion (see Chapter 13). But air quality is also affected somewhat during the feedstock growing process; for instance, farm machinery, which runs mainly on diesel, can cause a small amount of air pollution during planting and harvesting. In general, however, harvesting does not have a significant effect on air quality.

The only exception is when crops are intentionally burned to remove sharp tips and leaves that hinder access to the fields, as is standard practice in most sugar cane-growing countries.[115] This burning releases a host of gases and toxic compounds, including CO_2, carbon monoxide (CO), methane (CH_4), nitrous oxide (N_2O) and other nitrogen oxides (NO_x), which affect local and regional air quality.[116] Parts of Brazil, for example, have been blanketed with huge clouds of black smoke during harvest season.[117] Fires in cane fields have also occasionally spread to other areas, destroying native vegetation.[118]

To address such problems, the governments of Brazil (1998) and the Brazilian state of São Paulo (2002) have passed laws requiring a gradual phase-out of field burning and a transition to mechanical harvesting, though only São Paulo is currently enforcing the timetable.[119] In 20 per cent of Brazil's sugar cane fields, cane trash left in the field is no longer burned; instead, much of it is deposited in the field where it can help to maintain soil quality.[120]

Feedstock production also has a significant impact on the world's atmosphere due to the use of inorganic nitrogen fertilizer, which accounts for 60 per cent of anthropogenic nitrogen (N), as well as the cultivation of an increasing amount of leguminous crops such as soybeans, which account for about 25 per cent of anthropogenic N. Not only is nitrous oxide a potent greenhouse gas, but it also contributes to ozone depletion.[121]

Conclusion

The greatest environmental risks associated with biofuels result from the production of feedstocks, with associated impacts on habitat, biodiversity, and soil, air and water quality. Fortunately, many of the concerns about water use and quality can be mitigated by using water more efficiently and recycling more of it for fertilizer, or digesting it for biogas. Problems with water availability and use, however, may represent a limit on the production of biofuels. Air quality problems associated with feedstock production are relatively minor and can be reduced through measures such as shifting from petroleum diesel to biodiesel for farm machinery, and through regulations that limit or eliminate practices such as field burning.

Ultimately, it is the problems associated with the use of land, particularly virgin land, that will remain the most vexing and that deserve the most attention. As with agricultural crops grown for food, use of large-scale mono-crops grown for energy purposes could lead to significant devastation of tropical forests and other wild lands, a loss of biodiversity, soil erosion and nutrient leaching. Even varied and more sustainable energy crops could have negative impacts if they replace wild forests or grasslands. And eutrophication of water bodies, acidification of soils and surface waters, and ozone depletion, all associated with nitrogen compounds from agricultural production, are impacts of major concern. In fact, some studies have found that these three impacts are the greatest disadvantages of biofuels when compared with petroleum fuels.[122]

At the same time, new dedicated energy crops and sustainable resource management practices offer the potential for environmental improvements. Biofuels can have negative or positive effects on land use, soil and water quality and on biodiversity, depending upon the type of crop grown, what it is replacing, and methods of cultivation and harvest.

Feedstock selection is critical. Dedicated energy crops that are appropriate to the regions where they are planted – such as native perennial trees and grasses – can minimize the need for chemical inputs, avoiding some of the pollution associated with feedstock production, while also reducing water needs and providing habitat for birds and other wildlife. Oilseed bushes such as jatropha and pongamia can help to reduce desertification and restore degraded lands in countries from Mali to India. In the future, next-generation technologies that rely on agriculture and forest residue, or other forms of waste, could significantly reduce land requirements for biofuels production. More research is needed to determine how much residue can be safely removed to avoid degrading soil quality and reducing yields.

In addition, sound agricultural methods can achieve increases in productivity with neutral or even positive impacts on the surrounding environment, depending upon the feedstock choice and what it is replacing. A variety of management practices, such as the use of intercropping, crop rotation, double-cropping and conservation tillage, can reduce soil erosion, improve soil quality, reduce water consumption and reduce susceptibility of crops to pests and disease – thereby reducing the need for chemical fertilizers and pesticides. Using perennial crops as protective buffers or wildlife corridors can reduce chemical run-off and provide habitat for a variety of mammals and birds.

While there are great challenges to address, models do exist for ways of mitigating many of the risks associated with feedstock production. For example, to address concerns about biodiversity loss, the Brazilian state of São Paulo requires that sugar cane producers set aside 20 per cent of their total planted area as natural reserves.[123] The palm oil industry in Southeast Asia has promoted wildlife sanctuaries and green corridors to enhance biodiversity.[124] These efforts are supported on the international level by the Roundtable on Sustainable Palm Oil, formed in 2004 to respond to rising concerns about the environmental impacts of oil palm plantations (see Chapter 18). In India, which has more than 300 species of oil-bearing trees, a multi-species biodiesel programme could help to ensure plant genetic diversity.[125]

There is still a dire need for environmental policies and regulations at the local, national and regional levels, particularly in developing countries, to ensure that impacts on land, wildlife, and water, air and soil quality are minimized. For example, payment systems for water should be enacted to encourage more efficient use and the growing of crops that minimize water consumption. Large-scale feedstock producers must be required to set aside a share of their land as natural reserve, as São Paulo has done, and other lands should be designated for low-intensity farming. Farmers need to be educated and given the proper resources and incentives to select crops appropriately and to manage them in the most sustainable ways possible, such that wildlife habitat is maintained or improved and the use and impacts of chemical inputs are minimized. Governments should also encourage the use of degraded lands for feedstock production and encourage the planting of crops that have longer rotation periods, such as perennial grasses, while discouraging (if not banning) the use of virgin lands.

To date, most studies looking at the impacts of feedstock production have been species and context specific; therefore, more research is needed to determine which management practices are most effective and least harmful to wildlife and surrounding ecosystems under different and broader circumstances. In addition, more research is needed in a variety of areas, including the potential for using natural pesticides and fertilizers on feedstock crops; the potential impacts of large-scale plantations of oil-bearing trees such as jatropha; the potential to increase crop yields while reducing inputs; the impacts of residue removal from cropland and forests, and how much can be safely harvested; and possible perennial feedstocks that are suitable for arid regions. It is also critical to conduct further research to determine if the benefits of GM crops can outweigh their costs.

Over time, it is likely that the advantages of biofuels relative to petroleum fuels will increase as new feedstocks and technologies are developed and crop yields increase. It is important to get to this future as soon as possible by moving forward quickly to commercialize next-generation technologies, such as cellulosic ethanol and Fischer-Tropsch diesel from gasified biomass, which rely on less resource-intensive feedstocks. Furthermore, careful land-use planning and appropriate farming practices can establish safeguards against harmful biofuel cropping. At the same time, it is critical to develop international standards and certification systems to ensure that biofuels around the world are produced using the most sustainable methods possible (see Chapter 18).

It will be impossible to avoid all of the negative impacts of biomass production at every location. Biofuels *per se* are not environmentally preferable to petroleum fuels – and, again, their relative benefit is determined primarily by feedstock choice and management practices. Ultimately, the choice may be largely subjective as decision-makers weigh the merits and drawbacks related to different desired environmental ends – mitigating global climate change (which will probably affect entire species of plants and wildlife, as well as water resources, agricultural productivity, etc.), stemming the loss of species and landscapes, or improving air and water quality.

13

Environmental Impacts of Processing, Transport and Use

INTRODUCTION

The refining, transport and combustion of biofuels can result in significant environmental costs, particularly on local water and air quality. Generally, however, these effects pale in comparison to those generated by the use of fossil fuels, where the main detrimental environmental effects originate from the vehicle exhaust pipe (tailpipe). Even so, these impacts could expand considerably as biofuel production increases to meet rapidly rising global demand. Here, too, more sustainable practices and new technologies offer the potential for environmental improvements.

This chapter examines the main environmental impacts associated with biofuels processing, transport and use. In order to provide comparison, it first describes some of the environmental costs resulting from the processing and use of petroleum transport fuels. It then compares these with the (non-climate) environmental costs and benefits of biofuels at parallel stages of their life cycle.[1]

ENVIRONMENTAL COSTS OF PETROLEUM REFINING AND USE

While the use of oil has brought incalculable benefits to modern industrialized society, it has also exacted great costs, particularly to the local and global environments. Most of these occur during oil refining and fuel combustion. Delucchi (1995) estimated that in the US the costs of environmental externalities associated with oil and motor vehicle use totalled between €45 billion and €192 billion (US$54 billion and $232 billion) in 1991 alone.[2] Human mortality and disease due to air pollution accounted for more than three-quarters of these costs, or as much as €152 billion (US$184 billion) per year, according to the Union of Concerned Scientists.[3] In Germany, it is estimated that the quantifiable costs of air pollution and carbon dioxide emissions associated with the transport sector in 1998 totalled about €12 billion (US$14.5 billion).[4]

Oil refining

Refining petroleum is an energy-intensive, water-hungry and highly polluting process. Every day, an average US refinery releases 41,640 litres (11,000 gallons) of oil and other chemicals into the air, soil and water.[5] People who live near these facilities have higher incidences of respiratory problems (including asthma, coughing, chest pain and bronchitis), skin irritations, nausea, eye problems, headaches, birth defects, leukaemia and cancers than the average population.

Crude oil, chemical inputs and refined products leak from storage tanks and spill during transfer points. Numerous toxins are likely to enter the groundwater, including benzene (a toxic carcinogen), toluene (which debilitates kidney function, among other effects), ethylbenzene (a skin and mucous membrane irritant) and xylene (which probably inhibits development and leads to early death in exposed animals).[6] Other chemicals leak into the air. Gases such as methane, and slightly heavier hydrocarbons such as those in gasoline, evaporate. Other chemicals enter the air as combustion products, the most significant of which are sulphur dioxide (SO_2), nitrogen dioxide (NO_2), carbon dioxide (CO_2), carbon monoxide (CO), dioxins, hydrogen fluoride, chlorine, benzene, large and small particulate matter (PM), and lead.[7] Oil refineries are the largest industrial source of volatile organic compounds (VOCs) and CO, which lead to ozone and smog; the second largest industrial source of SO_2, which contributes to PM and acid rain; and the third largest industrial source of nitrogen oxides (NO_x), which are also ozone precursors.[8]

The refining process requires a great deal of energy, primarily in the form of natural gas and electricity. Petroleum refining accounted for 8 per cent of total energy consumption in the US in 1998.[9] And refineries consume significant amounts of water (typically 1.8–2.5kg of process water for each kilogram of crude feedstock) and discharge between 1.7 and 3.1 times as much water as biofuel-processing facilities. Water consumption amounts to some 17,000–23,500 litres (4400–6200 gallons) per minute for a refinery that processes 100,000 barrels daily.[10] Water is also usually used to cool waste products distilled out from crude to return them to liquid form. This water either evaporates or is released into surrounding ecosystems, where it can harm or eliminate species in streams or rivers.[11]

As mentioned in Chapter 12, unconventional oil resources such as tar sands and oil shale are becoming more economical to extract and refine as oil prices rise. But their environmental costs are high as well. Once sands are extracted from the earth, they must be steam cooked to separate the tarry residue and to purify it. The heavy oil that is removed must be refined into lighter synthetic crude oil before it can be used.[12] The extraction and processing of tar sands requires about three barrels of water for every barrel of oil produced and results in large quantities of wastewater. Impacts on air quality are also large, with direct emissions including SO_2, CO, CO_2, NO_x, particulates, ozone, lead, silica, metals, ammonia, trace organics and trace elements. Indirect air emissions result from generating the electricity required for extraction and processing. It is estimated that Shell's *in situ* process for removing sands from deep within the ground and processing them would require as much energy as the end use of the oil shale itself. The solid waste is significant as well: it is estimated that for surface mining, 1.1 to 1.4 tonnes of spent shale must be disposed of for every barrel of oil produced.[13]

Oil transport

Most of the world's crude oil comes from fields far from where it is refined and must be transported great distances from field to refinery, and from refinery to the refuelling station. Large tanker vessels account for 68 per cent of crude delivery to refineries, covering an average of 6600km (4100 miles) per trip. Oil pipelines, used mainly in places where deliveries can be land based, account for 30 per cent, while trucks and trains transport the remainder.[14]

Invariably, oil spills occur along the journey. Although most tanker spillage is relatively minor (occurring during loading or unloading, or from slow leaks), even small amounts can damage ecosystems. The dramatic spill events that make the news, such as the *Exxon Valdez* spill off Alaska in 1989, account for only a small share of the crude oil spilled into marine environments; but they can have significant environmental impacts, including harming or killing local populations of coastal mammals, large numbers of water birds and inter-tidal plant communities.[15] The *Exxon Valdez* spill cost an estimated €5.8 billion (US$7 billion), including clean-up costs, with punitive fines representing €4.1 billion (US$5 billion) of this total.[16]

Pipeline spills, although typically smaller, can also be ecologically disruptive, polluting soils and seeping into groundwater. Such spills can be fairly common in regions where pipelines are not maintained adequately: in Russia, for example, nearly 20 per cent of oil leaks during transit, often into sensitive Siberian ecosystems where it is too cold for the oil to evaporate or seep into the ground.[17] In conflict-ridden countries such as Colombia and Iraq, opposition groups have repeatedly blown up pipelines, spewing oil into jungle and desert ecosystems.[18]

Oil is shipped over great distances to refineries, and from refineries, gasoline and diesel fuels travel via pipelines and trucks to fuel depots. Upon leaving the refinery, 59 per cent of refined petroleum fuels enter pipelines (which travel an average of 957km, or 595 miles) before being loaded into trucks. The other 41 per cent goes straight into trucks, which travel about 60km (100 miles) to a commercial depot.[19] Along the way, gasoline and diesel fuels spill, contributing to the more than 50 per cent of 'oil spills' attributable to difficult-to-pinpoint urban run-off. Gasoline and diesel are lighter hydrocarbons that tend to evaporate, participating in the complex reactions that form ozone. Benzene, another evaporative air pollutant, is a known carcinogen. The most significant hydrological pollutant is methyl tertiary-butyl ether (MTBE), a fuel additive derived from petroleum that seeps quickly into nearby groundwater and is a likely carcinogen.[20]

Combustion of petroleum fuels

Compared to biofuels, petroleum contains a much wider variety of chemical molecules, including far more sulphur. Most of these have been sequestered in the Earth for tens and even hundreds of millions of years. In addition to spurring global warming, the burning of gasoline and diesel fuels releases a host of pollutants and heavy metals that affect local and regional air quality. Transport-related air pollution leads to reduced visibility, damage to vegetation and buildings, and increased incidence of human illness and premature death (including tens of thousands of premature deaths annually in the US alone).[21] Road transport is also a growing contributor to air pollution in many developing country cities, particularly where diesel remains the predominant fuel.[22]

Table 13.1 summarizes some of the main environmental and health impacts associated with petroleum's primary combustion products, including CO_2, CO, unburned hydrocarbons (particularly benzene), NO_x, SO_x, particulates and, in some countries, lead.[23]

BIOFUEL IMPACTS: REFINING

Like petroleum fuels, biofuels can have environmental impacts at all stages of their production and use. Relative to fossil fuels, however, the impacts resulting from refining, transporting and using biofuels are generally significantly smaller. Moreover, there are ways to improve the resource efficiency and impacts of these activities.

Water use

Processing the biofuel feedstock into fuel can use large quantities of water. The primary uses of water for biodiesel refining are to wash plants and seeds for processing and then to remove the soap and catalyst from the oils before the final product is shipped out. Sheehan et al (1998) estimate that a typical US soybean crushing system requires just over 19kg of water per tonne of oil produced.[24] For each tonne of soybeans that go into the refining process, 170kg come out as crude de-gummed soybean oil, 760kg are soy meal and the remaining 70kg include air and solid (non-hazardous remains of beans) and liquid waste.[25] The primary contaminant in the wastewater is soybean oil.[26]

Production of ethanol, in particular, requires a tremendous amount of water – for processing and for evaporative cooling – to keep fermentation temperatures at the required level (32° to 33°C).[27] But some feedstocks are more water intensive than others: each tonne of sugar cane in Brazil, for example, requires as much as 3900 litres for processing.[28] The concentration of slurry fed into the hydrolysis

Table 13.1 *Environmental and health impacts of emissions from petroleum combustion*

Combustion product	*Impacts*
CO_2	Contributes to global warming and climate change.
CO	Results from incomplete burning. In the atmosphere, CO reacts with oxygen to form ozone (O_3), a highly reactive molecule that damages plant leaves and human and animal lungs. When breathed, CO prevents oxygen molecules from attaching to blood haemoglobin. It is hazardous to people with cardiac, respiratory and vascular disease.
Benzene	The smallest aromatic hydrocarbon and a highly toxic carcinogen. Larger aromatic and olefin hydrocarbon chains are precursors for ozone and particulates.
NO and NO_2	Ozone precursors. They also react with atmospheric water to form nitrogen acids, creating acid rain, which, among other things, corrodes buildings and plants, and depletes soil nutrients.
SO_2 and SO_3	Acid-rain precursors, they also tend to form sulphate-based particulate matter. SO_2 exposure can cause respiratory disease, breathing difficulties and premature death. Sulphur can render vehicle catalytic converters useless.
Particulates	Formed from SO_x, NO_x and hydrocarbons, particulates contribute to ozone formation and affect visibility. Once inhaled, they can pass deeply into the lungs and contribute to asthma, cancer and heart disease. The primary difference between gasoline and diesel combustion is that diesel emits more particulates. Particulate matter contributes to more than 15,000 premature deaths in the US each year.
Lead	Has been phased out of gasoline in most countries, but is still used as an octane enhancer in some countries, particularly in Africa. As a neurotoxin, lead can impair the nervous system, stunt growth and cause learning disabilities. Lead poisoning is more common among children where gasoline has a high lead content.

Source: see endnote 23 for this chapter

process must be similar for all feedstocks; therefore, more process water is needed for less starchy grains than for starch-rich grains such as corn.[29]

Ethanol processing also results in large volumes of nutrient-rich wastewater that, if not cleaned and recycled, can speed eutrophication of local rivers and streams by affecting the water's dissolved oxygen content.[30] In addition, sugar mills must be flushed every year, putting huge amounts of organic matter into local waterways.[31] In Brazil, 1 litre of ethanol produces about 10 to 15 litres of stillage (vinasse), which is very hot and corrosive, with a low pH and high mineral content.[32] In 1979, the volume of vinasse produced in Brazil exceeded the amount of sewage waste produced by the nation's entire population.[33] For years, the costs of pumping and storage were so high in the mountainous regions of the country's northeast that vinasse was released directly into rivers, causing enormous fish kills at every harvest.[34]

Today, however, wastewaters and vinasse are recycled and used for irrigation and fertilization of Brazil's sugar cane crops, with varying quantities of vinasse used under different conditions as regulated by law.[35] However, some experts caution that vinasse cannot be used where water tables are higher, such as in India.[36] Also, if used excessively, vinasse (like nitrogen fertilizer) can cause eutrophication of surface waters due to the increased nutrient load.[37] Filter cake, another waste stream from ethanol processing, is also recycled as a fertilizer. As a result, Brazil has been able to significantly reduce its use of petroleum fertilizers, saving money while creating value from waste products.[38]

Water pollution is not exclusively a Brazilian problem. In the US state of Iowa (where a large share of US ethanol is produced), ethanol processing plants have sent syrup, bad batches of ethanol and sewage (containing chloride, copper and other wastes) into nearby streams. At times, the pollution dumped from such facilities has been strong enough to kill fish. While there are laws to prevent this practice in the US, they are not always clear and thus are open to interpretation, and some plants are violating them. As a result, efforts are under way to develop new water quality standards to reduce pollution.[39]

Fortunately, the problem of nutrient loading in streams and the associated problem of increased biological oxygen demand can be resolved by installing various treatment systems for these organic wastes, including anaerobic digester systems.[40] And, at least in the US, standard wastewater treatment technologies can eliminate virtually all pollutants and about 95 per cent of the water can be reused, substantially reducing the amount of fresh water required for processing; the remainder is retreated and then released. Methane, which can be used as a fuel, is captured in the process, creating an economic incentive to treat the water; nevertheless, regulations are necessary to ensure the proper treatment of wastewater.[41] However, as the scale of production increases, there is some concern that, even in the US, some small municipal systems will be unable to deal with the large quantities of wastewater discharged from high-capacity plants (e.g. those producing 110,000 to 150,000 cubic metres of ethanol annually).[42]

The other production-related pollutant that affects water is waste heat, which is needed for processing and is generated by the fermentation process. Hot water released into local streams or rivers can kill fish and alter ecosystems. Alternatives exist, however, for removing excess heat through evaporation (wet cooling system); air may also be used in dry cooling systems or hot water can be used for heat and cogeneration.[43]

Other waste products that result from the refining process include bio-solids from wastewater treatment and the ash content of biomass. But it is possible that uses will be found for these wastes, as has been the case with the petroleum industry.[44] As mentioned earlier, ash is already used in some instances to return nutrients to the soil.

Air pollution

Among the pollutants that biorefineries emit into the air are NO_x, SO_x, VOCs, CO and particulate matter. Emissions from corn-ethanol plants, for example, include SO_x, NO_x, CO, mercury, particulates and CO_2.[45] Corn-ethanol plants in Iowa have polluted both water and air, emitting cancer-causing chemicals such as formaldehyde and toluene.[46]

Biodiesel production requires methanol, which has the same environmental costs as those associated with petroleum production. In addition, direct emissions from biodiesel processing plants can include air, steam and hexane, which can be used to extract oil from plants and seeds. Hexane is an air pollutant, and though as much as possible is recovered and recycled, some is emitted into the air as well. Sheehan et al (1998) estimate that the average US soybean crushing system releases just over 10kg of hexane per tonne of oil produced. Alternatives have been found so that hexane is no longer needed; but these options are more costly.[47] In addition, where renewable resources are not used to produce process energy, pollutants associated with the use of natural gas and the generation of steam and electricity are released into the air. An estimated 3.6 kilowatt hours of electricity are required per tonne of soybeans entering a soy biodiesel plant.[48] On the other hand, Fischer-Tropsch (F-T) biodiesel is gasification-based and therefore has minimal local air pollution problems.[49]

As plant size increases, concerns about pollution – including air emissions, odours released during the drying of distillers grain in corn-ethanol plants and wastewater discharges – have risen as well.[50] However, with appropriate regulation and pollution-control technologies, emissions associated with biofuels refining can be minimized significantly.[51, 52] For example, NO_x emissions from boilers can be reduced by installing new NO_x burner systems.[53] VOC emissions, which result primarily from the blending of ethanol with gasoline, can be reduced by mixing the fuels at locations where pollutants can be collected and treated.[54] In some cases, new and larger plants are incorporating such emissions control systems and are finding alternative options that enable them to reduce such emissions.[55]

In addition, much of the air pollution associated with biofuels refining results from the burning of fossil fuels for process heat and power – which in the US, Germany, China and many other countries is mainly coal. Thus, emissions can be reduced through traditional power plant control technology or the use of renewably generated power.[56]

In Brazil today, mills and distilleries meet most if not all of their own energy needs with bagasse (a by-product of sugar cane crushing), which can generate thermal, mechanical and electrical energy.[57] Some plants even sell surplus electricity into the grid.[58] Elsewhere, agricultural and forestry residues can be used to produce required power and heat; however, it is important to ensure that enough residues remain to maintain soil organic matter and nutrient levels.[59]

BIOFUEL TRANSPORT AND STORAGE

Water pollution

Pure ethanol and biodiesel fuels offer significant environmental benefits compared to petroleum fuels, making them highly suitable for marine or farm uses, among others. They result in dramatically reduced emissions of VOCs and are less toxic to handle than petroleum fuels.[60] One other significant advantage relates specifically to water: both ethanol and biodiesel are biodegradable and break down readily, reducing their potential impact on soil and water.[61]

Biodiesel is far more water soluble than petroleum diesel, enabling marine animals to survive in far higher concentrations of it than petroleum if fuel spills occur (due to lower risk of suffocation).[62] Such benefits are helping to drive biofuel-promotion policies in China, where vehicle spills have polluted water bodies and gasoline and diesel leakage from pipelines has polluted groundwater – affecting biodiversity, drinking water and soil resources.[63]

At least one study has shown that biodiesel made with rapeseed oil can biodegrade in half the time required for petroleum diesel. Biodiesel also speeds the rate at which biodiesel-petroleum blends can biodegrade, which is not the case with ethanol.[64] There is evidence that ethanol's rapid breakdown depletes the oxygen available in water and soil, actually slowing the breakdown of gasoline. This can increase gasoline's impact on the environment in two ways. First, the harmful chemicals in gasoline persist longer in the environment than they otherwise would; benzene, in particular, can last 10–150 per cent longer when gasoline is blended with ethanol. Second, because gasoline breaks down more slowly, it can travel farther (up to 2.5 times) in the marine environment, affecting a greater area.[65]

Additionally, if ethanol is spilled, it can remobilize gasoline in previously contaminated soils, intensifying the impacts of the initial spill. Since up to 85 per cent of such spills occur at gasoline terminals, this is where such a problem is most likely to happen.[66] The transition to high levels of ethanol needs to be planned with such impacts in mind and should include regulations for the handling of fuels.[67]

Air pollution

Most biomass is carried to processing plants by truck, and most biofuels are transported by truck as well, although some travel by train or, in Brazil, via pipelines. The environmental impacts associated with transport include the air emissions and other pollutants associated with the life cycle of the fuel used – in most cases, petroleum diesel. As demand for biofuels increases and as consumption exceeds production in some countries, it is likely that a rising amount of feedstocks and biofuel will be transported by ship. While shipping is a relatively energy-efficient means of transport, it is also a major source of pollution due primarily to a lack of regulations governing maritime emissions. Pollutants include NO_x, SO_2, CO_2, particulate matter and a number of highly toxic substances, such as formaldehyde and polyaromatic hydrocarbons.[68] Emissions from diesel marine engines represent an ever-increasing share of air pollution, and most of these pollutants are released near coastlines, where they can easily be transported over land.[69]

The other potential concern associated with biofuels transport is the possibility for spills and evaporation. Biofuels can leak at the production facility, spill while being transported and leak from above- or below-ground tanks. They can also evaporate during fuelling and storage and from a vehicle's fuelling system.

In general, 'neat' biofuels are distinctly less toxic than spills of petroleum fuels. For biodiesel, evaporative emissions are not a particular concern since biodiesel fuel does not have a higher vapour pressure. Neat ethanol has a low Reid vapour pressure (RVP), and when stored as a pure fuel (or even as an E-85 blend), it has a lower vapour pressure than gasoline and thus will have fewer evaporative emissions.[70]

The primary concern regarding emissions from biofuel transport has to do with lower-level blends of ethanol in gasoline, which tend to raise the vapour pressure of the base gasoline to which ethanol is added. When ethanol is blended up to about 40 per cent with gasoline, the two fuels combined have higher evaporative emissions than either does on its own. The fuels are mixed via splash blending at the petroleum supply 'rack', so there is a potential for increased evaporative emissions from these lower-level blends at the point in the distribution chain and 'downstream' – mainly during vehicle refuelling and from use in the vehicles. These evaporative emissions from a vehicle's fuelling system can increase ozone pollution.

Adding the first few per cent of ethanol generally causes the biggest increase in volatility, so increasing the blend level to 2.5 or even 10 per cent will have similar results.[71,72] Evaporative emissions peak at blend levels between 5 and 10 per cent and then start to decline. Once ethanol's share exceeds 40 per cent, evaporative VOC emissions from the blend are lower than those from gasoline alone.[73]

Most International Energy Agency (IEA) countries have emissions standards requiring that VOC emissions, and thus RVP, be controlled.[74] Emissions resulting from higher vapour pressure can be controlled by requiring refiners to use base gasoline stock with a lower vapour pressure when blending it with ethanol, although this increases costs and reduces production levels. The US state of California and US federal reformulated gasoline programmes have set caps on vapour pressure that take effect during high ozone seasons in areas that do not meet ambient air quality standards for ozone.[75] As a result, the addition of ethanol does not increase the vapour pressure of the gasoline available during summer months.[76] Emissions from permeation are more difficult to control in the on-road fleet, although experts believe that most can be controlled in new vehicles that must meet stricter evaporative emission control standards (such as California LEV 2 and US Federal Tier II), with higher-quality tubes, hoses and other connectors (see Chapter 15 for more on vehicle and engine technologies).[77]

BIOFUEL COMBUSTION

The level of exhaust emissions that results from the burning of ethanol and biodiesel depends upon the fuel (e.g. feedstock and blend), vehicle technology, vehicle tuning and driving cycle.[78] Most studies agree that using biofuels can significantly reduce most pollutants compared to petroleum fuels, including reductions in controlled pollutants, as well as toxic emissions.[79] NO_x emissions have been found to increase slightly as blend levels rise, although the level of emissions differs from study to study.

Ethanol

Ethanol contains no sulphur, olefins, benzene or other aromatics,[80] all of which are components of gasoline that can affect air quality and threaten human health.[81] Benzene is a carcinogen, while olefins and some aromatics are precursors to ground-level ozone (smog).[82] Ethanol-gasoline blends also reduce toxic emissions of 1,3-butadiene, toluene and xylene. While few studies have looked at the impacts on pollution levels from high blends, it appears that impacts are similar to those from low blends.[83]

With ethanol fuel combustion, emissions of the toxic air pollutants acetaldehyde, formaldehyde and peroxyacetyl nitrate (PAN) increase relative to straight gasoline.[84] Most is emitted as acetaldehyde, a less reactive and less toxic pollutant than formaldehyde.[85] Neither pollutant is present in the fuel; they are created as by-products of incomplete combustion.[86] PAN, an eye-irritant that is harmful to plants, is also formed as a by-product.[87] A US auto-oil industry study determined that combustion of E85 resulted in a slight increase in hydrocarbon emissions relative to California reformulated gasoline. It also found that toxic emissions rose as much as two to threefold compared to conventional gasoline, due mainly to an increase in aldehyde emissions.[88]

There is concern that aldehydes might be carcinogenic; but the pollutants that are reduced by blending with ethanol (including benzene, 1,3-butadiene, toluene and xylene) are considered more dangerous to human health.[89] A study done in California determined that acetaldehyde and PAN concentrations increase only slightly with ethanol blends, and a Canadian study concluded that the risks of increased aldehyde pollutants are negligible.[90] Because of the reactivity of aldehydes, emissions can generally be managed with emissions controls.[91] For example, three-way catalysts can efficiently minimize aldehyde emissions.[92]

Ethanol-blended gasoline increases fuel oxygen content, making hydrocarbons in the fuel burn more completely in older vehicles, in particular, thus reducing emissions of CO and hydrocarbons. Note that oxygen sensors in newer vehicles control engine combustion, reducing the benefit that ethanol can provide in reducing CO and hydrocarbon emissions.[93] Ethanol used as an additive or oxygenate (e.g. a 10 per cent blend) has been found to achieve CO reductions of 25 per cent or more in older vehicles.[94] In fact, one of the goals driving the use of ethanol in the US during the 1990s was to reduce hydrocarbon and CO emissions, particularly in winter when emissions of these pollutants tend to be higher.[95] Ethanol in higher blends will positively affect the efficiency of catalytic converters because of the dilution of sulphur.[96] Ethanol can also be used to make ethyl tertiary butyl ether (ETBE), which is less volatile than ethanol and widely used in the European Union (EU).[97, 98]

As a result of its national ethanol programme *Proálcool*, Brazil was one of the first countries in the world to eliminate lead entirely from its gasoline. According to the São Paulo State Environment Agency (CETESB), ambient lead concentrations in the São Paulo metropolitan region declined from 1.4 grams per cubic metre in 1978 to less than 0.1 grams per cubic metre in 1991.[99] Most other countries, however, have been able to eliminate lead through other means, including a reduction in unnecessarily high octane grades and the development of cheaper refining alternatives (e.g. reforming and isomerization).[100]

Ethanol use has resulted in significant reductions in other air pollutants, as well. Emissions of toxic hydrocarbons such as benzene have declined in Brazil, in addition to emissions of sulphur and CO. For example, Brazil's transport-related CO emissions declined from more than 50 grams per kilometre in 1980 to less than 1 gram per kilometre in 2000 due to ethanol use. CETESB estimates that urban air pollution in Brazil could be reduced an additional 20–40 per cent if the entire vehicle fleet were fuelled by alcohol.[101] In 1998, Denver, Colorado, became the first US city to require blending of gasoline with ethanol; it is used in winter to improve fuel combustion and to reduce CO emissions. As a result, it is estimated that CO levels have declined by 50 per cent.[102]

As noted above, there is some evidence that emissions reductions associated with using ethanol blends, compared to straight gasoline, are not as significant in the cleanest vehicles available today. Durbin et al (2006) tested vehicles that qualified as low emission and ultra-low emission in California, and found that emissions of non-methane hydrocarbons increased as engine temperatures rose and that benzene emissions increased with higher concentrations of ethanol, while fuel efficiency declined. However, CO emissions decreased somewhat with ethanol use.[103] Some of the findings were inconsistent with those of other studies, highlighting the need for further research.[104]

As discussed earlier, ethanol used as an oxygenate can reduce emissions of several pollutants, particularly in older vehicles. However, the use of oxygenates, such as ethanol (and biodiesel), to alter the fuel-to-oxygen ratio will not necessarily have a positive effect on emissions if a vehicle's air-to-fuel ratio is set low or if too much ethanol is added to gasoline in a vehicle with a fixed air-to-fuel ratio. If that is the case, oxygenates can increase NO_x emissions and cause 'lean misfire', increasing hydrocarbon emissions.[105] In fact, Tyson et al (1993) argue that ethanol has no emission-related advantages over reformulated gasoline other than the reduction of CO_2.[106]

Ethanol blended with diesel can provide substantial air quality benefits. Blends of 10–15 per cent ethanol (combined with a performance additive) result in significantly lower emissions compared with pure diesel fuel: exhaust emissions of PM, CO and NO_x decline. For high blends, the results are mixed. Some studies have found higher average CO and hydrocarbon emissions, and others have seen reductions in these pollutants. However, all studies, to date, have seen significant decreases in both PM and NO_x.[107]

Flexible-fuel vehicles (FFVs) – which can take virtually any ethanol–gasoline blend up to 85 per cent in the US and up to 100 per cent in Brazil – are widely used in Brazil and are becoming increasingly available in the US. However, tests to date have found that the use of FFVs results in higher air emissions than new gasoline vehicles.[108] Because it is not possible to tune the combustion controls of a vehicle so that it is optimized for all conditions, controls are compromised somewhat to allow for different mixes.[109] It is possible that vehicles dedicated to specific blends, and operated on those blend levels, would achieve lower emissions than conventional vehicles.[110]

Biodiesel

Biodiesel – whether pure or blended – results in lower emissions of most pollutants relative to diesel, including significantly lower emissions of particulates, sulphur, hydrocarbons, CO and toxins.[111] Emissions vary with engine design, condition of vehicle and quality of fuel. In biodiesel–diesel blends, potential reductions of most pollutants increase almost linearly as the share of biodiesel increases, with the exception of NO_x emissions.[112]

In one of the most comprehensive analyses to date, a US Environmental Protection Agency (EPA) study of biodiesel determined that the impacts on emissions vary depending upon type (feedstock) of biodiesel and the type of petroleum diesel that it is mixed with. Overall, animal-based biodiesel did better in the study than plant-based biodiesel with regard to reducing emissions of NO_x, CO and particulates. On average, the EPA determined that B20 (made with soybeans) increased NO_x emissions by 2 per cent and reduced emissions of particulates by 10 per cent, CO by 11 per cent and hydrocarbons by 21 per cent, while also reducing toxic emissions. Biodiesel made from animal fats increased NO_x emissions the least, followed by rapeseed biodiesel and then soybean-based biodiesel; the same relationship held true for CO reductions, as well. Reductions in particulate emissions were also greatest for animal-based biodiesel.[113]

Tests carried out by the EPA showed that, when compared with conventional diesel, pure biodiesel (produced with soybean oil) resulted in average reductions of particulate matter by 40 per cent, CO by 44 per cent, unburned hydrocarbons by

68 per cent, polycyclic aromatic hydrocarbons (PAHs) by 80 per cent, carcinogenic nitrated PAHs by 90 per cent, and sulphates by 100 per cent.[114]

During 2000, biodiesel became the first alternative fuel to successfully complete testing for tier 1 and tier 2 health effects under the US Clean Air Act. Tests determined that, with the exception of minor damage to lung tissue at high levels of exposure, animals observed in the study suffered no biologically significant short-term effects associated with biodiesel.[115]

A 1999 Swedish study by Pedersen et al found that biodiesel (rapeseed methyl ester, or RME) led to an up-to-tenfold increase in emissions of benzene and ozone precursors compared with Swedish low sulphur diesel fuel, called MK1.[116] However, this study was conducted using a very small reactor; many US and European researchers were sceptical about transferring results from this study to the real world for combustion in a diesel engine. Since then, other studies have produced different results. For example, Krahl et al (2001) compared 100 per cent RME to MK1, fossil diesel fuel and another low sulphur diesel fuel (with high aromatic compounds content and flatter boiling characteristics, known as DF05), using a modern DaimlerChrysler diesel engine such as those generally installed in light-duty transport vehicles. They concluded that RME led to significant reductions in CO, hydrocarbons (HCs), aromatic HCs (including benzene), and aldehydes and ketones (which contribute to the formation of summer smog) compared with the other fuels.[117]

However, Krahl et al (2001) did determine that RME resulted in more emissions of particles in the 10–40 nanometre diameter range, and fewer in the larger diameter range relative to fossil diesel (ultra-fine particles are considered to be more toxicologically relevant than larger particles), whereas MK1 led to a reduction over the entire measuring range and DF05 led to an increase across the whole range, compared with fossil diesel fuel. At the same time, RME resulted in a considerable decrease in total particle mass relative to the other fuels tested and produced almost no soot. And the mutagenic effects of particle extracts from RME were the lowest, indicating a reduced health risk from cancer associated with the use of RME relative to fossil diesel and the other fuels tested.[118]

Impacts on NO_x emissions

Most studies conclude that ethanol and biodiesel emit higher amounts of nitrogen oxides (NO_x) than do conventional fuels, even as other emissions decline (see Figure 13.1).[119] There are exceptions, however. When ethanol is blended with diesel, NO_x emissions decline relative to pure diesel fuel; and some tropical oils are saturated enough – and thus have a high enough cetane value – that they increase NO_x less (and in the case of highly saturated oils such as coconut, actually decrease NO_x) relative to diesel.[120] NO_x are a precursor to ground-level ozone (smog). In addition, NO_x emissions increase acid rain and are precursors to fine particulate emissions;

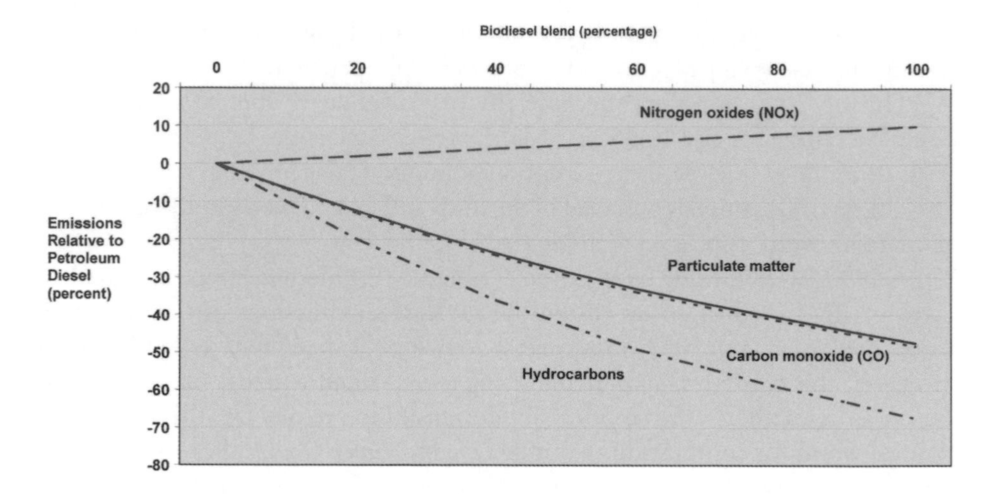

Figure 13.1 *Exhaust emissions from varying biodiesel blends relative to emissions from 100 per cent petroleum diesel fuel*

Source: Fulton et al (2004, p117)

associated health impacts include lung tissue damage, reduction in lung function and premature death.[121]

The level of NO_x emissions found varies significantly from study to study. Some cities, particularly in the US state of California, have complained that ethanol has increased local problems with NO_x and ozone.[122] California is using ethanol as an oxygenate to meet requirements under the US Clean Air Act because concern about water contamination led the state to ban MTBE. More recently, concerns about evaporative VOC emissions and combustion emissions of NO_x led California to sue the US EPA twice for a waiver; both times the waiver was denied.[123] But both the EPA and the California Air Resources Board agreed during the process that ethanol increases NO_x slightly in the on-road fleet.[124]

Fulton et al (2004), on the other hand, report that the impacts of biofuels on NO_x emissions levels are relatively minor and can actually be higher or lower than conventional fuels, depending upon conditions. In fact, there is evidence that NO_x levels from low ethanol blends range from a 10 per cent decrease to a 5 per cent increase relative to pure gasoline emissions.[125]

Studies by the US National Renewable Energy Laboratory (NREL) show inconsistent results with regard to biodiesel and NO_x, depending upon whether the vehicle is driven on the road or in the laboratory. According to McCormick (2005), they have seen 'NO_x reductions for testing of vehicles (chassis dyno) and NO_x increases for testing of engines (engine dyno)'. The former, which involves driving an entire car on rollers rather than testing emissions directly from an engine removed from the vehicle, is considered more realistic than the latter.[126]

NREL studies of in-use diesel buses have found a statistically significant reduction in NO_x emissions with biodiesel.[127] A US auto-oil industry six-year collaborative study examined the impact of E85 on exhaust emissions and found that NO_x emissions were reduced by up to 50 per cent relative to conventional gasoline.[128] But India's Central Pollution Control Board has determined that burning biodiesel in a conventional diesel engine increases NO_x emissions by about 13 per cent.[129]

Fortunately, newer vehicles designed to meet strict air standards, such as those in California, have very efficient catalyst systems that can reduce VOC, NO_x and CO emissions from ethanol–gasoline blends to very low levels.[130] With biodiesel, NO_x increases can be minimized by optimizing the vehicle engine for the specific blend that will be used.[131] Emissions can also be reduced with additives that enhance the cetane value or by using biodiesel made from feedstock with more saturated fats (e.g. tallow is better than canola, which is better than soy).[132]

It is also possible to control diesel exhaust using catalysts and particulate filters. High-efficiency diesel particulate filters (DPFs) remove particulate matter (PM) by filtering engine exhaust; such systems can reduce PM emissions by 80 per cent or more.[133] However, because of concerns about increased oil-film dilution during post-injections, German car manufacturers do not accept neat biodiesel in DPF-equipped vehicles.[134] There is also concern that the extra injection used to increase emission temperatures for regeneration of the particulate trap results in a dilution of engine oil when RME is used as a fuel, and this dilution can increase engine wear.[135] Rust particle filters, which are available in many new diesel automobiles and significantly reduce emissions of fine particulates, cannot operate with biodiesel.[136] According to some sources, biodiesel does not meet European air emissions standards that went into effect in January 2006,[137] although the Association of the German Biofuel Industry notes that biodiesel can meet updated European standards for trucks and commercial vehicles.[138]

Several groups are in the process of developing additives to address the issue of NO_x emissions associated with biodiesel blends, including NREL, the US National Biodiesel Board, the US Department of Agriculture and World Energy Alternatives.[139]

Advanced technologies

In general, the air quality benefits of biofuels are greater in developing countries, where vehicle emissions standards are non-existent or less stringent and where older, more polluting cars are more common.[140] For example, the use of ethanol can effectively reduce emissions of CO and hydrocarbons in old-technology vehicles today.[141] Less understood, however, are the impacts that biodiesel might have on exhaust emissions from vehicles that are underpowered, over-fuelled, overloaded and not well maintained – vehicles that are also most prevalent in the world's developing nations.[142]

Advances in pollution-control technologies for petroleum-fuelled vehicles will reduce, if not eliminate, the relative benefits of biofuels. Greene et al (2004) note that the main benefit of biofuels in such advanced vehicles may be to make it easier to comply with emissions standards in the future, thus reducing the cost of emissions-control technologies.[143]

At the same time, new technologies are on the horizon. For example, Volkswagen and DaimlerChrysler have invested in biomass-to-liquid (BTL) technologies that convert lignocellulosic fibres into synthetic biodiesel. This process enables them to produce a cleaner-burning biofuel. In the future, they hope to optimize fuels and vehicle engines in parallel.

CONCLUSION

The refining, transport and combustion of biofuels have environmental costs, particularly on local water and air quality, and these impacts could rise considerably as biofuel production increases to meet rapidly rising global demand. At the same time, more sustainable practices and new technologies offer the potential for environmental improvements.

Increasing efficiencies in water and energy use at refineries can help to reduce both air and water pollution. The UK-based biodiesel producer D1 Oils now recycles both water and methanol used in its refineries, and uses biodiesel to run its facilities.[144] Standards and regulations are also needed to minimize pollutants. In addition, encouraging smaller-scale distributed facilities will make it easier for communities to manage wastes, while possibly relying on local and more varied feedstocks for biofuel production and thereby benefiting local economies and farmers.

The combustion of biofuels – whether blended with conventional fuels, or pure – generally results in far lower emissions of CO, hydrocarbons, SO_2 and particulate matter (and, in some instances, lead) than does the combustion of petroleum fuels. Thus, the use of biofuels, particularly in older vehicles, can significantly reduce local and regional air pollution, acid deposition and associated health problems, such as asthma, heart and lung disease, and cancer.[145]

However, the air quality benefits of biofuels relative to petroleum fuels will diminish as fuel standards and vehicle technologies continue to improve in the industrialized and developing worlds. Even today, the newest vehicles available for purchase largely eliminate the release of air pollutants (aside from CO_2).[146] At the same time, concerns about higher levels of NO_x and VOC emissions from biofuels will probably diminish with improvements in vehicles and changes in fuel blends and additives. A combination of next-generation power trains (based on internal combustion engines) and next-generation biofuels can make a major contribution to reducing air pollution in the transport sector.

In the developing world, ethanol should be used to replace lead, benzene and other harmful additives required for older cars. And because high blends or pure biofuels pose minimal air emissions problems and are less harmful to water bodies than petroleum fuels, for all countries it is important to transition to these high blends as rapidly as possible, particularly for road transport in highly polluted urban areas and for water transport, wherever feasible.

Part V

Market Introduction and Technology Strategies

14

Infrastructure requirements

Introduction

A dramatic expansion in biofuel production capacity worldwide will require substantial new investments in biofuels and related infrastructure. Existing experience with current feedstocks and conversion technologies can provide useful insight into the infrastructure that will be needed for next-generation cellulose-based biofuel production in the coming years. The experience of the few countries where major biofuel developments have been under way for more than a decade will be particularly useful to those countries that are just beginning their own biofuel initiatives.

This chapter discusses the basic infrastructure considerations that need to be addressed for either of the two basic feedstock options, including degree of concentration of production (distributed versus centralized); transportation of feedstocks and of finished biofuel products (via truck, rail, barge, ship and, possibly, pipeline); investments in new conversion facilities; investments in biofuel storage capacity; and investments in vehicle refuelling facilities. The chapter focuses primarily on the experiences in ethanol infrastructure development in Brazil and the US, and on biodiesel infrastructure development in Germany. The emphasis is mainly on larger-scale biofuel production facilities, which are likely to dominate future production.

Centralized versus distributed production

Compared to petroleum refining, which is developed at a very large scale, biofuel production is lower volume and more decentralized. In the case of biodiesel, in particular, where a wide range of plant and animal feedstocks can be used, there has been a tendency for rather dispersed production facilities. Producers have the ability to extract raw vegetable oil at one site and send the oil to a different location for processing.

Ethanol fuel production, which is ten times greater than current biodiesel production, has tended to be more geographically concentrated, but is broadly distributed among different facilities within a specific production region. In the US, it is predominantly concentrated in Midwestern states that have abundant corn supplies, such as Iowa, Illinois, Minnesota, Nebraska and South Dakota. In Brazil, sugar cane and ethanol production is concentrated in the centre-south region, mainly in the state of São Paulo.

Compared to the oil industry, however, the ethanol production capacity in both the US and Brazil is significantly more decentralized. Petroleum refineries that produce gasoline and diesel fuel are considerably larger than the facilities that produce ethanol and biodiesel. Consequently, in the world's largest biofuel-producing regions, the average capacity of individual petroleum refineries is two orders of magnitude higher than the plants producing biofuels (see Figure 14.1).[1]

Despite the two countries' somewhat similar overall ethanol output, Brazil has three times more ethanol plants than the US. Accordingly, the average capacity of plants in the US is three times greater than the average capacity of those in Brazil. The largest plant in Brazil produces 328 million litres per year by crushing sugar cane, whereas in the US the largest corn dry-milling ethanol plant produces 416 million litres per year.[2, 3] There are various reasons for the differences in plant capacities. One key reason corn-to-ethanol plants can be larger is because substantial amounts of harvested corn can be stored for long periods of time, whereas sugar cane must be processed shortly after it is harvested (preferably within 24 to 48 hours to avoid deterioration of the sugar).

Despite being spatially concentrated in specific regions, ethanol production in Brazil and the US is decentralized among different plants due to feedstock transportation costs and handling logistics, which place economic limits on the size of processing plants. Since unprocessed biomass tends to be bulky (particularly compared to petroleum fuels), transport and logistics play an important role in facility siting and in overall biofuel economics.

INVESTMENT REQUIREMENTS FOR FEEDSTOCK TRANSPORT AND PROCESSING

Transporting and processing feedstock requires investments in a wide range of areas, including transportation infrastructure and facility construction, as well as overall operating costs. Since biofuel production technology is not homogeneous, these costs depend upon the region, technology used, feedstock type, labour costs, spatial distribution, existing transportation infrastructure and other factors.

Investment in feedstock transport

Feedstock transportation cost is a function of the total distance and time required for hauling the biomass and the time for loading and unloading.[4]

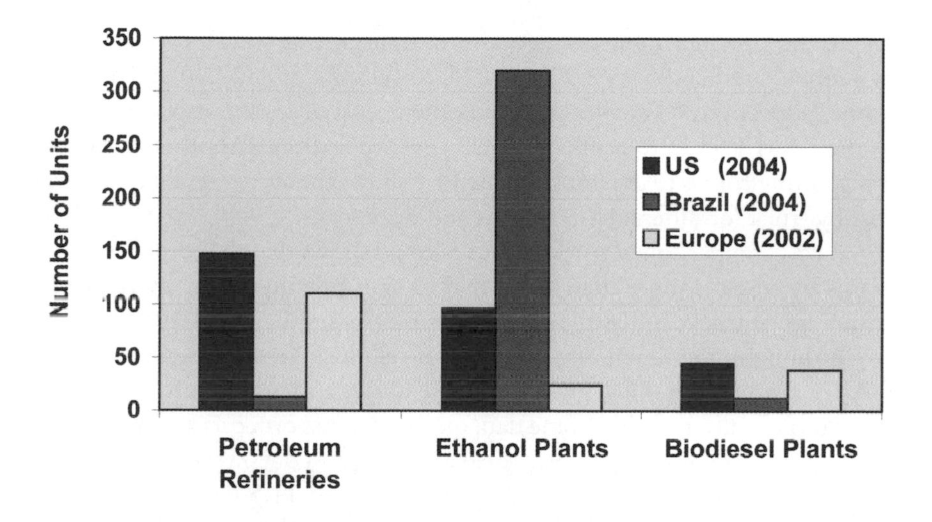

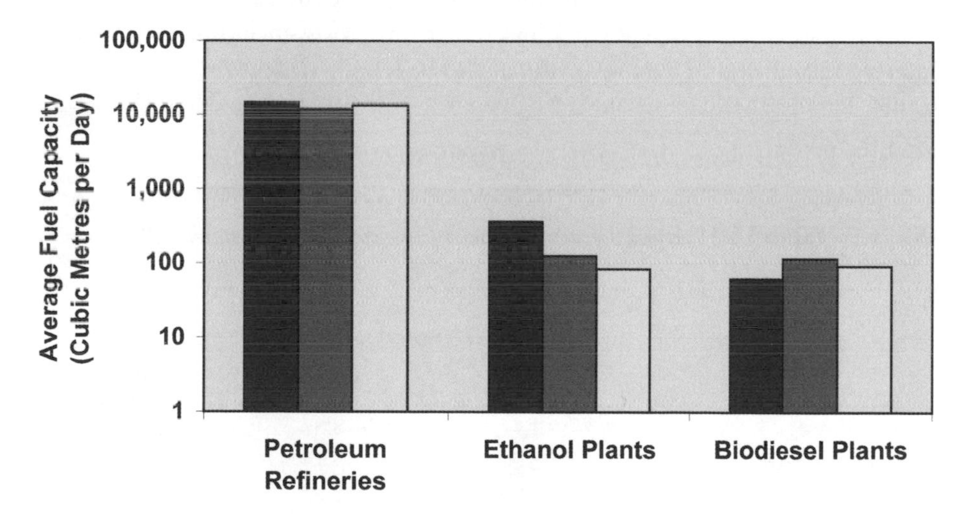

Figure 14.1 *Number of plants and their average fuel production in the US, Brazil and Europe*

Source: see endnote 1 for this chapter

In Brazil, the estimated costs of harvesting, loading and transporting 1 tonne of sugar cane over 18km are €3.07, €0.50 and €2.01 (US$3.71, $0.61 and $2.43), respectively, using medium-sized trucks; this corresponds to €0.11 (US$0.13) per tonne for every kilometre of hauling distance.[5] This is actually a conservatively high transport cost estimate since the sugar cane is often transported by special trucks with two to three trailers, rather than in single medium-sized trucks, which increases the fuel efficiency of cane transport and reduces the number of overall trips required.

In the US, the current average costs of transporting wheat, corn and soybeans by railroad are €26.95, €22.90 and €24.26 (US$32.59, $27.70 and $29.33) per tonne, respectively.[6] The average per kilometre cost of corn transported by railroad in the US is 1.58 Euro cents (1.91 US cents) per tonne. Another alternative for transporting corn is to use trucks. Table 14.1 shows the average grain transportation cost by truck for different US regions and distances.[7]

A cost comparison of trucks versus railroad shows that the cost of transporting grains by truck is more than twice that of transporting grains by railroad. The advantage of trucks is that they can access more sites than a railroad or waterway.

Although production of biofuels from cellulosic feedstocks such as crop and forestry residues, perennial grasses, organic wastes and other sources of fibre is not yet commercially available, some data on these transport costs are available. A US study of switchgrass transport estimated a hauling cost of €0.04 (US$.05) per tonne per kilometre, assuming three trailers per truck and 11.5 dry tonnes per trailer.[8] The hauling time cost, which includes time-dependent variables, such as the driver labour costs, as well as truck and trailer time costs, including depreciation, interest, insurance and fees, was estimated to be €3.65 (US$4.42) an hour per dry tonne. Finally, the loading/unloading cost was estimated to be €2.74 (US$3.32) per dry tonne.

Table 14.1 *Average grain transportation cost by truck in the US*

Region	*Transportation cost (Euro cents/tonnes per kilometre travelled)*		
	≤40km	*≤160km*	*≤320km*
National average	4.47	3.37	2.89
North-central region	3.99	3.14	2.80
Rocky Mountain	5.99	3.23	2.78
South-central	3.86	3.23	3.03
West	6.43	4.66	3.75

Note: Rates are based on trucks with a 36.3 tonne gross vehicle weight limit.

Source: see endnote 7 for this chapter

Investment in biofuel processing plants

Ethanol production

Table 14.2 compares the costs of producing ethanol from wheat and sugar beets in Germany with the costs of producing ethanol from corn in dry-milling facilities in the US.[9] In general, ethanol production costs in Germany are somewhat lower for wheat than for sugar beets. There are also savings related to plant economies

Table 14.2 *Engineering cost estimates for bioethanol plants in Germany versus the US*

Costs	Germany				US 53 million litre plant
	50 million litre plant		200 million litre plant		
	Wheat	Sugar beet	Wheat	Sugar beet	Corn
	(Euros per litre)				
Feedstock cost	0.23	0.29	0.23	0.29	0.17
Co-product credit	–0.06	–0.06	–0.06	–0.06	–0.06
Net feedstock cost	0.17	0.23	0.17	0.23	0.12
Labour cost	0.03	0.03	0.01	0.01	0.02
Other operating and energy costs	0.17	0.15	0.17	0.14	0.09
Annualized net investment cost	0.08	0.08	0.05	0.05	0.03
Total cost	0.45	0.49	0.40	0.34	0.26
Total cost per litre gasoline equivalent	0.67	0.73	0.59	0.64	0.40

Source: see endnote 9 for this chapter

of scale, with overall capital costs lower for a 200 million litre plant than for a 50 million litre plant.

In Brazil, the cost of an ethanol distillery with a 450,000 litre per day capacity is €28.12 million (US$34 million), whereas the production of 1100 tonnes of sugar cane per day (enough to feed an ethanol plant with this capacity using current technology) requires an investment of €14 million (US$17 million).[10] Such investments correspond respectively to 26 and 12 Euro cents (31 and 15 US cents) per litre, assuming an 8 per cent annual discount rate over ten years.[11]

Biodiesel production

Biodiesel investment costs in Brazil are estimated to range from €64 (US$77) per 1000 litres if methyl ester is used, to €83 (US$100) per 1000 litres if ethyl ester is used. In the US, the investment required for biodiesel production is about €109 (US$132) per 1000 litres of production capacity. Feedstock costs correspond to an additional 70–85 per cent of the final cost of biodiesel (see Table 14.3).[12]

A typical biodiesel plant is composed of two plants: the soybean processing plant and the transesterification plant. This separation of functions highlights a potential option of having soybean-processing facilities in completely different locations from the transesterification process. This separation could allow farmers to send unprocessed vegetable oil to a larger central transesterification facility, perhaps owned jointly by numerous farmers. It could also potentially be an

Table 14.3 *Cost estimates for investments in biodiesel production capacity*

	Facility cost per 1000 litres	*Total cost per 1000 litres*
Large facility: Brazil (>1 million litres)	€64 (US$77)	€427 (US$531)
Large facility: Brazil (ethyl esters)	€83 (US$100)	€553 (US$693)
Large facility: US	€109 (US$132)	€450 (US$563)
Small facility: US	€270 (US$334)	€1100 (US$1376)

Note: Assuming by-products are sold and feedstock costs correspond to 85 per cent of facility costs in Brazil and 75 per cent in the US.

Source: see endnote 12 for this chapter

approach used for international trade, where nations that produce vegetable oil could send their product to countries that would then produce finished biodiesel to their own specifications.

In terms of the capital costs involved in developing smaller-scale biodiesel plants, for a plant with an installed soybean processing capacity of approximately 1900 tonnes per year, paired with a transesterification plant with an installed capacity of 2.3 million litres per year, the costs would be about €414,000 (US$500,000) and €223,000 (US$270,000), respectively.[13] This corresponds to a capital cost of €0.27 (US$0.33) per litre of annual biodiesel production capacity.[14]

Malaysia, the world's largest producer of crude palm oil, is planning to expand its production by up to 25 per cent due to increasing demand for biodiesel. The Plantations Industries and Commodities Ministry aims to increase the yield of palm oil from current average levels of 4 tonnes per hectare (4300 litres) up to 5 tonnes (5400 litres) per hectare by 2010. The government is building three very large biodiesel plants, each with an annual capacity of 60,000 tonnes (65 million litres) and a cost of €26.4 million (US$32 million).[15] There are different feedstocks involved, different labour rates and different investment criteria for the two cases.

Cellulose conversion

The cost of a cellulosic biomass plant with a capacity of 542,000 litres per day is estimated to be €194 million (US$234 million), which implies an annualized capital cost of €0.15 (US$0.18) per litre (assuming an 8 per cent annual discount rate over ten years). This estimate is based on current technology and excludes operational costs, which on an annualized basis are similar to the facility's capital cost.[16]

In general, cellulosic conversion plants will require greater capital investments than plants that convert sugar or starch due to the extra steps needed to break down resistant cellulose plant fibres. The capital costs for gasification-based systems are expected to be somewhat higher than for enzymatic conversion systems; however, using wood in gasification systems should allow for larger plants that will benefit from economy-of-scale cost savings that should make gasification and enzymatic systems similarly competitive (since costs for both of these technologies are based on engineering estimates, there is clearly a need for real-world validation of costs before a definitive cost comparison can be made).

INVESTMENT REQUIREMENTS FOR BIOFUEL TRANSPORT, STORAGE AND DELIVERY

The widespread production and use of biofuels depends upon the existence of infrastructure for transport, storage, distribution and delivery of the fuels. The investments and costs required for these steps depend upon the type of fuel (gaseous versus liquid), the type of vehicle using the fuel and the existing transportation infrastructure.

Refuelling considerations for gaseous versus liquid biofuels

Since biofuels such as ethanol, ethyl tertiary butyl ether (ETBE), biodiesel and synthetic diesel (from various biomass-to-liquid options) are liquid fuels, their similarity to petroleum fuels allows for much lower refuelling station costs than for gaseous fuels, such as hydrogen or compressed natural gas (CNG).

It is possible to upgrade biogas from biomass digesters to natural gas quality for use in vehicles (similar to the CNG applications that are fairly common in many countries). However, the infrastructure costs for adding many new CNG refuelling stations will be fairly high. The cost for fast-fill CNG refuelling stations (i.e. ones that would allow for vehicle refuelling times similar to gasoline or diesel refuelling times) is on the order of €124,000 (US$150,000) for a small station, €331,000 (US$400,000) for a medium-sized station, and €830,000 (US$1 million) for a large station.[17] In addition, vehicle retrofit costs to allow for gaseous fuel use in existing vehicles must also be considered, compared to the convenience of using ethanol or biodiesel blends with essentially no vehicle modification costs in many cases (see Chapter 15). Since modern automobiles are often used for a 12-year time-frame, the turnover in a national vehicle fleet will be slow, and the ability to use existing vehicles in a biofuel implementation strategy will allow for a much more rapid impact on a nation's overall transportation fuel use.

In contrast with biogas, it is often possible to find at least one type of fuel option at a gasoline/diesel refuelling station that can be replaced with a biofuel pump, entailing relatively low costs for changing over an existing pump/storage tank to accommodate biofuel storage (costs for this type of retrofit are estimated to be around €830, or US$1000). For those refuelling stations where it is felt that an existing petroleum product cannot be dropped to accommodate a biofuel, costs will be distinctly higher if new underground storage tanks must be installed – for example, this cost is estimated to be around €18,000 (US$22,000) for an 11,000 litre underground storage tank.[18]

Refuelling infrastructure for flex-fuel and dedicated biofuel vehicles

In Brazil, 29,646 out of a total of 31,979 vehicle-fuelling stations sell 'neat' ethanol (actually hydrous ethanol that contains 4 per cent water) for use in flexible-fuel vehicles (FFVs) and pure-alcohol vehicles.[19] And in the US, approximately 590 refuelling stations out of 168,987 sell E85, a mix of 85 per cent ethanol and 15 per cent gasoline for use in FFVs. About 6 million FFVs on North American roads are able to use this fuel, although it is worth noting that few of these actually use E85.[20]

In terms of market applicability, pure ethanol can be used in FFVs in warm tropical and subtropical climates such as Brazil, whereas E85 is targeted to (and particularly appropriate for) markets with colder climates, such as the US and Europe. Ford Motor Company, Volvo and Saab have all announced plans to market E85 vehicles in Germany and elsewhere in Europe.

Germany has proven that pure biodiesel fuel (B100) can be used in existing diesel engines with some minor refitting, mostly concerning seals. B100 has received a fuel tax exemption in Germany, where more than 1500 refuelling stations now sell the fuel. Note that Europe's new EURO V regulations will require more stringent engine performance criteria for the European automotive industry. There are questions whether B100 will be compatible with the new particulate and nitrogen oxide (NO_x) standards, although B10 blends will probably comply.

Ethanol transportation infrastructure

Brazil

Brazil is increasingly interested in building infrastructure to export ethanol internationally, whereas in the US, ethanol infrastructure is driven almost exclusively by the domestic market. As a result, Brazil's ethanol sector has given some consideration to export issues and infrastructure requirements. This includes investments in the construction of larger maritime terminals or greater storage capacity (which allows for regulation of the supply), as well as the construction of pipelines to minimize transportation costs.

The infrastructure required to facilitate ethanol export demands the interconnection (via waterway) of producers from Brazil's southwest with storage facilities in São Paulo state, which then connect to ports in Rio de Janeiro and São Sebastiao. Upon completion of 550km of pipelines in 2010, the capacity of these two ports will reach 4 million cubic metres per year. All civil works required to put the infrastructure in place are contingent upon a €347 million (US$420 million) investment.[21] It is worth mentioning that Brazil's ethanol exports in the last two years have surpassed 2 billion litres, and only modest infrastructure has been added for ethanol handling.

Between 2008 and 2010, €132 million (US$160 million) will be invested in the waterway infrastructure in Brazil's midwest, as well as west of São Paulo state. This will entail the construction of four or five storage terminals and 90km of pipelines. Potential environmental concerns regarding these plans are addressed in Chapter 13.

The Brazilian oil company Petrobras planned to reach an export capacity of 2 billion litres per year of ethanol by 2007, with 1.2 billion litres per year in the southeast and 0.4 billion litres per year each in the south and northeast.[22] The goal is to reach 5.4 billion litres per year between 2008 and 2009, and 9.4 billion litres in 2010. It is expected that world ethanol production, which was estimated at 38.2 billion litres in 2004, will escalate to 60 billion litres in 2010.[23] Petrobras's expected investments do not include the acquisition of barges, rail cars and trucks, or the shipment fees associated with each transportation mode. On a volume basis, assuming 20 years of operation and solely considering the upfront investment costs (not future operational costs), the total investment equates to €2.50 (US$3) per 1000 litres of ethanol.

The transportation costs charged by Petrobras range from €3.56 to €13.65 (US$4.31 to $16.52) per 1000 litres, depending upon the pipeline used and distance. The oil company charges external users a monthly fee of €3.47 (US$4.20) per 1000 litres to store alcohol in its facilities.[24] This adds a significant cost to ethanol since, on average, it is stored for six months to guarantee the supply during the off-harvesting season. Thus, the storage cost of half the annual amount produced is around €21 (US$25) per 1000 litres, yielding an average cost of €10.30 (US$12.50) per 1000 litres for all ethanol commercialized. The producer's price of ethanol in São Paulo in 2005 was €288 (US$348) per 1000 litres, and its retail price was €456 (US$551) per 1000 litres. A rough estimate of distribution costs is obtained by subtracting 29.65 per cent in taxes plus the producer's cost from the retail price. Accordingly, €33 (US$40) per 1000 litres corresponds to freight costs in São Paulo.

Part of the ethanol consumed domestically in Brazil is mixed with gasoline, while the remainder is used as a 'neat' fuel. Therefore, the ethanol produced in sugar mills/distilleries needs to be transported to distribution terminals that distribute the gasoline–ethanol mix and the neat fuel ethanol to the retail market. Although railroad is more cost effective than road transport to transfer ethanol between producers and distribution terminals, this option is seldom available due to the limited availability of rail transport and the significant upfront investments required for this infrastructure. As a result, distribution between the terminals and the retailers relies primarily on tanker trucks.

Petrobras was responsible for around 20 per cent of Brazil's ethanol sales in 2003.[25] In 2003, the company owned 51 distribution terminals and shared nine other terminals with several users. Petrobras also used 11 third-party storage facilities and delivered the fuel to their own 7200 retail stations, which also sell gasoline–ethanol mix and diesel.

Petrobras has extensive experience with ethanol logistics, including pipeline transport, and is an example of a large-scale distribution biofuels infrastructure. Currently, the company operates eight storage facilities. Seven are located in the southeast region of Brazil, and five of these are in São Paulo state. These facilities contain tanks to collect and store ethanol and are interconnected by railroads to three maritime ports on the Atlantic Ocean (Rio de Janeiro, Santos and Paranaguá). The company also operates an ethanol distribution storage facility in the Brazilian midwest (Alto Taquari), which is connected to its transport network. Transport between the storage facility of Paulinia in São Paulo and Rio de Janeiro is done through a pipeline. In Rio de Janeiro, the product reaches a maritime terminal from where it can then be exported to international markets via ship.

Currently, most Brazilian ethanol is transported within the country by tanker trucks. Of the eleven facilities operated by Transpetro (the transportation branch of Petrobras), eight depend upon trucks, three are connected to pipelines and two are connected to railroads.[26]

US

In 2005, US ethanol consumption, including both E85 and ethanol mixed in gasoline as an oxygenate, was 14.8 billion litres.[27] By 2010, it is projected to reach 19.3 billion litres.[28] This increase is expected to occur through both the expansion of existing facilities and the construction of new facilities. However, the 2010 projection is conservative: over the past six years, growth in US ethanol consumption averaged 14 per cent annually, whereas the projection accounts for only 8 per cent annual growth. Due to this disparity, it is worth speculating about US infrastructure needs associated with a much larger supply of ethanol by 2010 – of as much as 37.8 billion litres per year.

The investment needed to put in place the equipment and convert existing storage tanks to ethanol is €127 million (US$154 million) for the low-end projection of 19.3 billion litres and €172 million (US$208 million) for the high-end projection of 37.8 billion litres. In addition, investments in the retail infrastructure to enable the mixing of ethanol and gasoline are needed. This would amount to €122 million (US$148 million) for the low-end scenario and €238 million (US$288 million) for the high-end scenario (see Table 14.4).[29]

After completion, the infrastructure associated with the low-end production scenario will rely on 59 per cent of the terminals installed in ethanol distribution centres. About 25 per cent of these terminals will have naval connections and 26 per cent will be connected through railroads. In comparison, the infrastructure

Table 14.4 *Investment requirements associated with expanding US ethanol production by 3.7 billion litres*

Investment requirements	*Cost (millions)*
Equipment and storage	€127 (US$154)
Retail infrastructure	€122 (US$148)
Operational costs	€323 (US$390)
Breakdown of operational costs	
Shipment by barges and ships	€98 (US$118)
Additional 54,832 rail cars	€128 (US$155)
Additional 399,375 truck shipments	€97 (US$117)

Source: see endnote 29 for this chapter

associated with the high-end projection will rely on 85 per cent of the terminals installed in ethanol distribution centres. About 19 per cent of these terminals will have naval connections and 20 per cent will be connected through railroads.

Operational costs are also part of the investments associated with the consumption of 19.3 billion litres of ethanol in 2010. An additional 3.7 billion litres produced in 2010 will cost €98 million (US$118 million) in freight charges associated with shipment by barges and ships.[30] With regard to ethanol transport by railroads in 2010, the expenses associated with freight for 54,832 rail cars amount to €128 million (US$155 million). Finally, 399,375 intra-regional truck shipments will cost €97 million (US$117 million). Thus, total freight charges in 2010 (for the low-end scenario) amount to €320 million (US$387 million).

Increasing the transport of ethanol also demands purchasing new transportation equipment. Table 14.5 shows freight and new equipment costs for the two ethanol production scenarios for 2010 (19.3 billion litres and 38.7 billion litres).[31]

The final use of ethanol also affects the transportation cost of the fuel. Although part of the ethanol is mixed with gasoline, it is possible to offer pure ethanol at the pumps. In the US, most of the ethanol consumed is mixed in gasoline. About 90 per cent of the ethanol is used as a 10 per cent blend in gasoline, and 10 per cent is used as an 85 per cent blend of ethanol in gasoline in E85 fuels.[32]

The world's two largest ethanol producers, Brazil and the US, have a spatially concentrated fuel production system and rely on inter-modal distribution to transport the biofuel. The costs of transportation are estimated in Table 14.6.[33] In Brazil, 85 per cent of ethanol production is located in the mid-south region, but 33 per cent of the consumption occurs in states that do not produce ethanol.[34] In the US, 88 per cent of ethanol production is located in the Midwest, while the major consumers are on the east and west coasts.[35] The fuel is transported by barge along the Mississippi River to New Orleans, then transported to the northeast and west coast by ship. Therefore, in both countries, a distribution infrastructure exists for the transportation of ethanol.

Table 14.5 *Transportation costs for two ethanol production scenarios in the US, 2010*

Transportation mode	*Freight*		*Equipment purchased*	
	Cost (€ millions per year)	*Description*	*Cost (€ millions)*	*Description*
Low-end scenario (19.3 billion litres)				
Barges and ships	98	3.7 million cubic metres; 726 barge trips	28	21 barges
Railroad	128	55,000 rail cars loaded	127	2549 tanker cars
Trucks	97	399,000 shipments	24	254 tractor trailers
High-end scenario (37.8 billion litres)				
Barges and ships	134	5.4 billion litres; 1559 barge trips	55	42 barges
Railroad	165	73,000 rail cars loaded	172	3472 tanker cars
Trucks	170	804,000 shipments	45	563 tractor trailers

Source: see endnote 31 for this chapter

Table 14.6 *Ethanol transportation costs in the US and Brazil*

Transportation mode (Cost-effective distance[a])	*Cost (Euros per cubic metre)*	
	US	*Brazil[b]*
Water (including ocean and river barge)	€8–€25	€10
Short trucking (less than 300km)	€8–€17	
Long distance trucking (more than 300km)	€17–€83	€26
Rail (more than 500km)	€17–€40	€17

Notes: [a] cost-effective distance is based on US estimates; [b] assuming that specific gravity is 0.789 grams per millilitre at 20°C and the exchange rate is 2.83 Brazilian reals to €1

Source: see endnote 33 for this chapter

Although pipeline transport is the cheapest way to transport liquid fuels, biofuels face several challenges in this regard. If ethanol–gasoline blends are to be transported in pipelines used routinely only for petroleum products, there is a concern that 'phase separation' can occur if water gets into the pipelines or fuel. Rather than being mixed with gasoline, the ethanol pulls itself out of the blend, creating a separated ethanol water stratum or layer (this is of greater concern in cold winter conditions since phase separation occurs more readily at cooler fluid temperatures). When 'neat' (pure) ethanol is to be transported, if it contains a small amount of water, the water component can contribute to corrosion inside the pipe since petroleum pipelines are generally made with steel that is not resistant to water corrosion (because petroleum does not contain water).

Biodiesel transportation infrastructure

Biodiesel fuel is easier to transport and store than ethanol because it can use the same infrastructure as diesel. However, because of the smaller scale of production, biodiesel is usually transported by trucks, which are not as cost competitive as pipelines that transport diesel. Transportation costs for biodiesel in the US can be as high as €330 (US$440) per 1000 litres.[36]

Currently, 1900 refuelling stations in Germany sell pure biodiesel in the form of rapeseed methyl ester (so-called RME100). Much of the fossil diesel sold in Germany contains a blend of 2 per cent RME, where the blending is done by the petroleum refineries. The goal is to increase this amount to 5 per cent RME over the next four years. This is likely to result in some competition in the German biodiesel distribution network between refiners who need biodiesel for use in low-level blends and the refuelling stations that need biodiesel for RME100.

With the phase-in of ultra-low sulphur fuel, one concern with biodiesel is that existing pipelines and storage tanks may have a build-up of sulphur residues on their internal surfaces that could be freed up by the solvent action of biodiesel, potentially raising the sulphur content of diesel–biodiesel blends above allowable levels. This may be only a temporary problem in a fuel changeover time period when new low-sulphur requirements take effect; but an assessment of this concern may be needed to determine the potential severity of the problem and possible solutions.

INTERNATIONAL TRANSPORT CONSIDERATIONS

Increased global trade in biomass and biofuels would necessitate consideration of the infrastructure and related requirements for international transport, particularly by sea. International bioenergy trade can include direct transport of biomass materials (chips, logs, bales, etc.), intermediate energy carriers (such as bio-oil or charcoal) or high-quality energy carriers (e.g. ethanol, methanol, Fischer-Tropsch liquids and hydrogen). In addition to factors such as the production method of biomass, the type of transport and the order and choice of pre-treatment operations are of importance.

Studies of intercontinental biofuels trade, and even of bulk transport of wood, have found that maritime transport of these commodities could be economically feasible since it does not appear that dramatic energy losses would be incurred. For example, exporting forest residues some 1500km from the Baltic region to The Netherlands – including inland transport and transfer, and using smaller-sized vessels – results in an overall energy use of 5 per cent of the energy content of the biomass transported. And exporting (cultivated) wood some 10,000km from Latin America to The Netherlands – accounting for inland transport and transfer, and using large-sized vessels – uses about 10 per cent of the energy content of the biomass.[37]

In a recent study, various options for transporting raw or processed biomass were evaluated, including direct transport of woody biomass (chips, logs or bales), an intermediate energy carrier (pyrolysis oil) and a high-quality energy carrier (methanol). In general, Latin-American biomass is cheaper than that in Europe. However, this financial advantage is counteracted by the higher costs for long-distance ship transport. The most favourable approaches are those with a high energy density, such as pellets, logs and liquid carriers. Long-distance transport of wood chips (e.g. by ship) is generally not desirable because chips have a relatively low bulk density and are vulnerable to fungi deterioration due to their moisture content and large specific surface area.[38]

International transport of biomass (or energy carriers from biomass) is feasible from both an energy and cost point of view. Such systems are, in fact, current practice: large paper and pulp complexes import wood from all over the world. Of course, when feedstocks such as wood are considered, trade-offs should be weighed between producing the biofuel where the feedstock is harvested (and then transporting or importing it to the country where it will be used) versus importing just the wood feedstock (and then converting it into biofuels or electricity in the country where the end products are to be consumed).

The cost implications for using international transport infrastructure to deliver biomass feedstocks for biofuel production are illustrated in Figure 14.2.[39] The charts compare the farm gate and international transport costs for delivering soybeans to Hamburg, Germany (for use in biodiesel) from the US versus Brazil. The average cost of transporting soybeans by trucks is 3.32 Euro cents (4.02 US cents) per kilometre in Brazil and 3.37 Euro cents (4.08 US cents) per kilometre in the US.[40] The final cost of producing and delivering soybeans to Germany is €231 (US$279) per tonne from the US and €248 (US$300) per tonne from Brazil. However, the farm-gate value is €187 (US$226) per tonne in the US, but only €145 (US$175) per tonne in Brazil. This illustrates the extent to which transportation

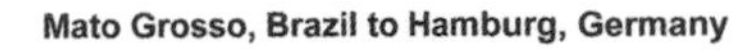

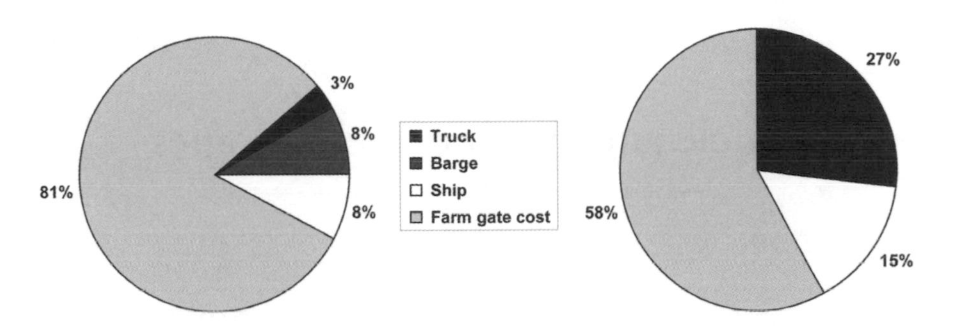

Figure 14.2 *Transport costs for delivering soybeans to Germany from the US versus Brazil*

Source: see endnote 39 for this chapter

costs versus feedstock production costs contribute to the final delivered cost of biofuel feedstocks.

Conclusion

Existing experience with first-generation feedstocks and conversion technologies provides useful insight into the infrastructure needed for next-generation cellulose-based biofuel production. Infrastructure will be needed for the transport of feedstocks and biofuels, as well as for feedstock conversion facilities, biofuel storage and vehicle refuelling.

Current infrastructure available for the use of agricultural and forestry resources needs to be evaluated to determine what expansion and refinements are required if renewable biomass resources are to play an expanding role in providing sustainable transportation fuel supplies. Some of the larger first-generation biofuel facilities require in the vicinity of 3000 tonnes per day of feedstock (such as 'dry mills' that produce ethanol from corn); and next-generation facilities are envisioned that would call for 6000 tonnes per day or more of feedstock (such as gasification/Fischer-Tropsch facilities that will convert wood to synthetic diesel). To enable the expansion of biofuel production in such facilities, as well as to provide for associated distribution requirements, it is clear that substantial infrastructure planning and development will be needed.

15

Vehicle and Engine Technologies

Introduction

This chapter provides an overview of the main technological issues related to the current and potential use of biofuels in motor vehicles. Although the primary focus here is on ethanol and biodiesel, use of straight vegetable oils (SVOs), dimethyl ether (DME), biomass-to-liquid (BTL) fuels and ethyl tertiary butyl ether (ETBE) is also discussed. Biogas, methanol and hydrogen are addressed very briefly as well.

Ethanol

Fuel-grade ethanol, produced from biomass, has been considered a suitable automotive fuel for nearly a century, particularly for vehicles equipped with spark-ignition engines (technically referred to as Otto cycle engines, but commonly known as gasoline engines). It is by far the most popular biofuel available commercially today. Brazil and the US are the world leaders in fuel ethanol use, although many other countries, including Australia, Canada, China, Colombia, India, Paraguay, South Africa, Sweden and Thailand, have also introduced it in the fuel market. Ethanol has long been regarded as a top fuel for car and motorcycle racing in Brazil, and, starting in 2007, it will be the standard fuel for the IndyCar series in the US as well.[1]

Corrosion problems have been avoided with the adoption of suitable fuel ethanol specifications such as the Brazilian Agência Nacional do Petróleo (ANP)[2] or the US American Society for Testing and Materials (ASTM)[3] standards, as well as care to avoid fuel contamination, particularly with water. Depending upon the characteristics of the ethanol, treatment with corrosion inhibitors has also been adopted.

Because ethanol has a solvent effect, it will clean existing deposits from a vehicle's fuel system when used either as a 'neat' fuel (in pure form) or as a blending agent in vehicles that have previously run on gasoline or diesel oil. In cases where ethanol blends are being introduced into existing vehicle fleets, replacing the fuel filter at shorter intervals than the standard service periods is recommended, particularly during the first few months of operation with these blends. The vehicle's spark plugs should also be checked and (if necessary) cleaned during the initial phase of operation with ethanol blends. This will avoid premature clogging of filters and the build-up of combustion chamber deposits on the spark plugs, thereby facilitating trouble-free engine operation.

Ethanol use in spark-ignition engines

Low ethanol content gasoline blends

Blending anhydrous (i.e. essentially, water-free) ethanol with gasoline at a ratio of 1:10 by volume (E10) has been the most popular and fastest way of introducing fuel ethanol in the marketplace. Extensive international experience demonstrates that, in general, such blends do not require engine tuning or vehicle modifications. And since most of the materials that have been used by the motor industry over the last two decades are E10 compatible, substitution of parts is not usually required.

The use of E10 has been covered under warranty by all manufacturers selling light-duty vehicles in the US and Canada for many years, and it is becoming common practice elsewhere. Because European Union (EU) fuel quality regulations have limited the ethanol content to 5 per cent (E5) or less, automakers have typically restricted the warranty coverage of vehicles sold in the EU to this level. An initiative launched by Volkswagen and DaimlerChrysler in February 2006 opts for E10 in new vehicles in Europe.

In general, blends up to E10 do not result in perceived changes in performance or driveability. Maintenance requirements also do not differ from those with gasoline use. Since the variation in fuel consumption is very small (0–2 per cent, on average), and use of these blends is considered environmentally friendly, public reaction tends to be positive.[4]

Gasoline blends beyond E10

In Brazil, all brands of automotive gasoline contain anhydrous ethanol in the range of 20–25 per cent (E20 to E25),[5] and manufacturers of cars and two-wheel vehicles have been customizing products to these blends for more than 25 years. Independent importers, meanwhile, have been adapting foreign vehicles by using ethanol-compatible materials in the fuel system and by tuning the engines (mainly the fuel delivery and ignition timing) for a mid-range point, usually at the 22 per cent ethanol level (E22). This customization has resulted in good driveability and performance, with fuel consumption comparable to gasoline operation.

Due to the availability of on-board electronic engine management systems capable of self-adjusting to different engine operation conditions, and to the standard use of ethanol-compatible materials, imported vehicles equipped with gasoline-only engines have been marketed in Brazil since 1998. According to an engineering consulting company specialized in conversion to E20/E25, these vehicles have presented trouble-free operation.[6] In all of the above cases, full warranty coverage for the vehicles has been provided by either the manufacturer or the import company.

Elsewhere, as in Australia and the US state of Minnesota, there have been initiatives to implement regular use of blends beyond E10, but generally limited to E20. Existing vehicles in regular use may eventually need to have parts of the fuel delivery system and engine changed, and the engine tuned for trouble-free operation and conformation with emissions standards. Because certain materials such as aluminium, magnesium, zinc, lead, brass, natural rubber, nylon and polyvinyl chloride (PVC) may degrade after long-term contact with high levels of ethanol, a careful evaluation of vehicle and engine customization needs for a particular blend is recommended before routine use is initiated. Ethanol-compatible parts should be used in the customization process. Depending upon the particular customization requirements, costs may run from a few Euros for substitution of fuel lines to more than €500 (US$605) if the fuel-supply system is fully upgraded (fuel lines, tank, pump, filter, etc.).

A number of automakers have been manufacturing E20-compatible versions of their vehicles, but few have declared this publicly. One exception is the Ford Motor Company, which announced in October 2005 that it would supply an E20-compatible Focus model for the Thai market.[7] Ford's position may well become a trend in the industry and foster additional E20 programmes.

Flexible-fuel vehicles

Gasoline blends containing 85 per cent anhydrous ethanol (E85) have been used in the US since 1992 – and, more recently, in Sweden and Canada – in so-called flexible-fuel vehicles (FFVs). These vehicles are specially designed to run on straight gasoline or any gasoline–ethanol blend up to E85 from a single tank. The technology is based on sensors in the fuel system that automatically recognize the ethanol level in the fuel. The engine's electronic control unit then self-calibrates for the best possible operation; if ethanol is not present, the engine will self-calibrate to gasoline-only operation. The process is instantaneous and undetectable by the vehicle driver. The main reason to limit ethanol content to 85 per cent is to enhance volatility conditions for cold start, particularly in cold climates, since the technology does not use any cold-start ancillary system.

As of February 2006, there were an estimated 6 million E85 FFVs on the road, with the vast majority in the US and a small share in Sweden and Canada. Ford, General Motors, DaimlerChrysler, Mazda and Nissan all offer E85 FFVs as standard vehicle versions.[8] In March 2006 Ford launched the UK's first E85 FFV, the Ford Focus Flexible Fuel Vehicle.[9]

In 2003, a variant of the E85 FFV technology was unveiled in Brazil. Instead of straight gasoline, the technology is capable of operating either within the E20/E25 range, with hydrous ethanol (E100) exclusively, or with any blend of E20/E25 and E100. In this technology, the ethanol sensors used in the E85 versions are replaced with an advanced software component in the engine's electronic control unit, which uses inputs from conventional oxygen sensors in the exhaust system (lambda sensors) and self-calibrates the engine to fuel requirements.

The E100 FFVs have become a sales phenomenon since their introduction in the Brazilian marketplace, in part because E100 is significantly less expensive than E20/E25 in much of the country. Unlike elsewhere in the world, both E100 and E20/E25 can be found easily at more than 29,000 retail stations throughout Brazil. As interest in importing the technology grows worldwide, a second generation is being developed that extends the operational range from straight gasoline to E100. Volkswagen, Fiat, General Motors, Ford, Renault, Citroen and Peugeot are all offering E100 FFVs as standard versions, and Toyota and Honda are expected to offer similar versions in 2006. As of December 2005, more than 70 per cent of new light-duty vehicle sales in Brazil were E100 FFVs, and cumulative sales of the vehicles have totalled more than 1.3 million since March 2003.[10]

This technology has proven feasible in Brazil in large part because the warm climate allows blending of hydrous ethanol to E20/E25 without the risk of phase separation. Yet, even at lower temperatures, cold start can be accomplished with automatic injection of E20/E25 stored in a small tank under the hood. International automotive suppliers Bosch, Delphi and Magneti Marelli are all developing new cold-start systems that will not require the auxiliary E20/E25 tank. This new technology set-up is based on sensors that monitor engine and ambient temperatures during cold start, as well as automatically heated fuel injectors equipped with multi-spray injection nozzles, new spark plugs and control software. Industry sources believe the new technology will be available commercially in 2007. The concept could also be applied in the E85 FFVs and allow for increased ethanol content in the blend.

FFVs are built with ethanol-compatible materials, have shown proven reliability, come with manufacturers' warranties, and cost roughly the same to maintain as gasoline vehicles. Because of ethanol's lower energy content and the tendency of car manufacturers not to take full advantage of the fuel's combustion properties, fuel consumption with either E85 or E100 is higher than with gasoline or E20/E25.[11] FFVs have demonstrated an average drop in fuel economy in the range of 25 to 30 per cent, depending upon vehicle/engine characteristics; however, it should be noted that these figures can be considerably improved at moderate costs with existing technology.

On the other hand, performance tends to improve with both E85 and E100, although the degree of improvement depends upon power train characteristics. For standard FFVs, a 5 per cent increase in power is not unusual; however, for a more advanced FFV concept, such as Saab's, which uses a 'smart' turbo-charging system to boost the turbo pressure as the ethanol content increases, the power increase can near 20 per cent.

Manufacturing FFVs does not add much to the cost of the vehicle. In both the US and Brazil, the price of FFVs has been very similar to that of baseline gasoline vehicles, and in some cases the same.

Dedicated ethanol vehicles

While FFV engines must retain dual-fuel capability, dedicated ethanol vehicles can take advantage of the combustion characteristics of ethanol and therefore perform better with lower fuel consumption. An increase in the engine compression ratio[12] of up to 13.5:1 (versus approximately 9:1 to 10:1 for conventional gasoline engines) allows for improved fuel combustion efficiency that partially offsets ethanol's lower energy content.

Most of the experience with dedicated ethanol technology comes from Brazil, where more than 5 million units have been sold since 1979.[13] Initially equipped with carburettors and old-style ignition advance systems, and upgraded with electronic fuel injection and mapped electronic ignition since 1991, ethanol vehicles have demonstrated proven reliability, good driveability and low maintenance costs. On average, fuel consumption has been 25 per cent lower than for equivalent E20/E25-fuelled versions. Volkswagen, Fiat, General Motors and Ford have all produced dedicated ethanol versions for more than 25 years, with full warranty coverage. Maintenance costs do not differ significantly from standard gasoline vehicles and, according to anecdotal reports, can actually be lower because of ethanol's ability to keep the fuel system and engine clean.

Growing interest in ethanol in the US, particularly since 2000, has stimulated research into high-efficiency engine technology. The US Environmental Protection Agency (EPA), through its National Vehicle and Emissions Laboratory, has conducted trials with a 'neat' ethanol port fuel-injected turbocharged engine, with a compression ratio of 19.5:1. The study concluded that high-combustion efficiencies can be achieved, yielding up to 20 per cent fuel economy improvement over baseline gasoline engines – results comparable to diesel engines.[14] This suggests that the concept could become a benchmark for dedicated ethanol engines whose use could also be extended to medium- and heavy-duty vehicles.

Ethanol use in compression-ignition engines

Diesel modified engines

Since the 1980s, attempts have been made to use ethanol in compression-ignition engines (diesel engines), although this application has been limited. The most successful experience has been in Sweden, where approximately 500 urban buses are operating on a mixture of 95 per cent hydrous ethanol and 5 per cent of an ignition additive known as 'Beraid'.[15] The additive is used to promote fuel ignition since ethanol is difficult to ignite in a compression ignition engine.

An initiative of the auto manufacturer Scania and the municipality of Stockholm, Sweden's first ethanol buses began operating in 1990 and were introduced as a way of meeting environmental requirements for cleaner fuels. The buses have been used mainly for inner-city service, where the environmental gains (in particular, the reduction in particulate emissions) are most noticeable. The power train is based on a diesel engine converted to ethanol through the following modifications: an increase in the compression ratio from 18:1 to 24:1; use of fuel injectors and a fuel pump with higher volumetric capacity; optimization of the ignition advance; increase in fuel tank volume; and use of ethanol-compatible materials in the fuel delivery system.

Scania reported a drop in fuel economy in the order of 40 to 50 per cent for the ethanol buses relative to diesel, while operational performance was considered adequate.[16] SL, the bus operating company, reported higher maintenance costs for the ethanol buses as well.

In the US, Detroit Diesel Company (DDC) has converted a two-stroke diesel engine to run on alcohol using modifications similar to Scania's, except for the substitution of a glow plug to start ignition since fuel ignition additives were not considered. A one-year trial in Peoria, Illinois, with two vehicles equipped with the ethanol engine showed results similar to those with the Scania buses. The incremental maintenance cost during this period was close to 5 per cent for one vehicle and up to 20 per cent for the second.[17] Both Scania and DDC have provided full warranty coverage for the vehicles.

Diesel–ethanol blends

Although ethanol's ability to blend with diesel oil is not as good as with gasoline, ethanol can be emulsified with diesel oil and the resulting blend can be used in a standard diesel engine. International experience has shown that although diesel–ethanol blends can contain up to 15 per cent ethanol, a good compromise in terms of fuel economy, performance, driveability and emissions can be achieved with about 7 per cent. In this case, an average fuel economy loss of approximately 2 per cent might be expected. Also, depending upon power train characteristics and in-service conditions, operational performance might be lowered slightly.

Diesel–ethanol blends and the particular additive package used to prepare the emulsion need to be evaluated carefully before use since some fuel delivery systems, such as the rotary-type fuel pump, may be very sensitive to the presence of ethanol and may suffer premature wear. Moreover, some parts, such as fuel filters and fuel lines, may need to be made ethanol compatible. Special care with water contamination should also be taken to avoid phase separation since the entry of dissociated water into the combustion chamber may damage the engine.

Dual-fuel operation

One approach that uses diesel and ethanol simultaneously, without having to blend the fuels, is 'fumigation', whereby a carburettor, fuel injector, heated vaporizer or mist generator is used to meter ethanol into the engine's air intake manifold. Fumigation of up to 50 per cent ethanol has been reported.[18] Injection of ethanol directly into the cylinder of an engine with an increased compression ratio and a glow plug to assist with ignition is another approach that has been tested, showing that up to 90 per cent diesel displacement could be achieved.[19] In both cases, an additive such as nitride glycol may be required to allow for lubrication of the mechanical moving parts. Both fumigation and direct injection require an additional fuel-handling system for ethanol that results in incremental hardware costs. An interesting characteristic of these technologies is that both permit running completely on diesel fuel in the event of a disruption in the ethanol supply.

Commercial success of these technologies has been limited, mainly due to the complexity of the existing systems. However, advances in on-board electronics, sensors and digital engine operation mapping, in addition to the possibility of precise ethanol metering at selected engine operating modes, could help to simplify the hardware, optimize the benefits of ethanol use and reduce costs.

Conversion to spark-ignition engine

Diesel engines can be converted to spark-ignition engines to enable them to run on ethanol, as has been experienced on several occasions. However, the standard conversion practice, which typically requires a significant reduction in the compression ratio, is not desirable because it significantly lowers the combustion efficiency. Nevertheless, as discussed earlier, a new approach to conversion with a high compression ratio and port injection typical of diesel applications would be advantageous and could foster commercial development of such conversions.

BIODIESEL

Fuel-grade biodiesel is usually defined as a methyl or ethyl ester derived from transesterified vegetable oil or animal fat that conforms to industry specifications. The European Union (EU) standard EN 14214 and the US standard ASTM D 6751 have become the international references for the fuel, though a considerable number of other national references exist, as well.

One characteristic of biodiesel that must be addressed adequately by users is the fact that it oxidizes much faster than ordinary diesel. Proper care is needed to avoid premature ageing during storage. Moisture can also be a problem, resulting in bacteria growth and the formation of corrosive, free fatty acids that could have a negative effect on the fuel injection system and the engine itself. Limiting exposure to airborne moisture and water deposits and using suitable additives improves biodiesel's ability to withstand long-term storage.

Biodiesel has been used in two ways in compression ignition engines: as a blend with ordinary diesel or as a straight fuel. Like ethanol, biodiesel has solvent properties that break down deposits in the fuel supply system; thus, fuel filters may face the risk of premature clogging. Preventive measures, such as those discussed earlier for ethanol, are worth considering.

Biodiesel as blending agent

Biodiesel mixes easily and completely with ordinary diesel fuel at any concentration. In the US, the 20 per cent blend (B20) has been a very popular option, while in France the 30 per cent blend (B30) is preferred because of its greater capability to reduce harmful emissions. Most diesel vehicles are able to run on blends of up to B20 with few or no modifications (e.g. substitution of parts containing certain plastics or rubber-like materials), particularly if the vehicle was manufactured after the mid 1990s. As with ethanol, blending has been the easiest and lowest cost way of introducing biodiesel in the marketplace.

The automotive industry prefers blends of up to 5 per cent biodiesel content (B5) for use in existing vehicle fleets because it enhances lubricity, especially of ultra-low sulphur diesel. In some countries, such as France, all diesel sold routinely contains up to B5.[20] Regarding warranty coverage, most original equipment manufacturers (OEMs) tell their customers that use of up to B5 is acceptable as long as the pure product conforms to an approved quality standard. Many OEMs fear that higher blend levels could degrade fuel lines, filters, o-rings and seals, and damage fuel-injector orifices, among other potential problems – resulting in leaks and faulty engine operation. Although some OEMs leave the risks of biodiesel use to the customer's discretion, others have threatened to void warranties in the event of problems attributable to biodiesel. OEM advice on using biodiesel blends varies widely, however. While some consider a 20 per cent blend (B20) acceptable, others will deem anything up to 100 per cent biodiesel (B100) acceptable.

Another concern expressed by the automotive industry is the higher viscosity of biodiesel, which at B20 or higher could affect fuel flow and fuel spray in the combustion chamber, particularly in colder conditions. If proper care of fuel handling and use is adopted, however, no problems should be experienced.

Biodiesel's high cetane value (ability to ignite under compression) is considered an advantage because this has a pronounced effect on combustion quality and, thus, on noise and emissions reduction. Mixing biodiesel with ordinary diesel adds cetane value to the resulting blend.

Despite industry worries and inconsistent warranty coverage, consumption of biodiesel blends has increased steadily, mainly in the EU (see Chapters 1 and 20). As a result, a growing number of manufacturers have been marketing vehicles equipped with biodiesel-compatible parts and engines.

Depending upon product characteristics, which vary based on the type and purity of the feedstock used and upon the production process itself, biodiesel shows up to 12 per cent lower energy content than ordinary diesel. Therefore, a slight drop in fuel economy and performance might be expected. A comparative study found that, on average, B20 would result in a decline in fuel economy in the range of 0–6 per cent, while a marginal loss of 2 per cent in performance was observed.[21] For B5, the change in fuel economy and performance is marginal, in the range of 0–2 per cent, and is usually not noticeable.

User feedback suggests that maintenance requirements for diesel engines operating on biodiesel blends of B20 or less are identical to those operating on standard diesel.[22]

Straight biodiesel

Straight biodiesel (B100) can be used in existing diesel vehicles; however, it may require modification of engine or fuel system components, as well as some fine tuning. Biodiesel-compatible materials such as Viton[23] are required for use with B100 unless specified, particularly in fuel hoses, pump seals and gaskets. Tuning of fuel injection timing may also be required, depending upon engine characteristics. Typically, timing needs to be retarded by 1 to 2 degrees to avoid rough engine operation and to minimize nitrogen oxide (NO_x) emissions.

Because of its high viscosity, B100 is better suited to warm climate conditions; however, the introduction of fuel tank heaters and anti-gel additives has made B100 use possible in very low temperatures.

For existing vehicle fleets, fuel economy with biodiesel may be lowered by as much as 15 per cent relative to ordinary diesel, while power loss may drop by 7 per cent.[24] It should be noted that these figures can be improved for new customized vehicles. Maintenance is generally similar to or less demanding than for ordinary diesel use. In fact, B100 use results in decreased soot deposits in the injectors and combustion chamber. Fuel filter life and oil changes can also be extended somewhat.

OTHER BIOFUELS

Straight vegetable oil (SVO)

SVO refers to either new or waste vegetable oil, both of which can be used in diesel engines. Due to its relatively high viscosity (approximately 12 times higher than ordinary diesel), using SVO in unmodified engines can result in poor atomization of the fuel in the combustion chamber, incomplete combustion, coking of the injectors, and accumulation of soot deposits in the piston crown, rings and lubricating oil.

If SVO is to be used in conjunction with diesel in a dual-fuel mode, necessary modifications include an additional fuel tank for SVO, a system to allow for switching between the two fuels, and a heating system for the SVO tank and lines if the vehicle operates in temperatures below 18°C. Under this configuration, the engine is started on diesel and switched over to SVO as soon as it is warmed up. It is then switched back to diesel shortly before being turned off to ensure that it contains no SVO when it is restarted.[25]

Another alternative is to use SVO exclusively. Modifications would include an electric pre-heating system for the fuel (including lines and filters), an upgraded injection system and the addition of glow plugs in the combustion chamber since vegetable oil is not highly flammable. The modification can be expensive, reaching a cost of €2000 (US$2420) or more.[26]

Dimethyl ether (DME)

Dimethyl ether (DME) is a gaseous fuel that can be produced from a variety of sources, including biomass feedstocks such as wood or black liquor synthesis gas. Interest in the fuel is rising in Japan and Europe since DME is considered a promising substitute for diesel or liquefied petroleum gas (LPG). The vapour pressure of DME is similar to LPG, and it can be contained in a gas cylinder at low pressure because of its ability to change into liquid at approximately 0.5 megaPascals (approximately 72 psi) and room temperature. Despite these physical similarities, however, DME differs significantly from LPG in its ability to dissolve most sealing materials used in normal automotive applications. Its application therefore requires use of metal-to-metal sealing or other materials such as Teflon or graphite.

Experience indicates that DME is an environmentally advantageous substitute for diesel, particularly in buses and urban delivery trucks. DME has a relatively high energy density (about 80 per cent that of diesel), a high cetane value (55–60) and near-zero sulphur and aromatics content. Combustion of DME in compression ignition engines reduces engine noise levels and emissions considerably, although further NO_x reduction may be necessary with after-treatment exhaust gas systems. Significant engineering efforts have been undertaken to customize engines to the fuel, particularly with regard to materials choice, modification of the fuel-injection system, and ignition advance. Due to its higher energy density, the driving range with DME can be expected to be higher than with compressed natural gas (CNG). Road trials are under way to evaluate the durability, performance and practicability of DME propulsion systems.

Large-scale DME production from biomass sources has yet to be accomplished, particularly since the costs are high relative to natural gas.[27] But this may change in the future. On the fossil fuel side, French oil company Total is working with a consortium of nine Japanese companies towards large-scale DME production (6000 tonnes per day) from natural gas by 2010.[28]

Biomass-to-liquid (BTL) fuels

Biomass-to-liquid (BTL) fuels can be produced through gasification of biomass and the Fischer-Tropsch (F-T) chemical reaction process (see Chapter 5). A suitable alternative to the use of solid biomass feedstocks would be to use biogas (produced from the anaerobic digestion of wet biomass feedstocks, such as animal manure) following a similar route of natural gas processing known as gas to liquid (GTL). The by-products of the process include naphta, diesel and chemical feedstocks. The resulting BTL/GTL diesel can be used as a straight fuel or blended with ordinary diesel or biodiesel.

European companies are particularly interested in this concept since diesel oil accounts for a rising share of the continent's vehicle fleet. A cooperative effort is already in place in Germany involving Volkswagen, DaimlerChrysler and Choren Industries GmbH, and has resulted in an experimental plant producing BTL diesel from forest waste wood under the brand names SunDiesel and SunFuel. Industry sources believe BTL diesel could become a viable commercial fuel by 2010 (see Chapter 5).

Because BTL/GTL diesel is considered a premium quality fuel due to its very low sulphur concentration (less than 1 part per million), near-zero content of aromatics and toxic substances, and high cetane value (above 70), it could boost the technological prospects for high-efficiency and ultra-low emission engines. With BTL, some companies see the possibility of combining the Otto (gasoline) and Diesel thermodynamic cycles into an optimized combined combustion system, although this concept is still at the research level.

Biogas

Biogas is usually obtained from local waste streams such as sewage treatment plants, landfills, organic industrial waste streams and digested organic waste. Due to its relatively low methane content (typically 60–70 per cent) and high concentration of contaminants, biogas is normally unsuitable for use in vehicles. To be fit for automotive use, its quality needs to be upgraded to as close to CNG as possible. Treatment generally includes removal of carbon dioxide, water and minor contaminants in order to achieve a minimum methane content, preferably 95 per cent.[29]

Treated biogas (TB) can be used in both heavy- and light-duty engines suitable for natural gas. A number of vehicle options can therefore be considered: dedicated CNG or liquefied natural gas (LNG) vehicles, bi-fuel (gasoline/CNG) vehicles, and dual-fuel (LNG/diesel or CNG/diesel) vehicles. Sweden is currently the leading user of biogas in transportation. Its CNG/biogas urban bus fleet comprises approximately 4500 vehicles, with 45 per cent of fuel consumption supplied as biogas.[30] Other countries, such as Denmark, Germany, The Netherlands, Sweden and the UK, have also developed experience in this field.

Because TB is so similar to natural gas, engine performance, driveability, emissions and maintenance are considered equivalent. Moreover, no differentiation in warranty coverage is required as long as the TB characteristics fulfil the vehicle manufacturer's requirements.

Methanol

Gasifying biomass is a known method of producing methanol, a product that has already proven itself as a neat fuel or blending agent. Since it has many similarities to ethanol, the technological alternatives for methanol use are similar.

However, due to its higher aggressiveness to materials, higher toxicity and lower energy content, use of fuel-grade methanol has been considerably inferior. Blending has been limited to low levels and, in some cases, is not allowed. Aside from dedicated methanol vehicles produced by car manufacturers, there is only one known case where original equipment manufacturers would provide warranty coverage for methanol use.

During the early 1990s, Brazil witnessed an innovative use of methanol for about three years when a localized shortage of E100 led to the introduction of a blend of 60 per cent E100, 33 per cent methanol and 7 per cent gasoline for use in dedicated ethanol vehicles. The blend has many properties that closely resemble E100 and has resulted in similar performance, driveability, emissions and maintenance.[31]

Ethyl tertiary butyl ether (ETBE)

Ethyl tertiary butyl ether (ETBE) is derived from a chemical process that uses approximately 40 per cent ethanol and 60 per cent isobutylene. It has been used mainly in Europe as a gasoline octane booster and as a partially renewable oxygenated fuel additive. ETBE content in gasoline has generally ranged from 5 to 15 per cent. It is fully compatible with existing engine technologies and gasoline, and has no special requirements for automotive use. Although it has lower energy content than gasoline (about 13 per cent), the impact of its use on fuel economy has been generally marginal. The most notable benefit of ETBE use has been its ability to avoid an increase of gasoline volatility and, therefore, an increase in evaporative emissions.

BIOFUELS AND ADVANCED PROPULSION SYSTEMS

Biofuels can be used efficiently in any of the advanced propulsion systems now reaching the market. These include mainly hybrids and fuel cells. Hybrid vehicle technology that currently relies on fossil fuels (i.e. gasoline, diesel, natural gas and liquefied petroleum gas) may well take advantage of biofuels' growing availability. For instance, a plug-in hybrid vehicle with a high-efficiency dedicated ethanol engine could present superior performance compared with current commercial versions. Depending upon technology characteristics, fossil fuel substitution could reach 100 per cent on a volume basis.

Fuel cell technology could also benefit. The availability of multi-fuel on-board reformers that could continuously generate hydrogen out of methanol, ethanol, DME or TB would enable vehicles to use a combination of conventional and lower-cost fuelling systems. Alternatively, commercial-size multi-fuel reformers could generate hydrogen from biofuels on-site at retail stations, avoiding costly hydrogen distribution infrastructure.

CONCLUSION

Biofuels have a proven record of technical feasibility. Although ethanol and biodiesel have been the front-runners in this emerging market, other alternatives are showing significant potential for fossil fuel substitution. Existing automotive technology is in many ways compatible with biofuels, resulting in affordable and efficient solutions for the use of clean and renewable energy. Hybrids and fuel cells, considered to be the ultimate motor vehicle technologies, will have the opportunity to take advantage of the availability and diversity of biofuels. An extraordinary synergy between biofuels and automotive technology is likely to be seen in the coming years, resulting in more sustainable transportation alternatives.

What combination of technology and biofuel will prevail in the future? At present, there is no clear answer to this question, as both automotive technologies and biofuels production processes are developing at a fast pace. However, one might expect to see conventional and advanced technologies co-habiting for many years, with biofuels increasing their share in the global market. Nevertheless, significant issues still need to be addressed, including educating people about biofuels, promoting incentives for technological progress, and creating the political will to invest in non-conventional fuel sources.

16

Transfer of Technology and Expertise

INTRODUCTION

The development of more efficient and reliable biofuel technologies, and the transfer of this technology and expertise around the world are crucial to the expansion of the global market for these fuels. This chapter describes the process of technological change and transfer, with specific application to biofuels. It discusses the main areas where technologies can help to boost the development of larger biofuel markets, the role of technology in the development of national biofuels strategies and the roles of various stakeholders, including governments and the private sector, in the technology transfer process. It concludes with a case study of Brazil's *Proálcool* programme.

THE TECHNOLOGY CHANGE AND TRANSFER PROCESS

The process of transferring biofuel technology and expertise can be best understood as a process of managing technological change. It involves the flows of knowledge, experience and equipment among different stakeholders, including governments, private-sector entities, financial institutions, non-governmental organizations (NGOs), research and educational institutions and labour unions. It encompasses technological cooperation and the diffusion of technologies both within countries, as well as between them. And it involves the process of learning to understand, utilize and replicate existing biofuel technologies – including the capacity to select and adapt them for local conditions and even to sell them back to the original source as improved technologies.

Technology transfer reflects specific actions taken by individuals and organizations. These include:

- acquisition of knowledge and skills by individuals through formal education and on-the-job training;
- assimilation of publicly available knowledge;
- investment and trade decisions made by firms;

- purchase of patent rights and licences; and
- migration of skilled personnel with knowledge of particular technologies.

Technology flows are also influenced by government policies and by financial aid and development programmes. The rate of such flows is affected by the motivations of the relevant stakeholders and by the barriers that impede them – both of which are influenced by government policies, including environmental and climate change policies.

Most technology flows occur in, or are driven by, the private sector (between commercial parties), although they can also involve the government or community. In most instances, managing technological change and transfer follows five basic stages: assessment and identification of needs; agreement; implementation; evaluation and adjustment; and replication (see Figure 16.1).[1]

BIOFUEL TECHNOLOGY CHANGE AND TRANSFER

As biofuels attract greater commercial and political interest on a global scale, the pace of technological change is accelerating. There is ample evidence that learning

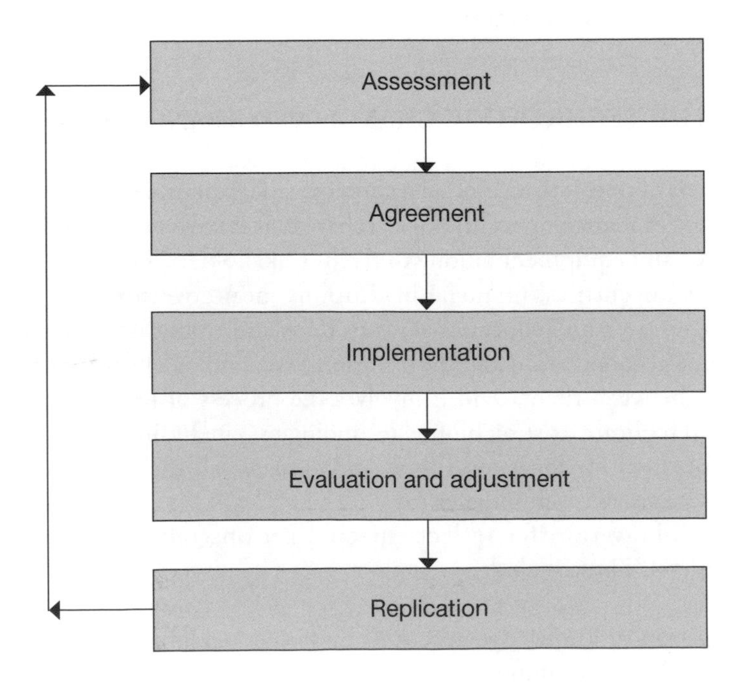

Figure 16.1 *The five basic stages of technology transfer*

Source: see endnote 1 for this chapter

curves for these fuels are evolving, driven by expanding markets, changes in design and development, and new information and communication technologies – resulting in sharply lower operational costs and investment requirements.[2]

Nevertheless, ongoing and future developments could substantially improve the energy and cost efficiency of biofuels, based on the relative efficiency of different farming, harvesting and processing approaches, fuel compositions, and engine technologies. In 2003, Towler et al identified three main areas where additional research would yield the greatest benefits for biofuels to become competitive with conventional energy sources:[3]

1 *Primary productivity.* Overall, improved productivity helps to boost the viability of biomass fuels by lowering crop prices (see Chapter 7).
2 *Crop modification* (increasing processing ability or the quantity of biomass for processing). Use of biotechnology is a possibility, although this raises concerns about the cost of seeds and the impacts of releasing genetically modified organisms into the environment (see Chapter 12).
3 *Conversion process.* New conversion technologies may make the spread of smaller-scale units feasible, thus avoiding the feedstock transportation costs of large plants. The efficiency of biomass gasification and liquid fuels synthesis via Fischer-Tropsch (F-T) processes can increase; but the overall viability would depend upon lower feedstock costs. In all cases, the prospect of co-production of other value-added goods would increase the feasibility of biofuels (see Chapter 5).

Biofuels technological change has led to a change in the role of key stakeholders. The relative importance of farmers and foresters has increased, for instance, because biofuels derive from agriculture and forestry and the cost of these feedstocks is key to the economics of the fuels.

The transfer of biofuels technology is now a global process. It does not just occur from North to South, but also, increasingly, within the developing world, and even from South to North. The technology embedded in Brazilian ethanol distilleries, for example, has been transferred to Costa Rica, Kenya and Paraguay, among other countries. Similarly, technology from Indian distilleries has been brought to Colombia.

Development of a National Biofuels Strategy

Developing a clear national strategy for biofuels implementation provides continuity to the overall biofuels learning curve, including the technology learning curve. For governments deciding to implement national biofuels programmes, it is useful to first consider what drives the key stakeholders in the biofuels economy. It is then necessary to review, study, research and analyse technological and other information that has accumulated over the years. These steps pave the way for getting the relevant stakeholders to reach a consensus and to commit to implementing a biofuels programme, allowing for adjustment and corrections as implementation proceeds.

Figure 16.2 describes the basic elements of a national biofuels programme implementation strategy.[4] In one way or another, these processes have unfolded in every country or region that has embarked on a biofuels programme, such as Brazil, the US, the European Union (EU) and Thailand.

Brazil's *Proálcool* programme, launched in response to the oil crises of the 1970s, provides a good example of the implementation process. Brazil had been blending ethanol into gasoline for the previous half century, in part to protect the sugar industry from the vagaries of the international market. Prior to and during the *Proálcool* launch, stock was taken of both Brazilian and international fuel ethanol technologies. Brazilian consultants and government research institutes were mobilized, as was the country's foreign automobile industry.

In the Brazilian case, the major driver for technology development and transfer was scaling up ethanol production and use. Ethanol technology had to respond

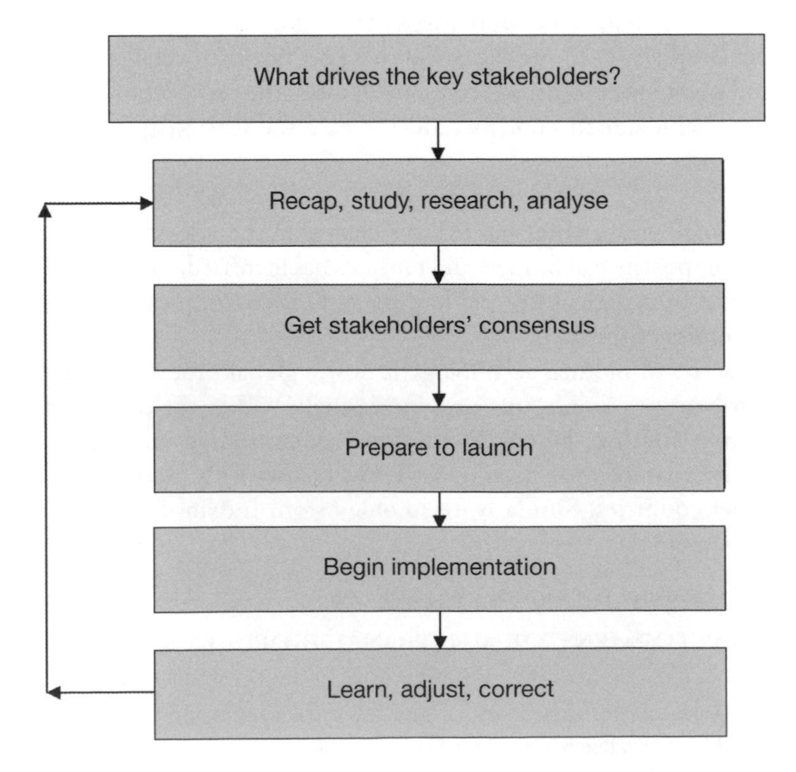

Figure 16.2 *National biofuels programme implementation strategy*

Source: see endnote 4 for this chapter

to a change of market, from the beverage and industrial markets to the much larger fuel markets. Ethanol distillery technologies had to resemble oil-refining technologies. Sugar cane agriculture had to expand to meet scaled-up demand for ethanol, via additional land use and, particularly, through increased yields. And conversion technologies had to be upgraded via technology transfer by working with organizations from numerous countries and within Brazil to improve conversion efficiencies and performance, and to lower ethanol production costs.

The Brazil sugar cane case (described later in this chapter) exemplifies the drivers for technology transfer with respect to oil, rural development and energy diversity. The importance of revisiting the stakeholder consensus, in particular, is evidenced by the disarray of the *Proálcool* programme after the sudden decrease in oil prices during the mid 1980s.

ROLE OF GOVERNMENT AND THE PRIVATE SECTOR IN BIOFUEL TECHNOLOGY TRANSFER

Domestic policies, as well as bilateral and multilateral governmental agreements, can play a key role in promoting change and transfer in biofuels technology. The development of Brazil's sugar cane ethanol technology, for example, relied on agreements between Brazil and a host of other developing countries. The rate of technological change is also closely linked to biofuel trade and investment, an area influenced by both governments and the private sector.

Depending upon their domestic policy choices, governments can either favour or inhibit the development and availability of particular biofuels resources, carriers or end uses. In Brazil, for instance, political action over time helped to rapidly increase biofuels' share in supplying the country's growing energy needs. (Note that during the 1970s and early 1980s, when Brazil established its nationwide network of ethanol stations and vehicles, it was a ruled by a military dictatorship. Military dictatorships have the ability to implement mandates quickly. In 1985, when the price of oil dropped sharply, a democratic government regained power and struggled for about ten years to adjust the national ethanol programme/mandate in a context of low petroleum fuel prices.)

In the US, the new energy law signed in July 2005 has provided impetus to a growing biofuels industry. Likewise, the EU directives on biofuels for transportation and on transportation fuels taxation (both from 2003) provided a boost to the biofuel market for transport in EU member states.[5] The EU has not met the target level for biofuels' proportion of fuel energy outlined in 2003; however, the *Biomass Action Plan* released in early 2006 encourages more demand and supply creation, as well as high priority for research into the biorefinery concept and into next-generation biofuels and continued encouragement of an industry-led biofuels technology. Over the past few years, similar political actions have begun to provide enabling environments for biofuel technologies in Canada, China, Colombia, India, Thailand and many other countries (see Chapter 17).

Governments also play an important role in promoting technological change and transfer through multilateral trade agreements. The Hong Kong meeting of the World Trade Organization's (WTO's) Doha round, concluded in December 2005, if fully implemented, is likely to result in the full liberalization, by 2013, of the EU trade in agricultural-related commodities, including sugar, fuel ethanol and biodiesel. This is likely to promote technology transfer within the EU and to other countries engaged in international trade of fuel ethanol and biodiesel with the EU. Another government initiative that can foster biofuel technology transfer is the International Energy Agency's (IEA's) Task 40, which deals specifically with the international trade of biofuels (see Chapter 9).

Similarly, multilateral environmental agreements highlight the importance of technology transfer to address climate change. Article 4.5 of the United Nations Framework Convention on Climate Change (UNFCCC), for instance, states that participating developed country parties 'shall take all practicable steps to promote, facilitate and finance, as appropriate, the transfer of, or access to, environmentally sound technologies and know-how to other Parties, to enable them to implement the provisions of the Convention'. Article 10(c) of the Kyoto Protocol to the UNFCCC, coupled with the Protocol's Clean Development Mechanism (CDM) is already promoting technology transfer in biofuels, with one example being support for the development of combined-cycle bagasse-fuelled steam and electricity generation in Brazil's sugar and ethanol mills.

Targeted bilateral and multilateral information sharing between countries can also play an important role in information dissemination. For example, Germany and China sponsored a joint workshop in 2003 to share information on new gasification and pyrolysis technologies for producing liquid fuels from biomass.[6]

Notwithstanding the key role of governments, in practice, the actual flow of biofuel technologies takes place in the private sector, particularly in the transportation and power generation sectors. While markets can be very imperfect mechanisms, once suitable macro-economic policies are in place to provide an enabling environment for innovation, markets may determine the choice of technology and the modes of transfer. In the biofuel context, there is ample room for ideas, such as diffusing technologies via transnational commercial networks; business charters for biofuel technological cooperation; biofuel technology investment corporations; and funds financed by private sources in the long term.

One way that government policy can provide an enabling environment for biofuels is by subsidizing biofuel programmes, at least initially. Brazil's *Proálcool* programme was launched with the help of lavish subsidies, which were gradually removed over the course of 20 years. In the US, the use of subsidies has played a key role in fostering ethanol and biodiesel use; these subsidies remain in force today and were extended by the new energy law of 2005. In the EU, the Fuel Taxation Directive allowed member states to remove excise taxes for biofuels for a period of years. In Germany, Spain and Sweden, they were totally removed (for more on the role of subsidies, see Chapter 7).

Other initiatives that can help to bring governments and the private sector together are public–private partnerships. At the community level, the main stakeholders are individual citizens, small-scale enterprises and NGOs concerned with agriculture and forestry.

As the world biofuel market develops, some countries may become either large or surplus producers of these fuels, while others may resort to imports. This could lead to movement of biofuel capital and technologies to the larger producing countries. There is already evidence that this is happening. Novozymes, a Danish company with a strong US presence, is investing in China as that country develops its starch-based ethanol industry, despite concerns over the protection of intellectual property rights. And the London-based biodiesel company D1 Oils Plc is promoting the use of jatropha oil as a biodiesel feedstock in India, the Philippines and parts of Africa. Brazilian companies, meanwhile, are engaged in extensive technology transfer from a variety of countries while developing the *Proálcool* programme. Many of these companies are now selling their own fuel ethanol technologies abroad, getting support from Brazil's government in promoting bilateral cooperation.

LARGE-SCALE BIOFUELS TECHNOLOGICAL CHANGE AND TRANSFER: THE CASE OF *PROÁLCOOL*

Brazil's fuel ethanol programme, *Proálcool*, provides a useful example of large-scale and long-duration biofuel technology change and transfer.[7]

Portuguese colonists introduced sugar cane to Brazil some 500 years ago, with the original goal of producing sugar for domestic consumption and export. Technology transfer was crucial to the crop's development from the very beginning, and imported cultivars were transferred widely both within mills and throughout Brazil. Until as recently as 1980, all of the main varieties originated from other countries: in the central-south, for example, the most common variety was NA56-79 from northern Argentina.

During the 1980s and 1990s, Brazilian agronomists brought in a wide range of technologies from abroad and adapted them for local use. At Copersucar, a major private cooperative of mills in Brazil, the main sources were the South African Sugar Technologists' Association (SASTA) and Texas A&M University in the US. These institutions were crucial to the developments that took place in Brazil, primarily in the areas of agronomy and agricultural engineering.

The work of the Agronomic Institute of Campinas (IAC) and the research and educational services of the University of São Paulo's Agricultural School Luiz de Queiroz (ESALQ) were also important in the agronomy of sugar cane production.

The scientific base provided by these institutions established a knowledge platform from which to absorb previously transferred technologies and to move to the next stage of technology transfer, driven by the emergence of *Proálcool*.

Proálcool's emergence leads to genetic improvements

The emergence of *Proálcool* during the 1970s drove two large varietal improvement programmes that, from 1980 on, led to the annual penetration of new Brazilian sugar cane varieties that surpassed earlier imports. One of these programmes, run by Copersucar through its research centre, Centro de Tecnologia Copersucar (CTC), focused on the needs of sugar cane cultivation in São Paulo (SP varieties). The other programme, Planalsucar (RB varieties), was carried out at experimentation stations of the government's Institute of Sugar and Alcohol (IAA) and was aimed at the north-eastern and southern regions. As a result, there are now no imported commercial sugar cane varieties in use in Brazil since they would not match the quality of domestic cultivars. Imported varieties are used only as potential progenitors for cross-breeding programmes.

The penetration of Brazilian varieties was a key factor in the technological leap in sugar cane cultivation in Brazil. The original germplasm bank was all imported; but the use of transferred technologies to improve these cultivars, including selection and improvement techniques, was crucial. During the 1970s and 1980s, the Hawaiian Sugar Cane Planters Association provided planning support to Copersucar's varietal improvement programme. Copersucar also exchanged information on a regular basis with the Bureau of Sugar Experiment Stations (BSES) in Australia and with US programmes. For example, all the main cultivars of the germplasm bank of Canal Point, Florida, were transferred to Copersucar. As part of the transfer process, workshops were developed with the specific objective of evaluating the Copersucar programme, engaging with research groups from Australia, South Africa and the US.

With regard to genetic changes, Brazilian researchers, trained in basic techniques in the US, implemented the Brazilian genetic improvement programme, initially at Copersucar. Copersucar also led an international consortium of groups from Australia, South Africa, the US and other countries. In Brazil, these developments led, for the first time ever, to the genetic mapping of the sugar cane plant in 2001. This effort was financed by the Research Support Foundation of the State of São Paulo (FAPESP), among others.

Today, Brazil is home to more than 500 transformed sugar cane varieties not yet in commercial use. Roughly 80 per cent of the country's sugar cane fields are planted with up to 20 of these varieties, according to the São Paulo Association of Sugar and Alcohol Manufacturers (UNICA). Today, this programme engages many industrial and university partners and promotes intensive exchanges among them.

Foreign and domestic technology transfer

Over the years, Brazil's sugar factories improved their technology through the use of imported equipment, as well as by taking advantage of the smaller incremental adaptations of Brazilian developers. The local capital goods industry also added to improved knowledge on key processes and equipment.

An interesting example of technology transfer within Brazil, established during the 1980s and still ongoing, is the Centre for Absorption and Transfer of Technology (NATT) in the north-eastern state of Alagoas. As Copersucar's CTC research centre expanded the scope of its work in response to the needs of central-south Brazil, a group of Alagoas sugar cane producers created NATT primarily (but not exclusively) to absorb CTC technology and to transfer it to Alagoas mills.

Since the launch of *Proálcool* in 1975, relevant stakeholders in Brazil have sought to obtain knowledge to increase the productivity of the sugar cane economy – about everything from the sugar cane plant itself to ethanol-making and end uses. During the late 1970s, with financial support from the Research Financing Foundation of Banco do Brazil (FIPEC), Copersucar sent a dozen Brazilians to Mauritius for one year to learn sugar and ethanol production. Upon their return, this group became the core of CTC's industrial unit during the 1980s. The integration of this group with CTC agricultural groups, adopting the biorefinery approach, was a key factor in CTC's success.

CTC's industrial unit contributed substantially to the absorption of foreign technologies via contracts with foreign and Brazilian companies, consultants, research centres and universities. In the area of ethanol fermentation, for instance, extensive work on yeasts, biocides and fermentation protocols was carried out with the Tropical Foundation of Technological Research André Tosello (FTPT) in São Paulo and with the Gulbenkian Foundation of Portugal. Training took place in the laboratories of the Centre National de la Recherche Scientifique (CNRS) in Toulouse, France. Consulting support was obtained from UNDGA, and research on the optimization of distillation systems was done in France. Other projects included consulting with Karlsruhe University in the area of infection-free fermentation and contracts with the German Research Centre for Biotechnology (GBF) in Braunschweig on new product fermentation.

Capital goods technologies

Brazil's sugar and ethanol capital goods industries incorporated knowledge from a variety of origins to develop equipment and machinery in all areas of sugar cane transformation, from sugar and ethanol production, to electricity generation, to milling. Sugar cane juice extraction technology, for instance, benefited substantially from foreign inputs, although this was always adapted in some critical way that made the resulting Brazilian version superior to the imported one. This process started with the migration of a group of South African consultants to Brazil during the late 1970s, which led to development of the Brazilian roller mill, and was later continued during the 1980s with the support of Australian consultants.

The technology for dehydrating hydrous ethanol, too, has evolved significantly in Brazil, progressing from a process of azeotropic distillation using benzene, to the use of cyclohexane, to the use of molecular sieves and ethylene glycol dehydration. In each case, Brazil's capital goods industry transferred and adapted new technologies to serve the Brazilian and export markets. Technologies brought in directly by the foreign capital goods industry (established in Brazil) also played a role in these developments.

End-use technologies

At the same time that technological developments were taking place in the agriculture and conversion stages of ethanol production, Brazil was developing and transferring end-use technologies, as well. The Air Force Technology Centre (CTA) and the country's private automobile industry developed new technologies that enabled the efficient use of ethanol in gasoline blends, hydrous ethanol and, more recently, blends of any composition of hydrous ethanol and gasoline (actually a 25 per cent ethanol blend by volume) in the country's new flexible-fuel vehicles (FFVs).

With support from CTA, Brazil's auto industry was able to successfully adapt the existing spark-ignition engine technology to use ethanol–gasoline blends. The partnership also led to development of the hydrous ('neat') ethanol engine technology in Brazil. This required developing an engine with a higher compression ratio to extract maximum benefit from the higher octane rating of ethanol compared to gasoline (see Chapter 15).

Flexible-fuel engine technology was first developed in the US and The Netherlands, and is based on the use of sensors that recognize the oxygen content of the fuel and trigger automatic adjustments in the air-to-fuel ratio. In the US and Europe, this led to development of the so-called E85 FFV engine, optimized to a blend containing 85 per cent anhydrous ethanol and 15 per cent gasoline. Unfortunately, due to a lack of infrastructure to distribute and store this E85 fuel, most of the 5 million or so E85 vehicles in the US and Canada run on gasoline instead. In Europe, the E85 engine has made limited inroads in the Swedish market and is being proposed for other markets.

Brazil's FFV technology, in contrast, relies on sensors derived from Italian, German and US technologies, and can use any blend of hydrous ethanol and gasoline (actually a 25 per cent ethanol blend by volume). Since every gas station in Brazil has at least one pump dispensing hydrous ethanol (thanks to prior government initiatives), there is no infrastructure constraint to market growth of Brazilian FFVs. In fact, the Brazilian market has responded so favourably to this new technology, which was adapted by the auto industry to Brazilian conditions, that today more than 70 per cent of the incremental auto output in the country is comprised of FFVs. The downside of this success is that the fuel consumption per vehicle is not optimized for the varied compositions of the blend that the consumer makes when filling up at gas stations. As a result, excessive fuel consumption is taxing the hydrous domestic ethanol supply and limiting the prospects for exports.

Infrastructure technologies

Brazilian mills and distilleries, as well as the oil company Petrobras, played a key role in developing the storage and distribution infrastructure for ethanol throughout the country. With the drive for ethanol exports, mills and distilleries invested in expanding the storage and loading capacities at ports such as Santos, Paranaguá, and Maceió.

Petrobras and other oil distributors in Brazil developed and transferred blending technologies to enable the distribution, storage and use of ethanol–gasoline blends with up to 25 per cent ethanol content. Petrobras also had to invest in technological developments to transport ethanol in pipelines due to the presence of water in pipeline systems. Brazil is the only country in the world that uses gasoline and other clear product pipelines to move ethanol (see Chapter 14).

Environmental technologies

Many developments have helped to promote the environmental sustainability of Brazil's ethanol economy. With respect to air quality, government policies favoured ethanol because of the fuel's high octane value, which eliminated the need to use the pollutant tetra-ethyl lead (TEL). Brazil was one of the first countries in the world to completely phase out TEL during the early 1990s. This same substitution approach could be transferred to other countries that still use TEL, as well as to the aviation industry, which obtains its high octane from TEL.

In the agricultural arena, the technology of harvesting sugar cane without engaging in prior field burning, now being adopted on a large scale in the state of São Paulo, could be transferred to elsewhere in Brazil, as well as to other countries still using this practice. The practice of returning vinasse (the large-volume liquid residue from sugar cane processing) to the fields as a combined source of fertilizer and irrigation water could also be transferred.

Studies of the carbon-recycling capacity of sugar cane-based ethanol have demonstrated that, compared to gasoline, fuel ethanol contributes substantially fewer greenhouse gas emissions (see Chapter 11). The major reason for this is that bagasse is the only fuel required to convert sugar cane into products and into surplus electricity that can be sold to the grid (in contrast, fuel ethanol made from corn in the US, and from sugar beets and wheat in Europe, resorts to fossil fuels to provide the required process energy). There are still many sugar cane-based operations in the world that could benefit from the transfer of Brazilian bagasse utilization technology.

Government policies and international cooperation

Governmental policies – translated into legislation and regulations – have provided an important enabling environment for investments in ethanol production in Brazil. Over the years, policy tools have ranged from direct subsidies to mandates making the use of anhydrous ethanol blends mandatory (by specifying gasoline as an ethanol-containing blend). The use of government banks to finance the development of sugar cane and ethanol production was also crucial to the success of the *Proálcool* programme and provided a level of comfort for investment to take place (on the international level, the World Bank extended financing to *Proálcool* during the late 1970s and again during the early 1980s, with part of the loan earmarked for technology development and transfer).

Proálcool drove technology transfer and change as the ethanol industry moved from meeting the needs of the beverage and industrial markets to satisfying Brazil's growing demand for transportation fuels – a much larger and more competitive market that required new technologies. These technological developments have played a vital role in lowering the costs of ethanol production in Brazil and making it competitive with gasoline when the price of crude is higher than €25 to €29 (US$30 to $35) per barrel.[8]

As a result of Brazil's successful implementation of *Proálcool,* other countries have become interested in learning from this experience and engaging in bilateral cooperation. Brazil has signed memoranda of understanding on technology transfer with countries in Central America, as well as with China, the Dominican Republic, Haiti, India, Nigeria, Thailand and Venezuela. In some countries, Brazilian ethanol is being imported to develop domestic markets while local production ramps up.

CONCLUSION

Because technology is typically more than just a piece of hardware or a set of ideas, it is not always easy to replicate another country's experience with technological change and transfer. One of the sources of Brazil's biofuel success, for example, has been the country's strong foundation of research, education and training, a capacity platform that required sustained effort over time to establish and maintain. This situation may not be easily found in other countries (particularly developing countries), although it does point to policy initiatives that they could consider. The willingness of Brazil's private sector to bring in technology from abroad and to adapt it to local conditions was critical to the development of *Proálcool.*

The ability to then resell the imported technologies, with added adaptations and improvements, reflects the existence of a strong capacity platform.

Countries with the elements of such a system of innovation in place will be able to move more rapidly along the biofuel pathway. This includes, among other things, having well-developed research, education and training; standards and norms; quality control; engine repair services; financing; venture capital; marketing; capital goods industries; openness to technology transfer; and foreign investment and trade. However, even if a country's system of technological innovation is well developed, variations in climate, water availability and agricultural potential could still hamper replication.

One way in which Brazil (and other biofuel leaders) can stimulate biofuel technology transfer abroad is through bilateral technological cooperation, supported by government diplomacy and implemented by the private sector. A basic requirement for developing a sustainable international market for a biofuel is the promotion of sustainable domestic markets in countries where it makes sense. Thus, Brazil has promoted bilateral cooperation on ethanol – through memoranda of understanding with China, Colombia, Central America, the Dominican Republic, Haiti, India, Thailand and Venezuela, among others.

As Brazil's experience demonstrates, the transition to a bio-based energy future offers plenty of strategic opportunities for private, public and multilateral capital to finance development, not to mention carbon recycling.[9] Any country interested in biofuel development would do well to consider the strategic approach suggested in Figure 16.2 of this chapter, and to decide through the interactions of relevant stakeholders on the best way to move forward and manage technological change and transfer.

Part VI

The Policy Framework

17

Biofuel Policies around the World

INTRODUCTION

Policy instruments are vital to the development of strong biofuel industries. If governments and others wish to significantly expand production and use of these fuels at the domestic and global levels, they will need to have an effective 'toolbox' of wide-ranging policy strategies. The most common policies supporting biofuels today are blending mandates and exemptions from fuel taxes. Other policy instruments have included loan guarantees; tax incentives for agriculture and forestry, consumers and manufacturers; preferential government purchasing policies; and research, development and demonstration funding for current and next-generation biofuels and technologies.[1]

Although governments adopt biofuel policies for a variety of reasons, the main driver, to date, has been to advance economic development in rural areas and create jobs. Subsidies for these fuels have been justified as indirect aid to domestic agriculture, and farmers increasingly recognize the market potential of energy crops as added sources of income. In parallel, governments have been motivated by a desire to reduce dependence upon foreign oil and to minimize the associated security and economic costs. Governments that have ratified the Kyoto Protocol are also promoting biofuels as a way of meeting national or regional greenhouse gas (GHG) emissions reduction targets, as the transportation sector accounts for a growing share of energy-related emissions linked to global climate change (approximately 25 per cent today).[2]

As awareness of the potential of advanced biofuels grows, new policy instruments are emerging to facilitate their market development. Research investments sponsored by the US Department of Energy (DOE), for example, recently led to a 30-fold reduction in the cost of producing enzymes used in cellulosic ethanol production, a major advance towards commercializing this technology. Researchers in several countries are also working on 'co-product' development, using bio-based resources to produce biofuels, as well as additional marketable products. And many countries are moving towards more sustainable approaches in their biomass planning processes, including Brazil's gradual phase-out of burning in sugar cane harvesting and Malaysia's development of *Sustainable Palm Oil Principles* in response to environmental concerns about palm oil production.[3]

It is clear from existing experience that the policies that governments adopt, and the specific ways in which these policies are designed and implemented, will be critical to how the biofuel industry develops and what impacts (positive or negative) it will have. This chapter describes the range of policies that have been used, to date, to promote biofuels at the national and international levels. The emphasis of the chapter is on market creation, with a brief analysis of which policies have been most effective thus far. Further discussion of specific types of policies, including quality and sustainability standards and certification systems, are found in other chapters and in the final recommendations of this book (see Chapter 19).

REGIONAL, NATIONAL AND LOCAL POLICIES

Several regions and countries have implemented targets, policies, standards and action plans that aim to boost biofuel production and consumption substantially in the coming decade. Table 17.1 highlights selected national, regional and state fuel-blending targets and mandates for ethanol and biodiesel.[4] Greater detail on these programmes in specific regions or countries is provided later in this chapter.

Africa

Several African countries currently have biofuel policies in place, some of which date back to the 1970s (shortly after Brazil began its *Proálcool* programme). Three of the first countries to experiment with ethanol – Kenya, Malawi and Zimbabwe – have had very different experiences with developing the fuel. South Africa and a handful of other African countries are also expanding biofuel efforts.

Malawi is the only country outside of Brazil that has been blending ethanol continuously on a national basis for more than 20 years.[5] The price of the fuel has been pegged to gasoline, with an incentive of 5 per cent or more, depending upon the volume of ethanol blended.[6] Zimbabwe, in contrast, initially used a cost-plus basis formula, offering the national oil company a 5 per cent incentive over the cost of ethanol production, although later the price of ethanol was pegged to the price of gasoline. Periodic droughts caused Zimbabwe to halt its blending during the early 1990s.[7] Kenya had a single distillery that produced ethanol from molasses in the 1980s, providing fuel to Nairobi; but it suffered setbacks due to low government-controlled retail prices, inadequate plant maintenance and operation, resistance from local subsidiaries of multinational oil companies, and unfavourable exchange rates.[8]

In South Africa, where the Sasol company is a leading producer of synthetic ethanol from coal, there is now a movement towards ethanol production from crops.[9] In 2006, a biofuel pilot project was under development in the Eastern Cape.[10] The South African government has developed a national biodiesel standard, based on the European Union (EU) standard EN 14214, and is expected to enact a fuel-blending mandate (1–3 per cent biodiesel) by the end of 2006.[11]

Table 17.1 *Selected regional, national and state biofuel mandates or targets*

Country or region	*Fuel*	*Mandates or targets*
Australia	Biofuel	350 million litres by 2010
Brazil	Biodiesel	2% of diesel by 2008; 5% by 2013
	Ethanol[a]	20–25% of all gasoline (current)
Canada:		
Ontario	Ethanol	5% of gasoline by 2007
Saskatchewan	Ethanol	7% of gasoline as of April 2005
China:		
National	Ethanol (corn)	2.5% of gasoline by end of 2005[b]
Jilin	Ethanol (corn)	10% of gasoline from October 2005
Colombia	Ethanol	10% in all cities of more than 500,000 people
European Union	Biofuels	2% of motor fuel by 2005[c]; 5.75% by 2010 (targets)
Austria	Biofuels	2.5% of all motor fuel by October 2005; 5.75% by October 2008
France	Biofuels	7% of motor fuel by 2010; 10% by 2015
Germany	Biofuels	2% of gasoline from 2007 to 2009
	Biodiesel	4.4% of conventional diesel, from 2007
Sweden	Biofuels	Eliminate use of fossil fuels by 100% by 2020
India	Ethanol	10% ethanol blending (E10) in 9 of 28 states and 4 of 7 federal territories (all sugar cane-producing areas) starting in 2003[d]
	Biodiesel	5% of diesel fuels, no set date
Japan	Biofuels (or gas-to-liquid fuels)	20% by 2030 (target)
Malaysia	Biodiesel (palm oil)	5% of diesel by 2008
Philippines	Biodiesel (coconut methyl ester, or CME)	1% CME for all government vehicles (began in 2004); 1–5% of diesel from CME biodiesel blends by 2006–2014
Thailand	Ethanol	10% gasoline blend to replace conventional gasoline by 2007
	Biofuels	10% of all motor fuel by 2012
US:		
National	Ethanol	28 billion litres (7.5 gallons) of ethanol to be produced by 2012
Hawaii	Ethanol	At least 85% of gasoline must contain 10% ethanol by April 2006
Minnesota	Ethanol	20% of gasoline by 2013 (up from current 10%)
	Biodiesel	2% of diesel as of October 2005
Montana	Ethanol	10% of gasoline

Notes: [a] Here, ethanol feedstock is sugar cane unless otherwise noted.
[b] Chinese provinces have had to suspend blending mandates due to ethanol shortages.
[c] This target applies to all member states of the EU. However, member states may choose targets that go further than the European target. The actual share achieved as of February 2006 is approximately 1.4 per cent.
[d] Due to poor cane crop yields during 2003 to 2004, India had to import ethanol in order to meet state blending targets, and has had to postpone broader targets until sufficient supplies of domestic ethanol reappear on the market.

Source: see endnote 4 for this chapter

Several other countries across Africa have enacted (or are in the process of enacting) initiatives to expand the production and use of biofuels, including Ghana, Ethiopia and Benin.[12] A few African countries, including South Africa and the Democratic Republic of Congo, currently export ethanol to the EU under the General System of Preferences (GSP) and Everything but Arms (EBA) agreements (see Chapter 9). It is possible that future policies will be designed to meet not only domestic needs, but also the growing international demand for biofuels produced in Africa.

Asia and the Pacific

Rapid population and economic growth in Asia and the Pacific are resulting in higher energy demand. To meet rising transport fuel needs, several countries in the region are implementing policies to accelerate biofuel expansion. China, for example, has promoted ethanol on a pilot basis since 2001 in five cities in its central and north-eastern regions – Zhengzhou, Luoyang and Nanyang in Henan Province, and Harbin and Zhaodong in Heilongjiang Province.[13] The country's Renewable Energy Law, endorsed in February 2005, lays out a biofuel policy framework that increases biofuel targets from the present level of 3 per cent of renewable energy to 10 per cent by 2020.[14] The province of Jilin, home to the world's largest corn ethanol distillery, offers tax breaks, low-interest loans and subsidies to compensate for the price differential between gasoline and ethanol.[15] Jilin's Tianhe distillery is producing more than 900 million litres (240 million gallons) of ethanol annually, operating only at about 75 per cent capacity.[16]

India's government developed a Draft National Biofuel Policy to promote biofuels in 2003.[17] The national government has mandated the use of E5 in nine states since 2003 and enacted an excise duty exemption for ethanol; however, due to droughts and crop failures, it has been unable to keep pace with the mandate using domestically produced ethanol.[18] In addition, the Indian National Bank for Agriculture and Rural Development has provided refinancing (at 100 per cent) to banks at a concessional interest rate for development of wasteland, helping non-governmental organizations (NGOs) and research organizations spread awareness about biofuels through demonstration projects and supporting state government initiatives to cultivate biodiesel crops, such as pongamia.[19] India has significant potential for producing biodiesel from jatropha and is working to expand its production and use of biofuels, particularly in the poorest areas (see Chapter 21).[20]

Elsewhere in Asia, Thailand's government announced in May 2005 that it plans to spend €16.5 billion (US$20 billion) over the next four years on energy and conservation programmes. The government also plans to phase out the gasoline additive methyl tertiary-butyl ether (MTBE), which currently comprises 10 per cent of gasoline blends, and to replace it with ethanol.[21] In the Philippines, government vehicles are required to use a 1 per cent biodiesel blend, and the government is deliberating the passage of a 1 per cent national biodiesel requirement that will increase to 5 per cent.[22] Ethanol is also mandated under the National Bioethanol Programme, starting with the use of 5 per cent blends in gasoline from 2007 to 2010 and 10 per cent blends from 2010 to 2017; this is scheduled to displace a total of 3.7 billion litres of gasoline over a ten-year period.[23]

Japan has set an ambitious goal of replacing 20 per cent of its oil demand with biofuels or gas-to-liquid (GTL) fuels by 2030.[24] In the nearer term, Japan has proposed a target of 500 million litres (132 million gallons), or about 1 per cent of projected fuel use, by 2010. To facilitate market development of ethanol, the government proposed an E3 standard in 2004 as a lead-in to a national E10 blend standard by 2010 (this standard may be substituted with ETBE blending).[25]

Australia's government has supported ethanol since 2000 with a variety of tax incentives and production subsidies.[26] The Australian biofuel industry is provided a lower excise rate than petroleum fuels, a production subsidy for domestic biofuels and capital grants to help cover the investment costs for new production facilities.[27] In addition, the government has established a target to increase production and use of biofuels to 350 million litres by 2010, a level that might well be surpassed under a new action plan. As of early 2006, more than 400 service stations around the country were selling ethanol and biodiesel blends.[28]

European Union

The EU has had a regulatory framework in place to promote biofuels since the early 1990s. For example, the Common Agricultural Policy (CAP) included production quotas for oilseed food crops (the so-called Blair House Agreement) for EU member states, as well as exemptions from certain taxes, and explicitly granted permission to grow non-food crops on set-aside lands. In 2003, the EU issued a directive stating that all member states should set national targets for the use of biofuels in the transport sector of 2 per cent by 2005 and 5.75 per cent by 2010.[29] As a result, most member states have developed national biofuel plans, and several are providing substantial tax relief to promote biofuel production – a direct consequence of the 2003 Transportation Fuels Fiscal Directive.[30]

This directive provides certain fuel tax exemptions for biofuels to enhance their market competitiveness. For example, Sweden and Spain grant 100 per cent tax relief for biofuels.[31] However, this varies greatly in other EU countries, creating the need for greater harmonization of energy tax laws within the EU to facilitate the development of alternative fuels.[32]

Despite these efforts, it has appeared unlikely that the EU targets for 2010 would be met under the 2003 policy framework – the EU market share for biofuels reached only 1.4% by the end of 2005, according to the US Department of Agriculture (USDA).[33]

In December 2005, the European Commission (EC) issued a *Biomass Action Plan* that sets out measures to promote biomass for transportation, heating and electricity through cross-cutting policies that address supply, financing and research. The plan concentrates on balancing domestic production and imports, using ethanol to lower fuel demand and reducing technical barriers. It states the EC's intention to propose a strategy with an integrated approach to reducing carbon dioxide (CO_2) emissions associated with the transport sector, including the use of biofuels, fiscal incentives, congestion avoidance, consumer information and improvements in vehicle technology.[34] It also proposes to amend EU standard EN 14214 to facilitate the use of a wider range of vegetable oils as biodiesel feedstocks, and to ensure that only biofuels 'whose cultivation complies with minimum sustainability standards count towards [EU biofuels] targets'.[35] In addition, the plan discusses the need to maintain current preferential market access for developing nations, acknowledges sugar reforms and the need to help developing countries to advance their biofuel markets, and mentions the need to keep these objectives at the forefront of considerations during bilateral and multilateral trade negotiations.[36]

In February 2006, the EC adopted a new and ambitious *EU Strategy for Biofuels*, which builds on the *Biomass Action Plan* to boost production and use of biofuels. It sets out three primary goals:

> *... to promote biofuels in both the EU and developing countries; to prepare for large-scale use of biofuels by improving their cost-competitiveness and increasing research into 'second generation' fuels; [and] to support developing countries where biofuel production could stimulate sustainable economic growth.*[37]

Key policy tools will include stimulating demand, possibly through biofuel obligations, examining how biofuels can best contribute to greenhouse gas emissions targets, and directing research money towards developing the biorefinery concept and next-generation biofuels.

In addition to regional-level policies, several European countries have national programmes to promote biofuels. Austria has established mandatory targets for these fuels combined with tax exemptions, while France has enacted a tendering process that sets a maximum amount of biofuels for the market, with tax reductions for this amount of fuel.[38] Slovenia, the Czech Republic and The Netherlands reportedly have plans to introduce obligations in the 2006 to 2007 time-frame, as does Germany.[39] And the UK is considering a trading system for biofuel certificates, as well as a blending obligation and certification system.

Latin America

Latin America is experiencing tremendous biofuels growth, following the leadership of Brazil in this area. Brazil's success stems from a combination of policies enacted over the years, beginning with the *Proálcool* programme launched in 1975 to reduce dependence upon imported oil. A combination of tax breaks and blending mandates drove investment in ethanol production and use and brought about rapid progress in the nation's ethanol industry. Subsidies to increase sugar production and distillery construction, along with government promotion of all-ethanol cars and development of a distribution infrastructure, also helped to fuel development.[40]

In recent years, Brazil has begun to focus on biodiesel production and use, as well. The government has mandated the use of 2 per cent biodiesel by 2008 and 5 per cent by 2013.[41] In 2004, Brazil issued an executive order and law encouraging biodiesel producers to buy feedstocks from family farmers; the following year, it passed a law that, among other things, exempts from taxes any biodiesel produced by family farms.[42] The oil company Petrobras has begun tendering for biodiesel, facilitating development of the market.

Brazil has a National Agri-Energy Plan that addresses fuel ethanol, biodiesel, agro-forestry residues and cultivated energy forests.[43] In addition, to facilitate the development of a diverse international biofuels trade, Brazil recently signed multiple memoranda of understanding (MOUs) with the governments of Nigeria, Japan, Venezuela, China and India – as well as with private entities in these nations. These MOUs are intended to create frameworks for countries to share technology and to help the latter countries develop, market and trade ethanol-related technologies and expertise. Brazil's aims are twofold: to increase demand for Brazilian biofuels around the world, and to help guarantee reliability of supply in the global marketplace, enhancing private-sector development. For instance, if a drought resulted in lower production levels in Brazil, other countries such as South Africa and India could still supply the market, and vice versa.

Other countries in Latin America have begun to enact biofuel incentives, as well. An ethanol-blending mandate is now in force in some regions of Venezuela, and the government is considering enacting a 10 per cent national blending requirement.[44] Colombia currently requires a 10 per cent ethanol blend in cities with more than 500,000 people.[45] Several other countries in the region also have biofuel initiatives, including Argentina, Mexico, Paraguay and Peru.

North America

Like Brazil, the US first began to seriously promote ethanol in response to the oil crises in the 1970s, primarily through tax policies. More recently, the 2002 Farm Security and Rural Investment Act (the Farm Bill)[46] contains an energy title designed to promote energy efficiency and the development of clean energy from alternative resources that can be produced by the agricultural sector, including biofuels. The title authorizes support for biofuels through a variety of programmes, including biorefinery development grants, biomass research and development (R&D), and federal procurement requirements for bio-based products. However, funding for these programmes has been inconsistent, and several programmes have received reduced funding or none at all. Planning is now under way for a new five-year Farm Bill for 2007, and it remains to be seen whether the government will support such programmes with concrete funding commitments.

Most US federal biofuel incentives, to date, have focused on ethanol. However, the nation's first federal biodiesel tax incentive was enacted as part of the 2004 American Jobs Creation Act (Jobs Bill) to help reduce the price of biodiesel for consumers (the 2004 Jobs Bill also applies to ethanol, extending federal tax credits through to 2010 and expanding the flexibility of these credits so that they apply to any ethanol blend fraction up to 10 per cent).[47, 48] Biodiesel use is also being promoted in the military: as of June 2005, the US Navy and Marine Corps were required to operate non-tactical diesel vehicles on a 20 per cent (B20) biodiesel blend.[49]

In 2005, the US government enacted a new Energy Policy Act (EPAct), the first major energy law adopted in 13 years. It includes several incentives to spur expansion of a biofuel market, including a Renewable Fuels Standard (RFS) that requires the production of 28 billion litres of ethanol by 2012, tax incentives for E85 refuelling stations, biofuels tax and performance incentives, and authorizations for loan guarantees, a bioenergy R&D programme and biorefinery demonstration projects.[50]

In addition to these national provisions, a growing number of US states are enacting policies to encourage market expansion of biofuels, including RFS laws and tax incentives. North Dakota, for example, committed in 2005 to providing up to US$4.6 million (€3.8 million) over two years to facilitate ethanol production, creating a tax incentive for consumers who purchase E85 gasoline, establishing an investment tax credit for ethanol and biodiesel production facilities, and offering income tax credits and other benefits for biodiesel.[51] In the autumn of 2005, New York state launched a *Strategic Energy Action Plan* that includes tax credits up to US$10,000 (€8265) for alternatively fuelled vehicles, depending upon vehicle weight.[52] And in early 2006, New York Governor George Pataki announced an initiative to make renewable fuels tax free and available at service stations throughout the state.[53] Minnesota has enacted the most ambitious mandates in the US thus far, calling for ethanol to represent 20 per cent of gasoline by 2013; it also requires a 2 per cent biodiesel blend. As of early 2006, new initiatives were under way in several other states and at the federal level, as well.

Several provinces in Canada are also promoting the production and use of ethanol through subsidies, tax breaks and blending mandates. Ontario, for example, has enacted a renewable fuels standard of 5 per cent ethanol beginning in January 2007. Manitoba and Saskatchewan have also mandated the blending of ethanol into gasoline.[54] At the national level, Canada aims to replace 35 per cent of its gasoline with E10 blends by 2010 in order to meet commitments under the Kyoto Protocol; this would require the production of 1.2 billion litres (350 million gallons) of ethanol.[55]

POLICY LESSONS TO DATE AND REMAINING BARRIERS

The modern biofuel industry is still relatively young, with little long-term policy experience. Brazil, the US and Malawi have the longest record of support for biofuels, and the experiences in these countries and elsewhere provide valuable lessons on ways to support the nascent industry. It is also important to address remaining barriers – whether policy or institutional – that slow the advancement of biofuels.

Research and development are critical to the success of biofuels; but, as with other renewable fuels and technologies, market creation is the most important force for driving their production and use. There is no example in the world of a country that has established a biofuel market without the use of mandates and/or subsidies, and the combination of these two policy tools has been most effective. But a comprehensive approach to market development is essential. Thus, it is also important to enact policies that develop the necessary infrastructure for production and distribution.

Lessons from Brazil

Brazil is a case in point. After Brazil's *Proálcool* programme was enacted during the 1970s, the nation's ethanol industry made rapid progress, spurred by a combination of blending mandates and tax incentives that drove investment in the production and use of ethanol. The Brazilian government also promoted the manufacture and sale of 'neat' (pure) ethanol cars and provided subsidies to increase sugar production and distillery construction. In addition, infrastructure was developed to distribute neat ethanol to virtually all pumping stations around the country. Largely as a result of these policies that created demand and provided access to the marketplace, the ethanol industry grew quickly, such that neat ethanol-fuelled vehicles represented 96 per cent of total car sales by the mid 1980s.[56]

But falling oil prices and rising sugar prices in the mid to late 1980s and the 1990s provided oil companies an opportunity to take back a large share of the market lost to ethanol, and ethanol growth slowed dramatically. From 2003, the ethanol and auto industries promoted the production of flexible-fuel vehicles (FFVs) that can run on virtually any mixture of gasoline and ethanol.[57, 58] As a result, the market was changed almost overnight as Brazilian drivers no longer needed to worry about price or supply fluctuations, but could change their consumption decisions even faster than producers could adjust.[59]

While the ethanol industry in Brazil began with a host of subsidies, none of these remain in place today, other than the fact that biofuels are subject to lower taxes and that there exists a 20–25 per cent ethanol-blending mandate in all gasoline. In addition, at the end of 2005, neat ethanol sold for nearly 40 per cent less than the gasoline–ethanol blend, even accounting for the lower energy content of ethanol.[60]

Lessons from the US

While biofuels supply a far smaller share of total fuel in the US than in Brazil, US production has expanded steadily during recent years. Ethanol production, in particular, has been spurred primarily by state and federal incentives, stricter environmental legislation (most notably the 1990 Clean Air Act), falling production costs and rising demand. In its early years, however, the biofuels market and industry grew far more slowly in the US than in Brazil. This is probably explained, in great part, by the fact that there were no production or blending mandates in the US until recently.

Although it remains to be seen if the US production mandate will prove successful (and we may never really know because many experts predict the market to expand beyond the mandated production level by 2010, regardless of federal law), it seems logical that blending mandates are generally preferable. Blending mandates create a market for biofuels – rather than requiring specific production levels – while also indirectly requiring that an infrastructure be developed to get the product to that market.

Biodiesel production remains low relative to ethanol production and use in the US; but it is growing rapidly. The US military has recently become a significant demand driver for this fuel due to the requirement that all non-tactical military diesel vehicles use a 20 per cent biodiesel blend. Because governments are generally the largest single users of energy within a country, government purchase requirements can play a major role in creating large and consistent markets for biofuels.

Lessons from Europe

Europe's experience with biofuels offers some important lessons as well. As mentioned earlier, the EU failed to meet its biofuel target of 2 per cent of all gasoline and diesel for transport use by 2005 under the EU 2003 Biofuels Directive, and it will be unlikely to meet its target of 5.75 per cent by 2010. In fact, as of early 2006, the European Commission was in the process of taking Slovakia, Luxembourg, Italy and Portugal to the European Court of Justice for failing to actively contribute to biofuel development, and was considering replacing the EU target with a biofuel obligation.[61] Experience in Europe to date suggests that voluntary targets are hard to meet, at least when they are not supported with incentives to create a market and drive investment in the industry. Those countries in Europe that have seen the greatest success with biofuels – including Germany, Spain, France and the Czech Republic – all have domestic supplies of first-generation feedstocks (i.e. rapeseed in Germany, wheat in France and the Czech Republic, and grapes in France), as well as biofuel policies that support local farmers.

It is also important that direct payments to support the biofuels industry are gradually reduced and then discontinued when the biofuels in question become more financially viable, as they have in Brazil. Countries can follow the German example in the field of renewable energy (e.g. wind and solar power), phasing out subsidies according to the level of economic performance: once a market exceeds a predetermined level, incentives such as tax exemptions should end. Not only does this ensure that the public will not be subsidizing the industry in the very long term, but such policy design can also enable countries to encourage the development of next-generation technologies, which can continue to receive tax breaks or other incentives until they are cost competitive.

Lessons from Asia

Policies promoting biofuels in Asia have also been instructive. Thailand has been extremely successful in entering the biofuels market. Driven by a desire to reduce oil dependency and to create jobs in rural areas, Thailand's 30-year-old biofuels programme recently launched a €30 million (US$25 million) 'investment roadmap' to support the target of 8.5 million litres of biodiesel production by 2012. Also interested in producing ethanol to absorb surplus sugar cane and cassava, Thailand has launched incentive programmes for those wanting to establish new plants, and special purpose vehicle schemes to encourage the vehicle production to be compatible with biofuels.

In India, where the government has put both biodiesel and ethanol blending requirements in place in certain states, the government seeks to replicate the success of the country's National Dairy Development Board. This model encourages the formation of farmer cooperatives that span a cluster of villages, where each farmer buys a share in the society and sells his or her oilseeds through the cooperative in exchange for greater organization of financial capital, fertilizers, planting materials and other inputs.[62]

Overall policy lessons and remaining barriers

Based on experiences to date, it is clear that long-term governmental and stakeholder commitments to biofuels are critical, that a combination of policies is needed to drive the market and development of necessary infrastructure, and that policies must be consistent and flexible enough to tackle new challenges as they arise. In addition, in order to address concerns about fuel quality, possible social and environmental costs, and other issues, these policies may be combined with standards and certification schemes (see Chapter 18).

It is also critical to eliminate or alter those policies that work against the production and use of biofuels. Despite the many initiatives worldwide to advance biofuels, there remain a number of policy or institutional barriers that cause market distortions and slow market entry for these fuels. Greater development of coal-to-liquid (CTL) fuels could also compete with biofuels and slow market growth. In general, petroleum subsidies tend to hinder or preclude new fuels from entering the market by making market penetration more difficult (see Chapter 8). China provides significant direct subsidies for oil, perhaps as much as US$20 (€16.5) per barrel during times of high prices, although the actual impact on the economy is unclear.[63] Meanwhile, indirect subsidies, including the cost of defending oil supplies, to the US oil industry have been estimated at roughly US$111 billion (€92 billion) annually for light-vehicle petroleum fuels (see Chapter 7).[64]

For biofuel to play a greater role in the transport sector, the playing field must be levelled through the gradual elimination of subsidies for petroleum-based fuels. As the price of oil increases, biofuels will become more cost competitive, making it easier for them to compete with conventional fuels. However, the same can be said for unconventional liquid fossil fuel resources such as tar sands, which could become a large part of the future energy mix if the price of oil is high enough and associated costs of land use, water and other environmental resources are not taken into account. This real possibility highlights the importance of also incorporating external security, social and environmental costs (particularly climate change) within the price of energy (see Chapter 11).

Other barriers that continue to exist on the domestic and international levels must be addressed as well. For example, if the oil industry controls the supply and distribution of biofuels, this may lead to price manipulation. Unreliable supply and demand for biofuels can create uncertainty and impede market development. A lack of environmental oversight can result in backlash if rapid biofuel expansion leads to unsustainable production practices. The lack of a global commodity market for biofuels may slow supply growth. And the absence of international standards and/or certification schemes creates uncertainty for equipment and technology suppliers, and may impede international trade (see Chapter 9). Overcoming resource and production constraints requires policy and/or institutional reforms, as well.

RELEVANT INTERNATIONAL POLICY INSTRUMENTS AND INITIATIVES

At the 2002 United Nations World Summit on Sustainable Development in Johannesburg, South Africa, the Brazilian government proposed an energy initiative aimed at establishing targets and timetables for increasing the share of the world's energy derived from renewable sources. It proposed that by 2010, renewable energy sources represent 10 per cent of world energy consumption, with new renewables (not including traditional biomass or large-scale hydropower) representing 5 per cent of total energy use.[65] In response to Brazil's proposal, Latin America and the Caribbean agreed to a regional 10 per cent target by 2010 for renewable resources.[66] While these initiatives do not include a specific share for biofuels, these fuels are an increasingly important option for meeting such targets. Several international

Table 17.2 *Selected international biofuels initiatives*

Initiative and sponsor	*Year launched*	*Purpose*
Task 40: Sustainable International Bioenergy Trade, International Energy Agency (IEA)	2004	Focuses on bioenergy trade, particularly to develop 'commodity markets' at the local, regional and global levels in order to foster long-term sustainability and stability.
Sustainable Energy Finance Initiative (SEFI), United Nations Environment Programme (UNEP)*	2004	Aims to encourage investments in sustainable energy through partnerships and to bring these energy resources into the mainstream.
International Partnership on Bioenergy, led by the Italian Ministry for Forestry and Territory	Grew out of 2005 Group of 8 (G8) commitment to further develop biofuels for transport, heat and power	Aims to focus on alleviating barriers to bioenergy development, among other issues.
Biofuels Initiative, United Nations Conference on Trade and Development (UNCTAD)	June 2005	Will help poorest developing countries to increase production, use and trade of biofuel resources and technology in coordination with other groups.
International Bioenergy Programme, United Nations Food and Agriculture Organization (FAO)	2006	Aims to promote biomass as a means of addressing poverty and climate change by facilitating partnerships and working to reconcile any 'food versus fuel' issues that might arise.

Note: SEFI is a joint effort of the UNEP Finance Initiative, the UNEP Energy Branch and the Basel Agency for Sustainable Energy, with support from the United Nations Foundation.

Source: see endnote 67 for this chapter

initiatives have also been established to facilitate the global market expansion of biofuels specifically (see Table 17.2).[67]

While it is encouraging that these policies and programmes are under way, there is a need for leadership to facilitate international policy developments and to integrate disparate efforts related to biofuels. It is possible that the United Nations Food and Agriculture Organization (FAO) or the International Energy Agency (IEA) could play this role. The United Nations Conference on Trade and Development (UNCTAD) is forming an International Advisory Expert Group for its Biofuels Initiative, and it has been suggested that the IEA Bioenergy

Programme's Task 40 could serve as an information clearing house, playing a coordination/facilitation role among various organizations, helping to design trading and other schemes, and perhaps developing projects.[68]

Well-functioning international capital markets will also facilitate the development of a biofuels infrastructure.[69] Policies can be enacted to ensure that markets function properly, as well as to remove barriers. International financial institutions that direct development aid flows, such as the World Bank and the Global Environment Facility (GEF), also play a role in market function. Their role in biofuel development is already beginning to grow, with new World Bank and Inter-American Development Bank biofuel projects being planned for the near term.[70] These institutions are eager to invest in biofuel projects able to simultaneously address poverty alleviation, climate change and sustainable growth.

The GEF was designed to ensure that development projects are sustainable and provide incremental environmental benefits beyond that which could be achieved without GEF assistance.[71] Already, it provides small grants to meet some of the objectives in its climate change 'focal area', which will also help to promote biofuels. One existing programme, for example, aims to reduce implementing costs and remove barriers for jatropha oil and rapeseed-derived biofuel in one or more developing countries.[72] One outgrowth of the GEF, the Sustainable Energy Finance Initiative (SEFI), aims to promote and support increased investment in energy efficiency and renewable energy by informing and connecting investors, creating a stable environment to catalyse investment flows and to minimize risk and uncertainty. SEFI has established guidelines for institutions and individuals interested in investing in renewable energy technologies, including biomass.[73]

RELEVANT INTERNATIONAL ENVIRONMENTAL INSTRUMENTS

A major driver behind rising interest in biofuels is concern about global climate change, as well as the 2005 entry into force of the Kyoto Protocol. The European Commission, for example, is examining integrated ways in which alternative fuels and vehicle technologies could help the EU to achieve its carbon dioxide emissions-reduction targets under the international agreement.[74] Large developing countries such as China and India could soon become some of the largest emitters of greenhouse gases, and biofuels provide an option for them to leapfrog to more sustainable fuels and technologies.

Significant opportunities continue to emerge for biofuel market expansion as countries look to address a number of goals, including domestic environmental and human health regulations, the creation of markets for energy crops as a new means of helping the domestic agricultural sector, and meeting international commitments on climate change.

Kyoto Protocol flexible mechanisms

There are two 'flexible' financing provisions that provide opportunities to promote biofuel development under the Kyoto Protocol. The first is the Clean Development Mechanism (CDM), which grants 'emissions reduction credits' for investments in emissions-reducing projects in developing countries.[75] The second is Joint Implementation (JI), which grants credits for projects in transition economies, such as those in Central or Eastern Europe. In order for projects to qualify for CDM or JI funding, they must achieve emissions reductions that are additional to any that would have occurred without the project.[76] For example, Brazil's current ethanol programme would not qualify for CDM funding because it is well established. However, Brazil is proposing a new biodiesel project under the CDM that might qualify for additional reductions (excluding the share that is compulsory).

To date, there are few biofuel-related projects under consideration for the CDM or JI. This is because it is more difficult to determine baselines and measurement methodologies for biofuel projects than for other renewable energy or energy efficiency projects – in other words, it is difficult to determine to what extent biofuel projects will reduce GHG emissions below business-as-usual levels, particularly because life-cycle emissions vary according to feedstock and processing methods (see Chapter 11).[77] Furthermore, as with other CDM and JI projects, leakage (i.e. whether the project will result in higher emissions elsewhere) can be a problem with biofuel CDM and JI projects.[78] In spite of these challenges, CDM transportation projects are 'in the pipeline' for Thailand (ethanol) and India (biodiesel and ethanol).[79]

UNCTAD views the development of biofuels as a powerful poverty reduction tool, and hopes to utilize the CDM to help mitigate climate change through expanded production and use of biofuels. A UK-based group, Agrinergy, also recognizes the important role that the CDM can play in promoting biofuels. This is particularly true for the near future (i.e. until cellulosic technologies become widely available) because the potential GHG emissions reductions from sugar cane-based ethanol can be significantly greater than those from grain-based ethanol, and ethanol production from sugar cane occurs primarily in tropical developing countries that can benefit from the CDM.[80] Agrinergy also recognizes the complexities associated with developing biofuel projects that could qualify for the CDM. Therefore, the firm is working to develop a CDM methodology for transport-related liquid biofuel projects, such that 'only under clear and specific circumstances will biofuel use qualify as a CDM project'.[81]

In addition, the 11th Conference of the Parties (First Meeting of Parties) of the United Nations Framework Convention on Climate Change (UNFCCC), held in Montreal in December 2005, strengthened the CDM and expanded the scope to include policies and programmes, not just single projects.[82] The latter is relevant for developing countries because it enables a cluster of activities to be registered under one project design document, a step that will greatly reduce transaction costs.[83]

Carbon finance instruments and developing carbon markets

The development of carbon finance instruments and carbon markets – for example, by the World Bank's Carbon Finance Unit (CFU), the Chicago Climate Exchange and the European Union's Emissions Trading System (EU-ETS) – could create a market for carbon credits/emissions trading and facilitate biofuel market expansion as investors and countries seek to shift to low-carbon technologies and activities to reduce GHG emissions. The near-term efficacy of such instruments, the robustness of carbon markets and the extent to which they will facilitate investments in sustainable emissions-reducing projects – including in biofuels – remain unclear.

Initial forays into carbon finance have indicated that the price of carbon may be too low to spur the necessary behaviour changes or investments in clean energy, and, in this case, to facilitate more robust biofuel development. Another major issue will be how much carbon credit to give biofuels – and how to achieve agreed-upon metrics to reach such a determination.

The World Bank CFU uses funds contributed by companies and governments in industrialized nations to purchase project-based GHG emissions reductions in developing countries and countries in transition. The emissions reductions are purchased through one of the CFU's eight carbon funds on behalf of the contributor, and within the framework of the Kyoto Protocol's CDM or JI. The role of the Carbon Finance Unit is to catalyse a global carbon market that reduces transaction costs, supports sustainable development, and helps to alleviate poverty and facilitate growth in the world's poorest nations.[84]

The Prototype Carbon Fund (PCF) was the first carbon-financing mechanism at the World Bank. It is a public–private partnership consisting of 17 companies and 6 governments. It became operational in April 2000, with a mission of pioneering the market for project-based GHG emissions reductions, while promoting sustainable development and offering a 'hands-on' educational opportunity to its stakeholders. As of early 2006, 21 projects had been funded through the PCF and 7 more were under development.[85]

The selling of emissions reductions – or carbon finance – has been shown to increase the bankability of projects by adding an additional hard currency revenue stream, which reduces the risks associated with lending. In this way, carbon finance provides a means of leveraging new private and public investments into projects that reduce GHG emissions, thereby mitigating climate change while contributing to sustainable development.[86]

CONCLUSION

Government-initiated policies, ushered in by key stakeholders, have been, and will continue to be, the main driver of the global biofuel industry. Biofuel subsidies have been justified as indirect aid to domestic agriculture; as a means of reducing dependence upon foreign oil and its associated security and economic costs, improving local air quality; and, more recently, as a way of meeting national or regional greenhouse gas emissions reduction targets. These goals will continue to be central, although the relative weight of these drivers may vary over time, and they require that future biofuel policies be designed to foster sustainable production for both industrialized and developing countries.

For all countries, policies that guarantee demand and encourage technological innovation are essential. Blending mandates, government purchasing policies, and support of biofuel-compatible infrastructure and technologies have been most successful in developing a reliable market for biofuels in the countries and regions where they are being implemented. Other policies, such as loan guarantees, tax incentives for consumers and manufacturers, and direct industry subsidies, will probably continue in some countries, but should be scheduled to be phased out when the fuels that they support approach commercial viability, as has occurred in Brazil and Germany.

It is also critical that small-scale biofuel development occurs alongside larger-scale production for export. Policies should be mindful of the benefits of both large- and small-scale biofuel production and not 'pick winners' – rather, they should support the most environmentally and socially beneficial production processes available, especially as increases in population and economic growth result in increasingly high energy demand. Technology transfer and investment from industrialized countries to facilitate market development will also be essential for biofuel development.

In general, development of the biofuel industry will require long-term and flexible commitments to these fuels. Successful use of biofuels to meet the growing liquid-fuel demand will necessitate a comprehensive approach to market development. Demands from industrialized countries are likely to exceed economical biofuel production capacity and lead to trade, and policies will need to be developed to address such elements as fuel quality and social and environmental concerns. Furthermore, reducing petroleum subsidies, and subsidies to other fossil fuels, will increase the competitiveness of biofuels worldwide.

In order to secure policies and programmes for sustainable biofuel development, there is a real need for leadership to facilitate international policy developments and to integrate disparate efforts related to biofuels. Although many international actors are currently involved in different aspects of biofuels policy, there remains a need for coordination to maximize the benefits of these many initiatives.

18

Standards and Certification Schemes

Introduction

As discussed throughout this book, the methods for producing different biofuels vary considerably, with some being more environmentally and socially sound than others. As international trade in biofuels increases, mechanisms are needed to assure both domestic consumers and importers that the fuels they use were produced using the most sustainable methods and equitable labour practices possible. One of the most pressing concerns is the potential impacts of feedstock production, including deforestation and competition with food uses.

While current biofuel trade is minimal, the call for such safeguards has already been raised, particularly in Europe. In 2005, consumers and activist groups in The Netherlands and the UK expressed rising concern about the environmental impacts of palm oil plantations on forests and wildlife in Southeast Asia – production driven, in part, by increased European demand for biodiesel.[1] UK imports of palm oil alone doubled between 1995 and 2004 to 914,000 tonnes, representing 23 per cent of the European Union (EU) total (most of this is for food uses, although plans for palm-based biodiesel are unfolding).[2] Such outcries have generated wider calls for the development of certification systems. Most initiatives call for voluntary systems; but a few governments intend to initiate mandatory schemes.

Setting standards and establishing certification schemes are strategies that can help to ensure that biofuels are produced in a responsible manner. They can also enable consumers to discriminate between fuels based on the sustainability of production. These schemes will be particularly important for countries that are actively striving to achieve reductions in greenhouse gas (GHG) emissions and that have included biofuels in their emissions reduction plans. Where they exist or could be created, strong national policies can act as an example and alternative to international certification schemes. Technical assistance from industrialized nations can also help countries developing biofuel markets to adopt sustainable practices.

This chapter discusses some of the main strategies being considered to guide biofuel production. It describes existing standards and certification schemes in the agricultural and forestry industries that could provide useful models for biofuels, as well as some of the initial efforts to design schemes specifically for these fuels. It concludes with a consideration of barriers and outstanding questions related to the creation of safeguards, and how assurance mechanisms might be pursued to facilitate biofuel trade that maximizes both environmental and rural development benefits.

THE NEED FOR SUSTAINABILITY STANDARDS AND CERTIFICATION

Today, a variety of sustainability standards and certification schemes exist or are under development in the areas of agriculture and forestry; however, no such system exists specifically for biofuels. Such initiatives are needed to help establish minimum social and ecological standards for these fuels and to guarantee responsible use of biomass from the raw material stage through to its final application.

Such standards would aim to address some of the dominant environmental and social concerns related to biofuels and their feedstocks, particularly when these are developed on a large scale. From an environmental perspective, the primary concerns include ecological impacts of monoculture crop plantations; damage to water and soil from the application of pesticides and fertilizers; soil erosion; nutrient leaching; increased use of freshwater resources; and the loss of biodiversity and wildlife habitat, particularly if the cropland area is expanded into previously undisturbed sites (for a more detailed discussion of these concerns, see Chapters 12 and 13).

Relevant social issues include potential impacts on agricultural and rural incomes; access to biofuel markets by small landholders and indigenous groups; job availability and quality (which could increase or decrease, depending upon the level of mechanization, local conditions, etc.); potential use of child labour; and access to education and health care for workers (for more information on these social issues, see Chapter 8).

It is important to note that these problems are not necessarily bigger or worse with biomass production than with similar agricultural activities, such as large-scale food and feed production. Nevertheless, establishing a certification programme could help to minimize the potential negative impacts of biomass production, while also working to promote sustainable biofuel trade.

RELEVANT STANDARDS AND CERTIFICATION SCHEMES

Biofuels, especially those produced from current-generation feedstocks, are produced primarily from traditional agricultural crops; however, unlike the majority of food crops traded, they are not consumed and therefore have a different significance for consumers. People concerned about the impacts of conventional agriculture on human health and food quality might be less concerned about crops raised for energy. Nevertheless, large-scale agricultural production of biofuels can have equally important impacts on the environment in terms of water and air quality, and especially biodiversity (see Chapters 12 and 13).

The rapid scaling-up of biofuel production in many countries is making certification an immediate imperative. 'Piggy-backing' on existing schemes such as those advocated by the International Federation of Organic Agriculture Movements (IFOAM) could be an excellent first step towards establishing a verified sustainable biofuels industry. Collaborative certification schemes could be a starting point, setting minimum standards for cultivation and harvesting practices for biofuel producers. As biofuel trade increases in volume and complexity, a more advanced and innovative certification scheme may build off earlier efforts.

Among the existing schemes that could act as models for biofuel standards and certification are the Rainforest Alliance's Standard for Sustainable Agriculture in Latin America; organic certification and labelling schemes in more than 100 countries; the Forest Stewardship Council's international forest certification system; a UK environmental assurance programme linked to the country's renewable fuels obligation; and the newly established Roundtable on Sustainable Palm Oil. Each of these is described in detail in the following sections (for an extensive list of other relevant standards and certification schemes, see Appendix 8).

Standard for Sustainable Agriculture

The Standard for Sustainable Agriculture (SSA) was developed during the 1990s by the New York-based Rainforest Alliance and collaborating organizations in Latin America. Through a coalition of independent conservation groups, known as the Sustainable Agriculture Network (SAN), these organizations promote the social and environmental sustainability of production in several key commodity areas, including coffee, bananas, cocoa, citrus, ferns and cut flowers.[3] The network aims to improve the social and environmental conditions of tropical agriculture by:

- certifying sustainable practices on farms and awarding a credible seal of approval (the Rainforest Alliance certified eco-label) to farms that comply with the SSA;
- changing the paradigm of farm owners, retailers and consumers to encourage all stakeholders in the agricultural industry to take greater responsibility for their activities;
- establishing contact between conservationists in the North and the South and offering them a way to work together;
- increasing public awareness about the effects of consumer purchases on tropical peoples and ecosystems, and offering the choice of more responsible certified products; and
- creating a forum for discussing the environmental and social impacts of agriculture.

In November 2005, after extensive public consultation, SAN approved the final version of the SSA, which includes principles in ten key areas:

1. social and environmental management systems;
2. ecosystem conservation;
3. wildlife protection;
4. water conservation;
5. fair treatment and good working conditions for workers;
6. occupational health and safety;
7. community relations;
8. integrated crop management;
9. soil management and conservation; and
10. integrated waste management.[4]

These areas are very similar to those that would need to be covered in sustainability standards developed for biomass and biofuels.

Organic certification and labelling

Organic certification is the most well-stablished system of certification, established over 30 years ago as a result of grass-roots momentum by farmers interested in cultivating agricultural products in a more environmentally sustainable way. As the movement grew, certification evolved from self-evaluating farmer-defined criteria to more sophisticated systems of third-party monitoring implemented by governments and non-governmental organizations (NGOs).

The proliferation of organic certification schemes led to a movement to establish consistent standards and to provide model laws and voluntary standards. The International Federation of Organic Agriculture Movements has advocated a decentralized system to support organic practices, and it now represents some 750 agricultural organizations in 108 countries. It has been instrumental in adopting formalized standards for organic agriculture, such as the Codex Alimentarius, as well as EU and United Nations Food and Agriculture Organization (FAO) regulations.[5] Thus, 'piggy-backing' on existing schemes such as IFOAM could help to launch the international movement towards a more sustainable biofuel trade.

The Forest Stewardship Council's forest certification system

Another promising model for a variety of biofuel feedstock areas is the Forest Stewardship Council's (FSC's) system for forest certification. The FSC, headquartered in Bonn, Germany, is an international network of non-profit organizations, businesses and governments that works to promote responsible management of the world's forests by granting certification to forestry-related operations that meet strict FSC standards.[6] The FSC accredits third-party certifiers, who then conduct field audits of candidate forests.

FSC standards are established via consultation with all three chambers of its general assembly (economic, environmental and social) and an elected board, which oversees the system. The standards consist of ten principles, covering such areas as indigenous peoples' rights, environmental impacts and maintenance of high conservation value forests, and a set of criteria to measure implementation.[7] Specific indicators can be tailored to each country or region.

The FSC framework works to ensure that all policies and standards development processes are:

- *transparent*, meaning that the process for policy and standards development is clear and accessible;
- *independent*, meaning that standards are developed in a way that balances the interests of all stakeholders, ensuring that no one interest dominates; and
- *participatory*, meaning that the FSC strives to involve all interested parties in the development of FSC policies and standards.[8]

Under the FSC system, any timber company, furniture maker, paper manufacturer, printer or retail outlet can undergo a voluntary third-party audit; if their operations pass muster, they will receive a certificate and licence to label their products with the FSC logo. The FSC's chain-of-custody control system guarantees the sustainable production of all FSC-labelled items along the path from forest to consumer, including forest management, timber operations, processing, transformation, manufacturing and distribution.

Over the past decade, the FSC has indirectly certified more than 50 million hectares of forest in over 60 countries, and several thousand unique products now carry the FSC logo.[9] Certification has caught on much faster in industrialized countries than in developing countries, however, due in large part to the expense. To address this problem, the FSC has established a mechanism that reduces the costs of certification by allowing small producers to be certified in groups. In addition, a special mechanism for small and low-intensity managed forests streamlines the assessment process and reduces some of the evaluation and monitoring costs. This has simplified the certification of non-timber forest products (such as fruits, nuts and ferns), and could ease the way for certification of ecosystem services such as clean water and carbon sequestration.

Linking assurance schemes to a renewable fuels standard

Another system that might provide a useful model for biofuels is now under consideration in the UK. The UK's Renewable Transport Fuel Obligation (RTFO) sets a target for 5 per cent biofuel use in the UK by 2008, and policy-makers have stressed the importance of sustainably produced fuels in meeting this target.[10] To this end, a feasibility study was conducted in 2005 assessing the merits and drawbacks of using the RTFO as an instrument to increase the use of renewable fuels in the transport sector.[11]

A key focus of the evaluation was if and how assurance schemes covering greenhouse gases, as well as broader environmental and social issues, should be linked to an RTFO. The study concluded that a 'simple, transparent and verifiable GHG certification scheme' could be developed at a low cost to the government and fuel suppliers, and negligible cost to the consumer. It was estimated that such a scheme could be developed and piloted within 18 months, a timescale consistent with the introduction of an RTFO.

The Roundtable on Sustainable Palm Oil

Another recent effort to establish sustainability standards, the Roundtable on Sustainable Palm Oil (RSPO), could prove particularly relevant to biofuel feedstock certification discussions. The World Wide Fund for Nature (WWF), in partnership with several companies and other interested parties, established the RSPO in 2004 in response to rising concerns about the environmental and social impacts of palm oil plantations. The group now has at least 95 members, representing the palm oil industry and other organizations working in and around the supply chain for palm.

In late 2005, all RSPO members voluntarily adopted the newly created Principles and Criteria for Sustainable Palm Oil, which set sustainability standards for palm oil production. Since November of that year, RSPO has been undergoing a two-year trial implementation period to field test these principles and to advance guidance on how the criteria should be interpreted and enforced.[12] Members are expected to support the roundtable's standards and to actively promote the use of sustainably produced palm oil.[13]

The principles cover eight categories of operation, including commitment to transparency; compliance with applicable laws and regulations; commitment to long-term economic and financial viability; use of appropriate best practices by growers and millers; environmental responsibility and conservation of natural resources and biodiversity; responsible consideration of employees and of individuals and communities affected by growers and mills; responsible development of new plantings; and commitment to continuous improvements in key areas of activity. Specifics range from using integrated pest management practices, to developing and implementing plans to reduce pollution (including GHG emissions), to allowing workers to join trade unions.[14]

Efforts are also under way to create a Roundtable on Sustainable Soy, modelled on the RSPO. Like the RSPO, this multi-stakeholder group aims to provide stakeholders and interested parties with the opportunity to jointly develop solutions for sustainable soy production. The process was initiated by the WWF and is now being managed by an organizing committee that includes several soy producers and retailers, as well as non-governmental groups.[15]

Implications of the World Trade Organization (WTO) Policy Framework for Biofuel Certification: International trade and equity

Although the issue of bioenergy trade is not explicitly one of the topics dealt with by the World Trade Organization (WTO), the intersection of energy and agriculture markets for trade make biofuel trade a very interesting and relevant issue. Biofuel trade is inextricably linked to food and forestry commodities and markets, sustainable development and climate change discussions, all of which are potentially pivotal issues for WTO legislation.

Barriers to trade

Because most current-generation biofuel feedstocks are agricultural commodities, the discussion of barriers to biofuel trade must begin with trade barriers in agriculture, a key area of WTO engagement. Harmonization of agricultural subsidies is being pursued; however, a number of approaches allow countries to subsidize the agricultural sector, such as support for research and development (R&D); infrastructure (so-called 'general benefits') that, in general, should not significantly and clearly affect export volumes; support for 'production limiting programmes' (e.g. set-aside regulations and 'green box subsidies'); and support for replacing undesired crops, such as narcotics.[16] Conservation efforts such as the EU's subsidies for transferring agricultural land to forest plantations are an example of allowed support that could affect developing biofuel markets.[17]

As a result of the latest round of trade negotiations, developing countries can maintain a support level of 10 per cent of the total value of agricultural production, while for the EU, the US and other industrialized nations, this is limited to 5 per cent, which may be lowered in future negotiations.[18] Under current rules, direct export subsidies are not allowed, although subsidizing transport costs is permitted. In addition, countries have made specific commodity-tied commitments to reduce subsidies during agricultural negotiations; each of these separate agreements must be considered.

Further complicating biofuels trade will be its definition as a 'product' by the WTO. Distinctions are made between 'old' products, for which agreements already exist, and 'new' products that have to comply with the most recent WTO rules. Additional complexities result because biofuels can fall under tariff schedules for agricultural commodities, industrial chemicals and energy carriers. This distinction could be of major relevance for bioenergy trade. If biomass-derived energy carriers are recognized as 'new' products, pathways around current protection measures from the EU for biomass-derived ethanol could be devised.[19]

The issue of trade barriers for biofuels was brought to light in the case of Brazilian ethanol exports to Europe, which has tariffs in place for commodities derived from sugar.[20] However, the issue was not pursued by Brazil. In the future, it is likely that Brazil or other WTO member countries will bring biofuels into negotiation rounds as part of a portfolio of topics under discussion.[21]

WTO policies and certification

Technical barriers to certification, such as specifying product and content definitions, are relevant for the wide variety of commodities and energy carriers that comprise an international bioenergy trade.[22] The WTO Technical Barriers to Trade (TBT) Agreement requires that domestic technical regulations use international standards as a basis where they exist, and where they are not ineffective or inappropriate.[23] The TBT Agreement also requires that such technical regulations are not an unnecessary obstacle to trade.

Regarding the general WTO rules on internal regulations (contained in the General Agreement on Tariffs and Trade, or GATT), the regulation of process and production methods (PPMs) is politically controversial at the WTO. The setting of PPM-based regulatory requirements is not, as such, inconsistent with existing legal rules, as interpreted by the Appellate Body in Asbestos.[24] According to the appellate body, consumer tastes and habits must be considered in determining whether two products are 'like' and, thus, are entitled to 'no less favourable treatment'.[25] If consumers differentiate products based on their production methods, or would do so if they had the information, then these products may well be 'unlike' in accord with the appellate body's jurisprudence.[26]

Moreover, for a regulatory measure to violate national treatment, it must not only treat 'like' products differently, but afford 'less favourable treatment' to the group of imported products when compared with the entire group of 'like' domestic products.[27] In the case of PPMs, the Appellate Body in Asbestos has emphasized that regulatory distinctions may be drawn even between products found to be 'like', provided that the distinctions in question do not systemically disadvantage imports over domestic products. For an emerging biofuel market, it is therefore possible to design measures that specify process and production methods desirable to importing countries so long as they do not systematically disadvantage imports and favour domestically produced fuels. These measures must also conform to the most favoured nation (MFN) requirement in Article I of the GATT, and not discriminate on the basis of national origin between the like products of different exporting WTO members.[28]

Current initiatives for biofuel certification

As the possibility of expanded biofuel trade becomes a greater reality, the push for certification in this area is gaining momentum. The Dutch government, for example, has passed legislation requiring that a certification system be developed to ensure the sustainability of imported biomass fuels. Using existing quality-control models, such as the EuroGAP criteria for agricultural products or FSC wood certification, the Dutch company Essent has developed the Green Gold Label. To qualify for the label, biomass has to be sustainable and traceable through the entire supply chain, from the plantations or forest to the consumer. Minimum conditions include that the biomass must be renewable (i.e. replanting must occur after harvesting); however, other environmental and social criteria are not currently included in the system. The monitoring process includes annual audits of biomass producers and suppliers, as well as quality-control inspections.[29]

Following the Dutch example, the EU is currently considering environmental standards for imported biofuels, as well as working towards bilateral agreements in this area.[30] The EU is responding to both consumer and producer concerns: increasingly, biofuel producers in Europe are asking for safeguards that require exporting countries to meet minimum labour, environmental and other standards on a par with their own. At the same time, the FAO is working with the Global Environment Facility (GEF) to develop criteria for evaluating proposals for such schemes. The FAO's suggestions, to date, have been based on a literature review, rather than on field trials; but the organization aims to field test them soon.

For their part, several biofuel exporting countries, as well as the FAO, have expressed concern about the trade implications of a rigorous biofuel certification scheme. A key worry is that certification schemes (or environmental standards, more generally) will create trade barriers for developing country exports and will be used as a way for importing countries (usually industrialized countries) to protect domestic biofuel industries. This raises a variety of other equity-related issues, such as who will set the standards and who will accredit the certifiers.

Given these concerns, it is critical to establish standards that both exporting and importing countries agree on, and to ensure that these standards are applied consistently and transparently so that they are not viewed as discriminatory (in all likelihood, some developing countries would, in fact, perform better than many industrialized countries on a range of sustainability criteria, including the greenhouse gas balance and fossil energy input, because they tend to experience higher crop yields and use fewer chemical inputs). It is also important to avoid creating a double standard between importing and exporting nations; the sustainability standards must apply equally for domestic production and use and for biofuels traded internationally.

KEY OBSERVATIONS ON BIOFUEL STANDARDS AND CERTIFICATION

Based on the experience, to date, in implementing sustainability-related certification schemes, several key lessons can be learned, with specific relevance to the development of effective biofuel certification:[31]

- *Select the most appropriate standards or certification approach possible.* Safeguards and assurances can be applied through a variety of means, including laws or regulations, voluntary certification schemes, or criteria to qualify for subsidies or incentive programmes. Relying on existing certification systems should be approached with caution since they may represent (or be perceived to represent) only some of the stakeholder interests.
- *Achieve consensus among diverse stakeholders about basic underlying principles.* Broad consultation and participation in the process is required for any voluntary system to be credible in the marketplace.
- *Design and adopt specific quantifiable criteria for sustainability indicators.* Despite their specificity, these should be flexible enough to be adapted to the particular requirements of a region. Where strict, specific criteria and indicators are difficult to establish due to differing opinions of stakeholders, the use of so-called 'process indicators' that show continuous improvement may help to facilitate progress in moving forward.
- *Ensure that compliance with the criteria is enforceable in practice, without generating high additional costs.* Issues of cost and who pays are critical to the success of a certification programme, particularly when seeking participation of smaller-scale producers who may have fewer resources.
- *Avoid leakage effects*, through which benefits gained in one location 'leak away' when damage occurs in another. In the context of biomass trade, leakage could occur when crop production activities are expanded into previously undisturbed natural habitats, leading to increased greenhouse gas emissions from soil or other environmental impacts.
- *Establish a system for monitoring and reporting.* In addition to assessing the sustainability of the biofuel feedstock supply and any possible leakage effects, it is critical to devise a system for measuring and reporting on energy efficiency and on the carbon and energy balances of the resulting fuels.

OUTSTANDING ISSUES TO BE ADDRESSED

A variety of issues specific to biomass and biofuels still need to be addressed when considering standards and certification systems. At this time, there may be more difficult questions than definitive answers. Developing appropriate responses to these questions is a task left to key stakeholders in the international community.

Key questions include:

- *Should biofuels be held to a higher standard than agricultural food products or petroleum-based fuels?* This is a particularly contentious issue in countries where the same crops are used for both food and fuel, or where two end products are processed in the same refineries (as in Brazil, where sugar cane refineries shift between sugar and ethanol, depending upon global markets). A potential double standard also exists with regard to petroleum fuels; rarely do consumers go to the refuelling station to request sustainably produced petroleum gasoline or diesel. On the other hand, it is not uncommon for oil companies to experience boycotts and other public backlash in response to real or perceived human rights or environmental abuses. Moreover, one of the main drivers behind the push for increased production and use of biofuels is their potential environmental benefit – justifying the importance of ensuring that these fuels are truly sustainable.
- *Will standards and certification schemes slow or speed market development of biofuels?* Because biofuel production is increasing at a rapid pace, taking steps towards certification that pick up on existing schemes is an important first move in establishing a sustainable biofuels industry. However, such schemes must not be an insurmountable hurdle for new market entrants. Establishing minimum standards between countries with established bilateral trade could be a starting point. As biofuel trade begins to include more actors, a more advanced and innovative certification scheme may expand on earlier efforts.
- *What would these standards include and how would less quantifiable targets be measured?* Would standards focus primarily on greenhouse gas emissions, or would they also include pesticide use, impacts on biodiversity, water and air quality, and labour practices? How would the baseline be determined for GHG reductions, and would a minimum reduction target be established? Or, would preferential treatment be provided to specific feedstocks or production systems (i.e. integrated agro-energy systems)? And how would impacts on habitat and wildlife be measured in economic terms? How would a certification system ensure that biomass production does not crowd out the production of much-needed local food sources?
- *Should standards and certification schemes be established at the national, regional or international level – or all of these?* The EU is currently in the process of considering environmental standards for imported biofuels and is also working towards bilateral agreements in this area. But is this the best option? How can we avoid a proliferation of standards that differ from one country or region to another? Similarly, how can turf battles among competing certification schemes and consumer confusion be avoided (a problem that has hampered efforts to develop meaningful certification in eco-tourism and organic foods)? It will be essential for standards to be either consistent or highly flexible. But would increased flexibility of standards minimize their potential to ensure sustainability?

- *At what stage of production or distribution should certification occur?* Because biofuel generally comes from many diverse sources, it would be difficult to certify it only at the fuel pump. Should certification, then, be done at the farm gate, possibly even linking incentives for farmer certification to carbon credits? Or would it be best to certify biofuels at the distillery gate?
- *Should standards and certification be voluntary or mandatory?* Would it be possible to first develop and adopt mandatory standards in a few existing markets, and then publicize these more widely to encourage buy-in as other countries enter the global biofuels arena? The list of standards could be relatively simple, with specifics defined at the national level. Or perhaps a 'club' of governments, companies and other interested parties could voluntarily adopt standards and certification schemes, as the Roundtables on Sustainable Palm Oil and Soy have done. Such a reciprocal concept would probably not pose a problem with the WTO. Some experts believe that as part of a voluntary certification scheme, it would be possible to develop an eco-label for those biofuels that meet standards higher than those mandated by law; the fuels could be tracked and identified as a percentage of blends available at the pump. But others are sceptical of this approach.
- *What is the best approach for developing standards and certification schemes?* Should actors work quickly to establish a few basic minimum standards that can then be improved over time? Or is it more important to develop a thorough certification system, with broad stakeholder participation, even if this is a longer-term, more time-consuming process? Is it preferable to set up bilateral agreements first, while the most significant holes in the knowledge base are still being filled?
- *Who will pay for the process of establishing such a scheme, or for certification itself, or for enforcement?* Will such schemes increase the costs of biofuels at the pump, or put additional burdens on farmers and producers? One exploratory study has demonstrated that existing social and environmental standards for natural products do not necessarily result in high additional costs.[32] It is essential to ensure that standards and certification programmes do not hurt small farmers and biofuel producers, or developing nations.

Conclusion

While there are many possible targets and outcomes associated with an increase in bioenergy trade, several are particularly critical. Trade in biomass and biofuels should, among other things, foster a stable and reliable demand for the services of rural communities; provide a source of additional income and employment for exporting countries; contribute to the sustainable management of natural resources; fulfil GHG emissions reduction targets in a cost-effective manner; and diversify the world's fuel mix. Achieving these diverse goals – particularly in a sustainable manner – may best be done through implementation of a sound standards and certification framework.

Many potential sticking points remain, however – not least the fact that environmental and social safeguards could be perceived as unfair trade barriers to biofuels exports. Yet, at the same time, such standard-setting could become a critical driver to facilitating development of sustainable trade in biofuels. Thus, a compromise must be reached between developing complicated certification schemes and ensuring both growing markets and long-term sustainable biomass trade.

Technical standards for sustainability could probably be produced in a reasonably short time; but these alone may not be enough. To guarantee consumer confidence in the sustainability of the biofuel products themselves – particularly as the industry grows larger – a certification scheme will probably be needed to back up the standards and oversee their application. The incremental development of such a certification scheme is probably the most feasible option, allowing for gradual learning and expansion over time. While not all biomass types may fulfil the entire set of sustainability criteria initially, the emphasis should be on the continuous improvement of sustainability benchmarks.

While a certification scheme should be thorough, comprehensive and reliable, it should also not create a significant hurdle for nascent biofuel industries. Criteria and indicators must be adopted according to the requirements of each region and be mindful of the implementation costs. For example, it will be important to pair any certification scheme with technical assistance, incentives and financing so that small- and medium-scale producers can qualify as readily as larger producers. Furthermore, it is important to ensure that any standards and certification schemes for biofuels address the issue of possible leakage effects, through which benefits gained in one location could 'leak away' when damage occurs in another.

Moving forward, additional research will be needed to determine whether an independent international certification body for sustainable biomass is feasible. This should be done by a consortium of stakeholders in the biomass-for-energy production chain. At this stage, and at later steps in the development process, public information dissemination and support will be critical.

Part VII

Recommendations

19

Recommendations for Decision-Makers

Introduction

Biofuels have the potential to help meet many of the challenges that the global community faces today – reducing the threat of climate change, reducing reliance on oil and improving international security, and alleviating poverty in some of the world's poorest nations. Alternatively, a massive scale-up in the production and use of biofuels could increase the concentration of economic wealth, while speeding deforestation and biodiversity loss and possibly accelerating climate change. The path taken will depend primarily upon policies put in place by leaders at national and international levels.

When thinking about biofuels, it is important for policy-makers to keep in mind that there are really two different biofuels 'worlds': large-scale, high-tech production, and smaller-scale, low-tech biofuel production focused primarily on poverty alleviation through rural energy provision and local agro-industry development (involving local ownership, employment, etc.). There is certainly overlap, and the two worlds can and should exist in parallel. But the appropriate technologies and policy orientations required to promote them both are quite different; thus, policy-makers need to clearly define their desired outcomes and to design policies accordingly. In many cases, multiple goals can be achieved; but the more high-tech and large-scale biofuel industries become – and the more involved the huge energy, auto, chemical, finance and other companies become – the greater the policy effort required to fulfil the social and environmental aims.

This book focuses almost exclusively on high-tech biofuels since small-scale biofuel applications are often for non-transport purposes and, therefore, are outside the scope of this volume. Further research is needed into the potential for, and deployment of, appropriate technologies for small-scale and non-transport biomass energy applications.

In order to achieve their goals, it is critical that decision-makers take a comprehensive approach that encompasses all relevant sectors and stakeholders. The fuels, their production methods, the means of distributing them and the vehicle technologies appropriate for using them all need to be coordinated across industry segments and government agencies. This will be challenging; but it is absolutely necessary. The alternative will result in inefficient feedstock and fuel production, missed production targets, incompatibilities in the infrastructure, bottlenecks in the system, lost economic development opportunities and environmental degradation.

Furthermore, biofuel strategies must be developed within the context of a broader transformation of the global transport sector, with the goal of making it dramatically more efficient and diversified. As has been the case with renewable energy for electricity generation, biofuels will be able to meet a greater portion of total transport energy needs if the sector becomes more efficient.

Some governments have already enacted policies to support biofuels production, use and, increasingly, trade (see Chapters 9 and 17). While specific policy decisions will have to be made on a country (or regional) basis, according to unique natural resource and economic contexts, this chapter elaborates overarching recommendations to policy-makers and describes a number of policy options that governments should consider enacting in order to advance sustainable biofuel development. These recommendations are drawn from experiences, to date, with biofuels, with other fuels and with other renewable energy technologies, and are also based on the challenges that biofuels face today.

DEVELOPING THE BIOFUEL MARKET

The most efficient way to hasten a rapid expansion of biofuel production is for governments to create a policy environment that is conducive to private-sector investment in the development of these fuels. Policy-makers should focus on creating a predictable and growing market for biofuels. In turn, this market will draw in the substantial capital, entrepreneurial creativity and competitive spirit required to advance technologies, build production infrastructure and achieve the learning and the economies of scale that are necessary to drive down costs.

Policy actions that governments can take right away, at no (or low) net cost, to help develop the market include the following:

- *Enact tax incentives.* Tax incentives have been used effectively in Brazil, Germany, the US and other countries to spur biofuel production and reduce biofuel prices at the pump. They can also be used to encourage certain types of biofuels development (i.e. small scale, community oriented), and to speed the adoption of biofuel-compatible vehicles and other infrastructure. (Tax incentives for biofuels can be made revenue neutral in a number of ways – for example, by increasing taxes on petroleum-based fuels. Governments that subsidize fossil fuels can save revenues and reduce the need to subsidize alternative fuels by reducing direct and indirect subsidies for the petroleum sector.)
- *Establish mandates and enforcement mechanisms.* Blending mandates create consistent and expanding markets, which, in turn, attract private-sector

investment in technology advancement, infrastructure development, etc. Voluntary targets have been somewhat effective, but have not achieved the level of success provided by mandatory schemes coupled with credible enforcement mechanisms. Enforcement is important to ensure that targets are met. Mandates can be designed to steadily increase requirements for the share that must come from next-generation fuels. Mandates should also be tied to environmental and social standards (see below).

- *Use government purchasing power.* The enormous purchasing power of governments has been used successfully in a number of countries to expand the market for various products. Government purchasing of vehicles and fuels that are certified under sustainability schemes (which could eventually involve a greenhouse gas component) could provide a powerful market driver. Local governments can switch entire fleets to vehicles that run on biofuels, as many have already done. National governments could gradually increase the share of their fleets that are fuelled by biofuels and ramp up to 100 per cent; the one exception might be tactical military vehicles.
- *Collaborate to set international fuel quality standards.* While many nations have developed or adopted biofuel quality standards, others still need to take this step. In order to develop a significant international biofuel market, fuel quality standards must be agreed upon and enforced at the international level. This is necessary for consumer confidence and will gain increased importance as international trade in biofuels expands. Automakers need assurances of consistent fuel characteristics so that they can honour vehicle warranties.
- *Account for externalities.* Although it is extremely difficult, decision-makers should find ways of assigning monetary values to currently uncounted externalities, including local and regional pollution, health problems, climate change and other environmental costs, as well as potential benefits, such as job creation and rural revitalization. This can be done through tax increases or incentives. For example, in the case of climate change, this could be achieved through a carbon cap and trade system (note, however, that this would probably not benefit biofuels in the short term; see Chapter 11).
- *Facilitate public–private partnerships.* Public–private partnerships have resulted in important technological breakthroughs that have led to dramatic cost reductions (e.g. in the enzymes needed for the breakdown of cellulose via enzymatic hydrolysis), and will continue to play an important role in advancing next-generation technologies.
- *Increase public awareness.* Consumer demand could be a powerful driver of the renewable fuels market. Strategies to increase the public's awareness and comfort level with biofuels include various forms of public education, such as formal awareness campaigns, public announcements, university research and signage along highways. Typically outside the government sphere, but also potentially effective, informal methods include discussions on radio, blogs, podcasts and the use of biofuels in movies and television shows.

Mandates paired with subsidies have also proven to be an effective combination for biofuels industry promotion; however, subsidies should be phased out once a domestic industry has been established. Subsidies are often difficult to discontinue once created, so phase-outs should be strategically designed into the enabling legislation. For instance, subsidies for current-generation biofuels can be phased out first, while those for next-generation feedstocks and refineries continue.

Mandates and subsidies can be used together, or, as in the case of Germany, mandates can follow subsidies. As of early 2006, the German government was in the process of replacing subsidies for first-generation biofuels with a fuel-blending mandate, but intended to maintain the subsidy for next-generation biofuels to further their development. In the near term, the promotion of biomass generally for various bioenergy and materials uses will help to develop the biomass feedstock production sector, while the next-generation liquid fuel conversion technologies are developed.

Public concerns regarding possible environmental impacts of biofuel feedstock cultivation must also be addressed if biofuels are to gain broad public acceptance (see Chapter 18 for a discussion of certification and other proposed schemes to ensure the sustainable production of biofuels).

NATIONAL AND INTERNATIONAL RESEARCH, DEVELOPMENT AND DEMONSTRATION

To date, the world's engineering and scientific skills have not been focused coherently on the challenges associated with large-scale biofuel development and use. Thus, there is enormous potential for dramatic breakthroughs in feedstocks and technologies that could allow biofuels to play a major role in enhancing energy security, reducing greenhouse gas emissions and providing much of the world community with economical transport.

There has been a tremendous surge in private-sector investment in biofuels during recent years; but this investment tends to be oriented towards short-term and high-payoff research. There are many long-term research needs that governments are best suited to address; governments and international organizations should help to coordinate public and private efforts by bringing together the best minds and resources in national research facilities, universities, civil society and industry. Because intermittent funding seriously hampers research efforts, funding for research, development and demonstration must be consistent, as well as long term. It is worth noting that much of this research will probably have applications across the broader agricultural sector.

Research is needed to develop feedstocks and sustainable management practices, as well as technologies for harvesting, processing, transporting and storing feedstocks and fuels. Research is also required to better understand the potential environmental and societal impacts of biofuels throughout the entire supply chain. Biofuels and bioenergy, as a whole, are a cross-sectoral topic, which can only be analysed in an integrated way. Some of the key areas for further research are provided below.

Feedstock production

- *Improve conventional feedstocks.* Improve energy yields of conventional biofuel feedstocks, while developing sustainable management systems that include minimizing the use of chemical inputs and water. This includes research into the potential for modifying food crops to maximize both food and cellulose (for energy) production.
- *Develop next-generation feedstocks.* Improve management techniques and develop high-yield perennial crops suited for biofuel applications that require low inputs, are location appropriate, and can improve soil and habitat quality while sequestering carbon (see Chapters 4 and 12).
- *Advance alternatives to chemical inputs.* Research the potential for integrated pest management and organic fertilizer development and use, including the use of mixed crops, rotations and other management techniques.
- *Assess the risks of genetic modification.* Potential risks and costs of developing and using genetically modified (GM) crops must be fully assessed to determine if benefits outweigh costs. It is also important to research and develop appropriate safeguards for the use of genetically modified industrial organisms' required biological conversion of cellulosic biomass to ethanol.
- *Supplement environmental life-cycle studies.* Research is needed to fill in gaps in the existing body of analyses with regard to global climate impacts and effects on local and regional air, soil and water quality, and habitat, including a better understanding of the impacts of land-use changes, and of the scale of nitrous oxide (N_2O) emissions from feedstock production, and their potential impact on the global climate (see Chapters 11 to 13).
- *Develop methodology for measuring life-cycle greenhouse gas (GHG) emissions.* There is need for consistent, internationally used methodologies and assumptions for measuring GHG emissions associated with the production and use of biofuels from various feedstocks, associated land-use changes, management strategies and processing practices (see Chapter 11).

Feedstock collection and handling

- *Improve equipment and harvesting practices.* Agricultural equipment and harvesting practices must be optimized for both crop and residue harvesting in order to maximize economic benefits for farmers, while minimizing soil compaction and minimizing interruption of primary food crop harvests.

- *Ascertain sustainable residue removal rates.* Conduct research to determine sustainable extraction levels of agriculture and forestry residues to maintain soil quality under varying conditions.
- *Improve waste-handling practices.* Develop optimal means for the safe handling and collection of various municipal waste resources (e.g. waste grease and cardboard).
- *Optimize feedstock storage and transport methods.* For example, improved methods are needed to prepare feedstocks for transport through reducing bulkiness and water content.

Processing

- *Maximize efficiency of input use.* Technologies and practices should be optimized to make the most efficient use possible of water, energy, chemicals and other inputs, and to minimize waste through recycling of wastewater, waste heat, etc.
- *Advance the biorefinery concept.* Continue support for the integration of a variety of related operations, including use of animal and crop residues as fuel feedstocks and/or for process energy, and co-products (such as wet distillers grain) as animal feed, bio-plastics, etc.

Fuel distribution and end use

- *Advance fuel and power train development.* Combine research and design needs to optimize engine designs/performance to take full advantage of the unique properties of biofuels (e.g. higher oxygen content, higher octane, etc.), and evaluate fuel specification criteria to identify potential fuel changes that could improve engine performance.
- *Optimize vehicles.* This includes fine-tuning control systems and engine designs to run on varying blends for maximum fuel efficiency and minimum emissions across the full range of potential blend mixes.
- *Develop materials.* Research materials for higher-quality tubes, hoses and other connectors to reduce evaporative emissions.
- *Develop fuel additives.* Additives are needed to reduce emissions of nitrogen oxides (NO_x) and other harmful emissions from blends of fossil and biofuels.

Demonstration and field trials

In addition to resource assessments, policy analyses, and applied crop and processing research, it will be critical to advance experience on the ground in varied settings. This will include field trials of new energy crops in different climate and soil conditions. Pilot conversion facilities, using cutting-edge technologies, should be funded and constructed in a wide range of settings in order to work out any related problems or challenges, and to develop and make use of *in situ* ingenuity and local adaptation of technologies, crops and crop management, and handling systems. This should involve well-organized and well-monitored efforts in several countries (with varying climates, soil conditions, social structures, etc., including heavily degraded and desert lands) in order to build a body of practical experience over the next decade.

Outreach/extension

On the national level, findings need to be disseminated to producers through demonstration projects, extension services (where they exist) and other farmer education mechanisms, including feedstock demonstration projects. In addition, farmers will need the appropriate know-how, capital and incentives to risk planting new crops and to follow best practices. Sustainable management and good crop choices should be tied to existing or newly created government incentives.

Information clearing house

On the international level, a clearing house is needed – such as the Renewable Energy Global Policy Network (REN 21) or a small international institution – to gather and make available to the global community information regarding relevant findings and experiences with biofuel research and policies from around the world. This could be a subset of REN 21 or a separate body focusing on biofuels and agriculture.

INCENTIVES FOR RAPID DEPLOYMENT OF ADVANCED LOW-IMPACT BIOFUELS AND TECHNOLOGIES

Policies are needed to expedite the transition to the next generation of feedstocks and technologies that will enable dramatically increased production at lower cost, combined with the real potential for significant reductions in environmental impacts. To date, high costs and risks associated with the construction of new conversion facilities have hampered the development of next-generation fuels. Governments and international financial institutions can play a critical role in reducing financial risks and providing low-cost capital, helping industry to move quickly through early commercialization barriers.

Specific actions that governments can take to expedite the transition include:

- *Provide incentives.* Create tax structures and other incentives that favour next-generation biofuels and integrated 'biorefineries' and bioprocessing.
- *Enact mandates.* Mandates could require that an increasing share of total fuel comes from advanced feedstocks and technologies.
- *Fund research and development (R&D).* More sustainable feedstocks and technologies are needed, including those that provide enhanced net reductions in GHG emissions and in fossil inputs.
- *Support farmers.* Farmers will need information, crop and equipment assistance, market access and other help to make the transition to producing new feedstocks.
- *Facilitate conversion of existing plants.* Retraining and retooling are important for converting existing plants to next-generation facilities.
- *Provide capital.* Low-interest, long-term loans and risk guarantees are required to facilitate the development of commercial cellulosic refineries and 'biorefineries'.
- *Encourage the development of new uses and demand for co-products.*
- *Encourage technology transfer.* Transfer of technology and capacity-building to countries with nascent industries (particularly those with great potential for producing sustainable feedstocks and fuels) will be of utmost importance.

INFRASTRUCTURE DEVELOPMENT

Ethanol use can increase to 10 per cent of non-diesel fuel, possibly more, with minimal changes to current car fleet or infrastructure; biodiesel blends can be higher. To go beyond this, however, governments need to address the 'chicken or the egg' dilemma: vehicles are needed that can run on high blends of biofuels; but consumers will not buy them without a distribution system that ensures access to these fuels. Such a distribution system is not likely to develop without the vehicles to demand/use it. This dilemma can be resolved with technologies such as flexible-fuel vehicles (FFVs) (see below and Chapter 15).

To enable the expansion of biofuels, infrastructure changes will also be required on the production side (especially for next-generation biofuel production). New crops and production methods, as well as associated distribution requirements, will necessitate substantial infrastructure planning and development. The existing infrastructure available for the use of agricultural and forestry resources should be evaluated to determine what expansion and refinements are required for renewable biomass resources to play an expanding role in providing sustainable transportation fuel supplies.

In order to encourage the necessary infrastructure transition, governments could:

- *Advance flexible-fuel vehicle technology.* Governments could advance the development and availability of flex-fuel vehicles, including those appropriate for high blends, through legislative mandates or softer incentives (such as targets; for example, governments could call for 100 per cent of new cars available in the domestic marketplace to be biofuel compatible within ten years). In promoting FFVs, governments should not allow trade-offs in fuel economy or air quality standards.
- *Promote the use of flexible-fuel vehicles.* In addition or instead, governments could establish incentives for consumers who buy such vehicles and use them with biofuels. Governments should also commit to transitioning to flex-fuel vehicles for non-diesel, non-strategic fleets.
- *Require fuel companies to provide biofuels.* Because of the control that the fossil fuel companies hold over fuel distribution and sale in most countries, most governments may have to require that these companies distribute and sell biofuels. Governments could, for example, require that all refuelling stations over a certain size convert at least one pump to biofuels (this would have to be phased in as fuel becomes available). This may not be appropriate in countries where blending mandates exist, and such a requirement could destroy market niches for smaller distributors.
- *Support small refuelling stations.* Smaller petroleum dealers and refuelling stations should be supported since they have a higher chance of success (as has occurred in Sweden).
- *Support development of new fuel standards.* As higher blends become more desirable, fuel standards will need to be modified. Because this is a lengthy process, this should start as soon as possible.

OPTIMIZING ECOLOGICAL IMPACTS

While many perceive biofuels as environmentally beneficial because they are 'renewable', these fuels have the potential to positively or negatively affect the natural world (everything from local soil and water quality, to biodiversity, to the global climate) and human health, depending upon factors such as feedstock selection and management practices used. Whether the impacts are largely positive or negative will be determined, in great part, by policy.

As described in detail in Chapters 11 and 12, the most significant potential impacts associated with biofuel production result from changes in land use, including natural habitat conversion. With regard to climate change, land-use changes (from razing of tropical forests to replacement of grasslands) for the production of biofuel feedstocks can result in large releases of carbon from soil and existing biomass, negating any benefits of biofuels for decades. Therefore, governments must prioritize the protection of virgin ecosystems and should adopt policies that compel the biofuel industry to maintain or improve current management practices of land, water and other resources.

Next-generation feedstocks and technologies offer the potential to improve soil and water quality, enhance local species diversity and sequester carbon if lands are managed sustainably. This provides governments with yet another reason to speed the transition.

In addition, national and international standards and certification schemes will be necessary to safeguard the resource base (see Chapter 18 and below). Standards and best management practices take time to develop properly; so it is critical to initiate practical step-by-step processes that entail consistent progress towards increased sustainability. Work on this has begun, but should be supported with more substantial resources and greater international coordination.

Some specific actions that governments should take to help safeguard the environment and human health, while ramping up biofuels production, are provided below.

Feedstock production

- *Conserve natural resources.* Local, national and regional policies and regulations should be enacted to ensure that impacts on wildlife and on water, air and soil quality are minimized. For example, payment systems for irrigation and processing water could be adopted to encourage more efficient use, and nutrient and water recycling should be encouraged.
- *Protect virgin and other high-value habitats.* Governments must find ways of protecting natural forests, wetlands and other ecosystems that provide air and water purification, soil stabilization, climate regulation and other vital services. Options include enforcing bans on wild land conversion for biofuel feedstock production, including strong penalties for non-compliance; using satellite and global imaging technology to track land-use changes; tying tax incentives, carbon credits, qualifications for government purchase, sustainable production certification, and so on to the maintenance of natural ecosystems; and requiring land reserves. Large-scale feedstock producers can be required to set aside a share of their land as a natural reserve, as the Brazilian state of São Paulo has done.
- *Encourage sustainable crops and management practices.* Extension services for farmers should provide them with the proper resources and incentives to select sustainable crops (particularly native species that reduce the need for water, fertilizers and pesticides), reduce the frequency of tilling and replanting, and provide habitat for wildlife. They should encourage sustainable management practices, including minimal use of inputs, buffer zones between waterways or wild lands and crops, intercropping, crop rotation, and adjusting harvest schedules to minimize conflicts with wildlife. Subsidies can be linked to meeting specific criteria.

- *Improve degraded lands.* Encourage the rehabilitation of degraded lands through appropriate perennial feedstock production.
- *Maximize GHG benefits.* Feedstocks should be selected to maximize GHG reductions (see Chapters 4 and 11).

Processing, distribution and end use

- *Develop licensing procedures.* Refineries should meet strict environmental standards, which include efficiency of water use and recycling, air and water pollution controls, etc.
- *Promote the use of renewable process energy.* Provide incentives to use biomass as process energy and guarantee fair access to the grid for sale of excess electricity.
- *Establish emissions standards for biofuels.* Just as regulations exist for conventional fuels, they are necessary for transport and combustion of biofuels. Regulations are needed to minimize spills and hydrocarbon emissions during transport and fuelling, and to minimize evaporative and combustion emissions from storage, handling and combustion stages of the supply chain.
- *Encourage rapid transition to high-blend fuels.* High blends with properly optimized vehicles can minimize a variety of harmful emissions. High biodiesel blends, particularly in urban areas of developing countries (where there may be weak emissions standards), can reduce public health risks, especially from particulate emissions. Cities can commit to shifting public buses and other government vehicles to 100 per cent biodiesel over a few years.
- *Encourage biofuels for a range of uses.* In developing countries where lead is still used as a transport fuel oxygenate (particularly in Africa), ethanol should be phased in rapidly to replace it. Biofuel (especially pure biodiesel) use for marine applications is particularly beneficial and should also be encouraged. Biofuel use for agricultural machinery (as in Germany), and construction and other heavy equipment, which is generally far more polluting and has a much slower turnover rate, should be encouraged as well.

MAXIMIZING RURAL DEVELOPMENT BENEFITS

If biofuels continue their rapid growth around the globe, the impact on the agricultural sector will be dramatic. Increased jobs and economic development for rural areas in both industrialized and developing countries are possible if governments put the appropriate policies in place and enforce them. The more involved farmers are in the production, processing and use of biofuels, the more likely they are to benefit from them. Enabling farmer (and forest material producer) ownership over more of the value-added chain will improve rural livelihoods. This not only helps to improve the well-being of farm families; it increases the positive effects as greater farm income is circulated in local economies and jobs are created in other sectors. As biofuel industries grow, this multiplier effect will have impacts on the regional, national and international levels. Greater farmer ownership will also help to prevent a repetition of the dynamics in the current global food industry, where very large processors are able to exert pressure on producers.

In regions where access to modern forms of energy is limited or absent, government and development agency support for small-scale biofuel production can help to provide clean, accessible energy that is vital for rural development and poverty alleviation.

Specific options for decision-makers include:

- *Cooperatives and small-scale ventures.* Governments can provide support for cooperatives and small-scale biofuel production facilities – for example, through tax structures that give preference to small-scale feedstock and fuel production, or preferential government purchasing from farmer/cooperative-owned facilities. Cooperatives allow small- and medium-sized producers to share more in the economic gains of the biofuel industry and to negotiate on a more equal footing.
- *Purchasing from small producers.* Governments can require fuel purchasers and distributors to buy a minimum share from farmer or cooperatively owned facilities.
- *International development funding.* National and international development institutions can provide financial and technical support for small-scale biofuel initiatives for rural energy provision and poverty alleviation.
- *Technical and materials assistance.* Governments, civil society and others can provide assistance to small landholders in obtaining materials (the seeds and seedlings for energy crops), know-how and market access.
- *Appropriate fiscal policies.* Governments can implement policies that allow for local approaches to be developed. Government action to ensure markets for biofuels and for energy crops (e.g. mandates and preferential purchasing) helps to give producers the confidence to adopt new crops and crop management systems. In addition to providing markets for their products, ensuring fair prices for farmers is also essential to improving rural livelihoods.

ENCOURAGING SUSTAINABLE TRADE IN BIOFUELS

For the dozens of nations that are just beginning to develop biofuel industries, many decisions will have to be made, including the type, scale and orientation (i.e. for domestic consumption, for export or both) of production. Policies will need to be designed, appropriately based on domestic economic and resource situations; and with the rapid pace of biofuels development, they will need to be put in place soon. Decision-makers will also need to factor in the impacts that the policies of other nations (e.g. the European Union Biofuels Initiative) and international trade policies (e.g. continuing trade liberalization negotiations) will have on their own biofuel and biofuel feedstock markets. In general, biofuels trade restrictions should be removed over time, respecting the fact that the countries with nascent industries will want to protect them.

Integrated planning is necessary at the national level so that short-term or sectoral interests do not take precedence over strategic national priorities. For instance, market incentives at the micro-economic level might encourage biofuel exports. But when other factors – such as national employment needs; domestic energy and security needs; trade balance; food security and land-use concerns; the condition of domestic transport and export infrastructure; and GHG reduction obligations – are taken into consideration, exports might not make sense at that point in time. In many nations where displacing a modest amount of petroleum could make a significant difference, production for domestic use should take precedence over export. Alternatively, the value of biofuels as an export commodity to earn foreign exchange may be preferable in other instances. National leaders will need to weigh these factors for their countries.

Well-established markets such as the US and the EU have enormous fuel needs and growing energy security concerns. Due to policy initiatives actively promoting the use of biofuels, markets in these countries are large enough to accommodate both domestic production and imports (and the more rapidly biofuel-compatible transport infrastructure is phased in, the faster their biofuels markets will grow). International trade may help to ease fuel supply issues, linking a larger number of producers in order to minimize the risk of supply disruption. Furthermore, as renewable fuel use becomes more widespread, opportunities for countries with more developed biofuel industries to export their technologies will expand.

Some agriculture incentive programmes in wealthy countries have been blamed for supporting food production in a way that harms competitors in developing countries. These could be transformed into programmes that, instead, support biofuel production, a process that has begun in Europe and is being discussed in the US. While this is a step in the right direction, replacing highly subsidized and protected commodity food production in rich countries with highly subsidized and protected biofuel production is not the aim. Biofuel support strategies must be planned with gradual phase-outs or other means of moving beyond the subsidies once they are no longer necessary.

Trade and the environment

Energy crops and biofuels may be categorized as agricultural goods under the World Trade Organization (WTO) Agreement on Agriculture. Industry proponents may seek an exemption from the agreement's restrictions on domestic price supports by including biofuels subsidies in the so-called 'Green Box'. To qualify for Green Box status, the incentives must be 'non-trade distorting', meaning that they do not affect global market prices. This will be a difficult test to meet if financial incentives for biofuels are tied to production levels, especially if the trade grows to a significant size. The more that incentives are clearly tied to producing public goods – such as clean water and air, wildlife habitat preservation, carbon sequestration and soil erosion control – unconnected to crop yields and refinery production levels, the more likely they are to pass muster.

Alternatively, if biofuels are categorized as industrial goods, they may qualify for treatment as 'environmental goods'. To be included in such a category, they should be required to meet strict environmental standards for their production.

Developing countries have traditionally fought attempts to differentiate among traded goods based on process and production methods (PPMs). However, some biofuel producers in developing countries could rank quite well in a scheme based on production standards. For example, the ethanol industry in Brazil has generally achieved very low net GHG emissions (for more information on trade and biofuels, see Chapter 9).

Standards and certification

There are increasing calls in Europe and elsewhere for traded biofuels to be certified based on social and environmental standards. This could provide a means of ensuring that the production of these fuels provides net positive impacts for the planet and for society. However, if not developed in a participatory and transparent way, such a certification scheme could be viewed as a means for industrialized countries to erect new trade barriers to protect their domestic biofuel producers.

A certification framework based on sound standards could become a critical driver to facilitating the development of sustainable trade in biofuels. A compromise must be reached between developing complicated certification schemes to ensure long-term sustainable biomass trade, on the one hand, and putting safeguards in place quickly to direct the rapidly growing market, on the other. The incremental development of such a certification scheme is probably the most feasible option, allowing for gradual learning and expansion over time. Existing certification schemes provide useful models. While not all biomass types may fulfil the entire set of sustainability criteria initially, the emphasis should be on the continuous improvement of sustainability benchmarks.

While a certification scheme should be thorough, comprehensive, transparent and reliable, it should also not create a significant hurdle for nascent biofuel industries. Criteria and indicators should be adaptable to the requirements of different regions and be mindful of the implementation costs. It will be important to pair any certification scheme with technical assistance, incentives and financing so that small- and medium-scale producers can qualify as readily as large-scale producers. Furthermore, it is important to ensure that any standards and certification schemes for biofuels address the issue of possible leakage effects, through which benefits gained in one location could 'leak away' to another (for more information see Chapter 18).

Moving forward, additional research will be needed to determine whether an independent international certification body for sustainable biomass is feasible. This should be done in collaboration with a consortium of all stakeholders in the biomass-for-energy production chain. At this stage, and at later steps in the development process, public information dissemination and support will be critical. It will be important to evaluate how likely broad participation by the petroleum industry, biofuel industry, importers and consumers will be. Their participation is necessary in order for such a scheme to be accepted in the market. Costs and benefits for the various participants need to be analysed.

KEY OVERARCHING RECOMMENDATIONS

- *Develop the market.* Biofuel policies should focus on market development. An enabling environment for renewable fuels industry development must be created in order to draw in entrepreneurial creativity, private capital and technical capacity.
- *Speed the transition to next-generation technologies.* Policies are needed to expedite the transition to the next generation of feedstocks and technologies that will enable dramatically increased production at lower cost, combined with the real potential for significant reductions in environmental impacts.
- *Protect the resource base.* Maintenance of soil productivity, water quality and the myriad other ecosystem services is essential. The establishment of national and international environmental sustainability principles and certification is important in protecting resources, as well as in maintaining public trust regarding the merits of biofuels.
- *Facilitate the sustainable international biofuel trade.* The geographical disparity in production potential and demand for biofuels will necessitate a reduction in barriers to biofuel trade. Freer movement of biofuels around the world should be coupled with social and environmental standards and a credible system to certify compliance.
- *Distribute benefits equitably.* This is necessary in order to gain the potential development benefits of biofuels. Enabling farmers to share ownership throughout the production chain is central to this objective.

To achieve a rapid scale-up in biofuels production that can be sustained over the long term, governments must enact a coordinated set of policies that are consistent, long term and informed by broad stakeholder participation.

Governments should promote biofuels within the context of a broader transformation of the transportation sector. Biofuels alone will not solve all of the world's transportation-related energy problems. Development of these fuels must occur within the context of a transition to a more efficient, less polluting and more diversified global transport sector. They must be part of a portfolio of options that includes dramatic improvements in vehicle fuel economy, investments in public transportation, better urban planning, and smarter and more creative means of moving around a village or across the globe.

In order to achieve their full potential to provide security, environmental and social benefits, biofuels need to represent an increasing share of total transport fuel relative to oil. In combination with improved vehicle efficiency, smart growth and other new fuel sources, such as biogas – and, eventually, even renewable hydrogen or electricity – biofuels can drive the world towards a far less vulnerable and less polluting transport system.

Part VIII

Country Studies

20

Biofuels for Transportation in China

China study team leader: Professor Wang Gehua, Institute of Nuclear and New Energy Technology, Tsinghua University, Beijing

China study authors: Professor Dr Zhao Lixin, Chinese Academy of Agricultural Engineering (CAAE), Beijing; Zhang Yanli, CAAE, Beijing; Dr Fu Yujie, Northeast Forestry University, Harbin; Elisabeth-Maria Huba, Consultant, Beijing; Liu Dongsheng, CAAE, Beijing; Professor Li Shizhong, Centre of Bioenergy, China Agricultural University, Beijing; Professor Liu Dehua, Department of Chemical Engineering, Tsinghua University, Beijing; Heinz-Peter Mang, Institute for Energy and Environmental Protection (IEEP), Beijing

Contracting, policy dialogues and editing of study: Hans-Joerg Mueller (GTZ), Elke Foerster (GTZ) and Dirk Peters (GTZ)

The Chinese government began promoting biofuels several years ago in the face of rising energy demands and a sizeable grain surplus. In the coming decades, energy consumption in China is expected to continue its dramatic climb. The aim of this study is to assess China's future role as a biofuel producer and importer, to identify potential impacts on global markets, and to point to related investment opportunities.

CURRENT SITUATION

Ethanol

In 2004, China produced 1.3 million tonnes of ethanol for use as fuel. The government currently provides a subsidy of €137 (US$166) per barrel of ethanol, and five provinces now mandate a blend of 10 per cent ethanol (E10) in gasoline. The corresponding regulations are being handled at the province level, although the Chinese government intends to create a national statutory framework for a nationwide E10 blending obligation by 2020. This would translate into a demand of approximately 8.5 million tonnes of ethanol.

Production costs for ethanol currently range between €0.23 and €0.38 (US$0.28 and $0.46) per litre, depending upon a raw material price of between €0.16 and €0.32 (US$0.19 and $0.39) per litre. In the near future, Chinese ethanol production will be based mainly on sweet sorghum and cassava (manioc); in the past, production was based on surplus wheat, which is no longer available.

Biodiesel

China currently produces 50,000 tonnes of biodiesel per year, primarily from used cooking oil (edible oils). Production costs range from €0.17 to €0.35 (US$0.21 to $0.42) per litre and are thus still relatively low compared with other regions (in Germany, for example, substantially higher costs of €0.68 or US$0.82 per litre, are common). Existing biodiesel facilities and those currently under construction will provide a total annual capacity of approximately 2 million tonnes by 2010, corresponding to about 3 per cent of China's predicted diesel consumption.

POTENTIAL

China's ethanol production target for 2020 is between 8 and 20 million tonnes, based on an expansion in the current area from 2.7 million to 7.6 million hectares in 2020. According to the National Development and Reform Commission, domestic production will be sufficient in 2020 to supply a 10 per cent blend (even with conservative assumptions). This assessment assumes that 25 per cent of the required biomass will be covered by foodstuffs.

The Chinese Ministry of Science and Technology aims to produce 12 million tonnes of biodiesel by 2020. It estimates that, optimistically, the area planted in oilseed crops (40 million hectares in 2004) can be expanded to a maximum of 67 million hectares. Cultivating jatropha and a species of pistachio is currently under discussion. Based on a 10 per cent blend, however, even a conservative estimate suggests that there will be a 7 million tonne biodiesel shortage in 2020.

According to forecasts, about 9 million jobs can be created in Chinese agriculture and industry through the production of biodiesel and ethanol.

OUTLOOK

Decentralizing Chinese biofuel production will require improvements in the technologies for converting biomass and suitable wastes into liquid biofuels. Such decentralization would enable savings in energy expenditure and reduced transportation costs. China is very interested in the development of biomass-to-liquid (BTL) technologies, particularly in light of the large quantities of agricultural and forestry waste products generated in the nation's rural regions. Yet, despite the fact that the first BTL conference was held in China in 2001, not a single BTL pilot plant is in operation today.

China's biodiesel market, meanwhile, would develop much more quickly if the government were to introduce standards for cultivation, processing technologies and distribution networks. Rather than national biodiesel standards, China currently applies inadequately defined diesel standards. For ethanol, the Chinese regulations follow US standards, which differ from European standards. In practical terms, this will be reflected in higher blending ratios, further increasing Chinese demand for ethanol.

Although the prospects for international ethanol trade are good, and production costs in China remain below the world market price, it cannot be assumed that China will become a global ethanol supplier. The reasons for this are limited production capacities and high domestic demand. In the future, China will probably continue to struggle to meet its rapidly rising demand for fuel with domestic biofuel production. It is therefore likely to emerge as a buyer, both regionally and globally. This will lead to corresponding price increases on biofuel markets. To facilitate greater international biofuel trade, China will require ports with suitable import and export capacities and associated investments.

Table 20.1 *Summary of the ethanol situation in China*

Parameter	*Current*	*2020*
Fuel ethanol output (tonnes)	1 million	8–28 million
Total ethanol (tonnes)	>3 million	
Fuel ethanol area (ha)	2.7 million	4.3 million (2010) 7.6 million, minimum (2020)
Subsidies for ethanol (per tonne)	€137 (US$166)	0
Mandatory blending	10% in five provinces	10%
Production costs for ethanol (per litre)	€0.23–€0.38 (US$0.28–$0.46)	
Feedstock price alone	€0.16–€0.32 (US$0.16–$0.32)	
Ethanol net energy balance (in:out)	1:1.1 (corn) 1:2.1 (sugar cane) 1:0.7 (cassava)	

Table 20.2 *Summary of the biodiesel situation in China*

Parameter	*Current*	*2020*
Area planted in oilseed crops (ha)	40 million	Up to 67 million
Cooking oil consumed (tonnes per year)	18 million*	70 million
Biodiesel production (tonnes per year)	50,000–60,000 (2004)	1.5–2 million (2010) 10.6–12 million (2020)
Biodiesel plant capacity (tonnes per year)	82,000 (2004) 241,500 (2006)	1.5–2.2 million (2010)
Biodiesel production costs (per litre)	€0.17–€0.35 (US$0.21–$0.42) depending upon feedstock	
Biodiesel energy (GJ per hectare)	120	130

Note: * According to the National Bureau of Statistics, China consumed a total of 18 million tonnes of edible oil in 2004. Of this, 4 to 5 million tonnes became waste oil, and from this 2 million tonnes was collectable.

Source: the full study, *Biofuels for Transportation in China*, is available at www.gtz.de/de/dokumente/en-biofuels-for-transportation-in-china-2005.pdf

21

Biofuels for Transportation in India

India study team: N. V. Linoj Kumar, The Energy and Resources Institute (TERI), New Delhi; Dr Sameer Maithel, TERI; K. S. Sethi, TERI; S. N. Srinivas, TERI; M. P. Ram Mohan, TERI; Dr K. V. Raju, Institute for Social and Economic Change (ISEC), Bangalore; Dr R. S. Deshpande, ISEC; Dr. Mohammed Osman, International Crop Research Institute for Semi-Arid Tropics (ICRISAT), Hyderabad

Contracting, policy dialogues and editing of study: Michael Glueck (GTZ), Ali Kaup (GTZ), Elke Foerster (GTZ) and Christine Clashausen (GTZ)

Like China, India is a very populous nation experiencing rapid economic growth and rising energy demands. The Indian government actively promotes the production of biofuels. This study assesses the current situation and future opportunities for biofuel production and use in India.

CURRENT SITUATION

Ethanol

India has been operating an ethanol programme for several years; however, its activities have been severely hampered since 2002 by crop failures due to drought.

India currently produces 665 million litres of fuel ethanol annually, derived primarily from molasses from sugar production. Subsidies for using sweet sorghum as a feedstock are also being considered. The nation's current sugar cane acreage is 4.4 million hectares, although the goal is to expand this slightly to 5 million hectares by 2007. Nine provinces currently have an official blending obligation of 5 per cent ethanol in gasoline. Production costs are currently around €0.36 (US$0.44) per litre, which is an average value from an international perspective.

Biodiesel

Biodiesel is a relatively new fuel in India and not yet commercially available, although it is already being produced in pilot projects – for example, public–private partnerships involving DaimlerChrysler, Hohenheim University and Deutschen Investitions- und Entwicklungsgesellschaft (DEG); and the German company Lurgi, the Indian company Southern Online Bio Technologies (SBT) and the Gesellschaft für Technische Zusammenarbeit (GTZ). Several states (Andhra Pradesh, Tamil Nadu, Chhattisgarh, Uttaranchal and Rajasthan) have established a policy framework to support biodiesel.

POTENTIAL

Over the medium term, India's ethanol programme aims to achieve a 5 per cent blend in gasoline nationwide. Based on close links with the sugar industry, this is regarded as ambitious, but achievable.

The targets for biodiesel production are less clearly defined since production has only just begun. According to information contained in the *Wasteland Atlas* (Government of India, 2003), 55.3 million hectares of so-called 'wasteland' are available in India alone. The government is very hopeful that this degraded land can be cultivated with oilseed crops, particularly jatropha and pongamia. To achieve a 20 per cent biodiesel blend (B20) by 2020, 38 million hectares of wasteland would have to be cultivated, and the current yield of 1 to 2 tonnes per hectare would have to increase to 5 tonnes per hectare. Availability of degraded land is limited to some extent due to unresolved ownership issues. In addition, the long-term economic viability of jatropha plantations on degraded soils has not yet been established.

OUTLOOK

Due to high domestic fuel demand, India is likely to emerge as a biofuel importer rather than an exporter. This is based on attempts to diversify the sources for increasing fuel demand (in 2003, India imported 90.4 million tonnes of crude oil, and it is projected to import 166 million tonnes in 2019). Achieving the government's proposed blending targets will require significant investments in national production and processing facilities.

Table 21.1 *Summary of the ethanol situation in India*

Parameter	*Current*	*2006–2007*
Fuel ethanol production (litres)	665 million	823 million
Total ethanol production (litres)	2 billion	
Total sugar cane area (ha)	4.4 million	5 million
Mandatory blending	5% in nine states	
Production costs for ethanol (per litre)	€0.36 (US$0.44)	

Source: The full study, *Biofuels for Transportation in India*, is available at www.gtz.de/de/dokumente/en-biofuels-for-transportation-in-india-2005.pdf

22

Biofuels for Transportation in Tanzania

Tanzania study team: Dr Rainer Janssen, WIP – Renewable Energies; Dr Jeremy Woods, Themba Technology; Gareth Brown, Themba Technology; Estomih N. Sawe, Tanzania Traditional Energy Development and Environment Organization (TaTEDO); Ralph Pförtner, Integration Umwelt und Energie GmbH

Contracting, policy dialogues and editing of study: Christine Clashausen (GTZ) and Elke Foerster (GTZ)

Tanzania is an African country that has great agricultural potential, and at the same time is affected strongly by changing sugar trade regulations. Currently, it does not have a policy framework for biofuels. This study assesses the resource base, policy environment and other factors that will influence the potential for future biofuel production in the country.

CURRENT SITUATION

Tanzania is completely dependent upon petroleum imports (1.2 million tonnes in 2003), which account for roughly 40 per cent of all imports and are responsible for a significant share of the country's foreign exchange spending. The transport sector consumes more than 40 per cent of imported refined petroleum products. In 2002, Tanzania consumed roughly 134,000 tonnes of gasoline and 390,000 tonnes of diesel, and the Tanzania Petroleum Development Corporation forecasts annual growth of 5 per cent in both gasoline and diesel demand to 2010.

During the summer of 2005, fuel prices for unleaded gasoline ranged between 1120 and 1195 Tanzanian shillings (€0.77 to €0.83, or US$0.93 to $1) per litre, and diesel prices ranged between 1075 and 1095 Tanzanian shillings (€0.75 to €0.76, or US$0.91 to $0.92) per litre. Tanzania has a complex taxation system for petroleum products consisting of three main taxes (excise duty, road toll and value-added tax), which together comprise about 40 per cent of the final fuel price charged to consumers.

Currently, there is no commercial biofuel production in Tanzania. However, several stakeholders are engaged in the development of biofuels. The main players with regard to commercial biodiesel production include the local organizations FELISA (palm oil), KAKUTE, and the foreign companies Diligent, PROKON and D1 Oils (jatropha oil). For sugar cane-based ethanol production, the key players are the country's four main sugar companies: Kilombero Sugar Company, Mtibwa Sugar Estates, Kagera Sugar Limited and Tanganyika Planting Company.

Current biofuel activities and opportunities in Tanzania can be roughly divided into large- and small-scale approaches. Large-scale biofuel production, such as ethanol production from sugar cane promoted by the sugar industry, will focus primarily on biofuels for transportation. Supportive policies and regulations will be required to secure the rather large investment required for start-up. Smaller-scale biofuel activities conducted by organizations such as FELISA (Farming for Energy for better Livelihoods in South Africa) and Kakute, in contrast, are concerned mainly with generating rural income and revenue opportunities from oilseed crops through production of either plant oils (for food and/or fuel) or commodities such as soap from jatropha oil.

POTENTIAL

A recent assessment by the United Nations Food and Agriculture Organization (FAO) found that 44.4 million hectares of land in Tanzania are potentially available for crop production (both food and non-food). While these figures present only a broad picture of land use in a very large and diverse country, they suggest that land availability is not likely to be a barrier to bioenergy production in Tanzania.

Table 22.1 estimates the potential for bioenergy production from 'potentially available land' (44.4 million hectares) and can be used to gauge the limits of the actual production potential.

Table 22.1 *The potential for bioenergy production in Tanzania*

Parameter	*Potential (estimated)*
Land area (ha)	44.4 million (30 million very suitable)
Sugar crops	570,000
Cereal crops	24 million
Root crops	14 million
Energy	3.3 exajoules (EJ)
Palm oil	186 gigajoules (GJ) per hectare
Jatropha oil	59 GJ per hectare
Ethanol from sugar cane	173 GJ per hectare
Ethanol from C-molasses (treacle)	20 GJ per hectare

Using a range of annual biomass production of 75 to 300 gigajoules (GJ) per hectare, the limits of bioenergy production in Tanzania would be in the range of 3.3 to 13.3 exajoules (EJ) per year. This compares with total annual primary energy consumption in the country of 0.602EJ.

For the introduction of a biofuel industry in Tanzania, the following expected energy yields for the production of different transport fuels in Tanzania are important – namely, biodiesel from palm oil (186GJ per hectare), ethanol from cane juice (173GJ per hectare), biodiesel from jatropha oil (59GJ per hectare) and ethanol from C-molasses (20GJ per hectare).

Sugar crops provide the simplest and most cost-effective feedstock options for ethanol production. The area under sugar cane has grown from 23,000 to 39,000 hectares between 2000 and 2005, suggesting increasing availability of suitable feedstocks for ethanol production. Current production of C-molasses by Tanzania's cane sugar industry (about 90,000 tonnes in 2004–2005) could be converted into more than 20 million litres of ethanol per year, enough for a 10 per cent blend of ethanol into gasoline (E10) – or nearly 7 per cent of national gasoline consumption on an energy equivalent basis.

In August 2005, the retail price of gasoline in Dar es Salaam was 1120 Tanzanian shillings per litre. At that price, ethanol would be competitive at a retail price of 729 Tanzanian shillings per litre, or about €0.53 (US$0.64) per litre. At current petroleum prices, therefore, production of ethanol in Tanzania is likely to be competitive with gasoline.

Current production of oilseed crops is much lower than existing demand, and a biodiesel programme of any real impact would require planting considerably more land with oil crops than is now the case. Oil palm and jatropha are the two oilseed crops most likely to be used as feedstocks for biodiesel in Tanzania. Of the oilseed crops available, oil palm has the highest potential yield of oil per hectare of land harvested. However, there is currently great demand for palm oil for food and other uses, and local production meets less than 5 per cent of this.

There is a current proposal for a palm oil biodiesel project in the Kigoma region. The project would involve cultivation of 8000 hectares of oil palm, first to produce palm oil to meet local food and soap production demands, then eventually to produce biodiesel. If the project achieves the target oil yield of 5000 litres per hectare, palm oil production could approach 40 million litres per year. This production would in itself not be enough to displace current imports (in 2002, Tanzania imported roughly 172 million litres of palm oil). Alternatively, 40 million litres of palm oil could be converted into about 39 million litres of biodiesel. Diesel fuel consumption in Tanzania is projected to be about 700 million litres in 2010. Thus, if all the projected palm oil production in the Kigoma project were to be converted to biodiesel, a national blend of 5.7 per cent would be possible (5.2 per cent on an energy equivalent basis).

The other favoured crop for biodiesel production is jatropha. Tanzania has had some experience cultivating jatropha for small-scale oil production, which has been particularly promising in demonstrating the potential for rural poverty alleviation and empowering women. Cultivation of jatropha around the world has tended to be small scale, so production and yield data for plantation-scale cultivation is limited. The oil yield from jatropha plantations is reported to be about 1600kg per hectare from the fifth year onwards, although some local experience in Tanzania suggests that actual domestic yields may be significantly less than this. On the basis of a yield of 1600kg of oil per hectare, 19,700 hectares of jatropha would need to be harvested to produce enough biodiesel for a 5 per cent national blend with petroleum diesel in 2010.

OUTLOOK

Exploiting the large resource potential for biofuel production in Tanzania is hampered mainly by lack of information. The absence of set policies and regulations makes investment in the biofuel sector difficult, as the prospective return on investment remains largely unclear. In the meantime, the Tanzanian government is well aware of the benefits offered by the introduction of biofuels for transport applications, and is seriously assessing the various options for developing policies and strategies for increased use of biofuels.

Activities towards implementation of biofuel policies in Tanzania are driven mainly by the Ministry of Energy and Minerals (MoE). At an expert workshop and policy discussion in Dar es Salaam, organized in the framework of this regional study, MoE representatives strongly supported the proposed establishment of a high-level Biofuels Task Force that would provide advice and recommendations for the elaboration of biofuel policies and regulations suitable for the Tanzanian context.

The production and use of biofuels in Tanzania has the potential to offer large opportunities for investors. For the time being, these opportunities must be carefully identified on a case-by-case basis (in close cooperation with local partners) until the Tanzanian government has committed itself to actively promoting development of the national biofuel sector and biofuel market.

Source: the full study, *Biofuels for Transportation in Tanzania*, is available at www.gtz.de/de/dokumente/en-biofuels-for-transportation-in-tanzania-2005.pdf

23

Biofuels for Transportation in Brazil

Brazil study team leader: Fundação Brasileira para o Desenvolvimento Sustentável (FBDS); Professor Eneas Salati, FBDS technical director; Agenor O. F. Mundim, FBDS energy coordinator

Brazil study authors: Franz J. Kaltner; Gil Floro P. Azevedo; Ivonice A. Campos; Agenor O. F. Mundim

Contracting, policy dialogues and editing of study: Detlev Ullrich (GTZ), Jens Giersdorf (GTZ), Dirk Assmann (GTZ) and Ricardo Kuehlheim (GTZ)

Brazil is currently the world's largest producer and exporter of biofuels worldwide. It has a long history of biofuel production and a well-established national ethanol programme.

CURRENT SITUATION

Ethanol

Over the last 30 years, Brazil has implemented a very successful renewable energy programme called *Proálcool*, which was also the world's first large-scale biofuel programme. As a response to the international oil crisis, the programme was launched in 1975 to reduce the country's imports of oil and to promote the production of ethanol using sugar cane as the feedstock. In a transition period from the mid 1990s to 2002, government price controls and subsidies for production and logistics were eliminated for sugar and ethanol. In 2003, the first flexible-fuel vehicles (FFVs) (able to run on any blend of ethanol and gasoline) were introduced to the Brazilian market. Their numbers have increased rapidly: in 2005, sales of flex-fuel cars totalled 855,000.

Proálcool has had a variety of positive effects, including the creation of roughly 625,000 direct jobs in harvesting and processing, the development of a national technology, and the emergence of a completely mature industry. In addition to contributing to better climate protection (i.e. the reduction of an estimated 46.6 million tonnes of carbon dioxide emissions per year), replacing gasoline with ethanol has led to important foreign earnings savings for Brazil. Between 1976 and 2004, avoided fuel imports represented a savings of €50.2 billion (US$60.7 billion in December 2004 dollars).

Today, ethanol from sugar cane grown in the centre-south region of Brazil is by far the cheapest biofuel, making the ethanol industry competitive. Production costs are estimated to be around €0.15 (US$0.18) per litre, considerably lower than in other countries. Sugar cane is currently cultivated on about 5.5 million hectares, with 52 per cent of the area cultivated for ethanol production and 48 per cent for sugar production in 2004. Currently, 48 per cent of Brazil's total gasoline needs are met by ethanol. Biofuels for transportation represent 22 per cent of the country's total fuel consumption (including gasoline, diesel, biodiesel and ethanol).

Biodiesel

Through its Biodiesel Production and Use Programme, the Brazilian government is following a similar strategy to mitigate dependence upon fossil fuels and to push socio-economic development in rural areas, through a graduated tax exemption based on the region and scale of production. Creation of a national biodiesel market is just beginning, however. The chain of production is being structured to foster sustainable development by enabling participation by smallholders in oil production. Currently, the acreage for oilseed plants (primarily for food and feed) – mainly soybean – totals some 23 million hectares.

With regard to international trade, opportunities exist for exporting ethanol to countries that use biofuel either directly (in blends of more than 70 per cent) or in blends of up to 10 per cent.

POTENTIAL

Projections for Brazil's sugar/ethanol sector suggest that rising internal and export market demands for sugar and ethanol can easily be met. It is assumed that the industry should be able to produce 33.7 million tonnes of sugar (12.8 tonnes for internal consumption and 20.9 tonnes for export) and 26.4 million cubic metres of ethanol (4.4 million for export) by 2015. This would require an increase in sugar cane production of about 230 million tonnes in 10 years – a doubling in ethanol production and a 44 per cent increase in sugar production.

Considering the potential for biodiesel in Brazil, soybean oil can play an important role in the first years of implementation of the national biodiesel programme since the country is already one of the world's major producers of soybean oil. However, the low oil content, comparatively poor energy balance and low employment-generation impacts of soybeans must be taken into account. Furthermore, soybean production could expand into sensitive ecosystems if not directed otherwise. Similar problems may occur with other crops relevant for biodiesel production, such as castor oil and palm oil.

OUTLOOK

Ethanol trade is likely to continue to expand internationally. Worldwide, sugar production increased by 50 per cent between 2004/2005 and 2005/2006 alone, with biofuels as a major driver. Ethanol trade has increased even more steeply. In the future, Brazil is likely to provide more than 50 per cent of international ethanol trade.

Brazil is a potential exporter of biodiesel, as well. Given the limited potential for increased biodiesel production in Europe, Brazil faces an unprecedented opportunity to build market share on the European continent. Because of restrictive specifications and national policies for biofuels around the world, however, the market for biodiesel exports remains rather dispersed, varied and impaired by various trade barriers.

Brazil's biodiesel programme set a 2 per cent blending target in January 2006, and there is a potential market of about 800 million litres of biodiesel (B2) per year. From 2013 on, a mandatory increase to a 5 per cent blend (B5) will be considered, which would create a firm market of 2.4 billion litres per year. Substituting 2 per cent of petroleum diesel with biodiesel would lead to gains in Brazil's foreign currency reserves from reduced fossil fuel imports of about €132 million (US$160 million) annually.

Furthermore, the substitution of 1 per cent of Brazil's diesel consumption through the harvesting of various oil crops could result in the creation of approximately 190,000 jobs in rural areas. Achieving this goal, however, requires an emphasis on the importance of combining the biodiesel production programme with national land-use reform.

With regard to the ethanol sector alone, some €8.3 billion (US$10 billion) in private investment will be needed by 2015 to meet the national and export market demands for sugar and ethanol mentioned above. And foreign investments will certainly be necessary to expand Brazil's biofuel sector overall. Germany, for example, has a long tradition of investment in Brazil and could provide processing technology and equipment, as well as know-how on planning and capacity-building, in addition to direct investments. In addition, a major factor in the development of the Brazilian biodiesel programme will be the definition of biodiesel standards (technical, environmental and social) and the formulation of a sustainable global trade strategy.

Table 23.1 *Summary of the biofuel situation in Brazil*

Parameter	*2004*	*Future*
Fuel ethanol production (litres)	14.91 billion	20.5 billion of alcohol plus export of 5.5 billion
Fuel ethanol area (ha)	5.5 million of sugar overall: 52% for ethanol; 48% for sugar	–
Production costs for ethanol (per litre)	€0.15 (US$0.18)	€0.088 (US$0.107)
Ethanol net energy balance (in:out)	1:8.3	1:10.2
Acreage for oilseed plants (ha)	Soy: 22 million Palm: nearly 60,000 Castor: 134,000 Sunflower: 52,800	Soy: 100 million Palm: 66 million Castor: 4 million
Biodiesel production (tonnes per year)	Still in experimental phase	Government target: 2 million (2013+)

Source: the full study, *Biofuels for Transportation in Brazil*, is available at www.gtz.de/de/dokumente/en-biofuels-for-transportation-in-brazil-2005.pdf

24

Biofuels for Transportation in Germany

Germany summary author: Uwe Fritsche, Öko-Institut, Germany

Germany is currently a major player in the European Union's biofuel strategy. The country's domestic policy framework provides strong support for increasing the share of biofuels in road transport. While the national focus has historically been on biodiesel, other biofuels such as ethanol, Fischer-Tropsch (F-T) diesel and biogas are gaining in importance. This summary discusses the current extent of German biofuel production and explores the potential for continued domestic production in the future, including implications for the agricultural sector and related industries.

CURRENT SITUATION

Today, there is broad consensus in Germany that biofuels are a key element for a sustainable future, as illustrated in key government documents. As of 2006, the main commercially available biofuels in Germany were biodiesel made from rapeseed methyl ester (RME), ethanol (both pure and as ethyl tertiary butyl ether, or ETBE) and straight vegetable oil (SVO) (in addition, a few litres of experimental 'Sundiesel' were produced by the manufacturer Choren to demonstrate the potential of F-T diesel from biomass, and the first German refuelling station for biogas was scheduled to open in mid 2006).

Biodiesel

Overall, market development of biodiesel in Germany has been impressive. Between 1991 and 2004, sales of RME increased 50-fold, and in 2006 a market volume of nearly 2 million tonnes was expected – representing a 100-fold increase over 1991. Of the 1.1 million tonnes produced in 2004, the majority was used directly for blending with fossil diesel, while some 0.4 million tonnes was sold at filling stations. About 60 per cent of the biodiesel sold at filling stations is purchased by truck and bus fleet operators.

Ethanol

Compared to biodiesel, ethanol contributes only minimally to the current biofuel market in Germany. The main feedstock for ethanol production is cereals. Since early 2005, ethanol from cereals has increased. As with biodiesel, a blend of 5 per cent ethanol to gasoline is currently allowed in Germany (so-called E5).

In terms of market share, overall domestic ethanol production has actually decreased in Germany over the past few years. In 2004, a total of 220 million litres was produced, some 12 per cent less than in 2003 (note that data on ethanol use for fuels and other applications is not currently separated). Thus, Germany accounted for only 10 per cent of European ethanol production in 2004.

During recent years, biofuels were exempted from the German fuel tax so that they could compete with taxed diesel fuel. With the rapid increase in biodiesel production, tax losses totalled €559 million (US$676 million) in 2005 and will rise further if the tax exemption is maintained as planned until 2009. In late 2005, the new German government called for a change so that only existing 'pure' biofuel capacities and next-generation biofuels (as well as biogas) will continue to benefit; in early 2006, it announced the introduction of a mandatory 2 per cent biofuel blend in gasoline and a 4.4 per cent blend in diesel, starting in 2007.

POTENTIAL

In addition to having adequate technologies for converting organic residues and wastes to biofuels, the key factor determining the potential to grow energy crops is having the available land, without competing with food, feed and fibre demands. Energy crop potentials for Germany have been derived via a complex modelling of future developments in food, farming practices and competing land uses from nature conservation, human settlements, infrastructure, etc.

Even assuming an increase in organic farming of up to 30 per cent of all food by 2030 (from roughly 4 per cent currently), Germany's net land balance is quite positive. Analysis indicates that nearly 4 million hectares of land could be used for energy crops out of some 17 million hectares of total agricultural area in the country.

In addition to having available land, Germany's future bioenergy potential depends upon the types of energy crops grown on this land. Germany is a pioneer in analysing the most suitable crops from both an environmental and nature protection perspective. Analysis shows clearly that crops such as sugar beets do not perform well and that perennial grasses and short-rotation woody crops (SRWC) are very favourable, as are double-cropping systems.

OUTLOOK

Today, some 2000 of Germany's 15,000 fuel filling stations supply RME100. It is expected, however, that use of RME100 in private cars will decrease since the automotive industry will not authorize the use of pure RME in future EURO IV/V engines (EURO IV is the European Union's existing air emission standard for automotive vehicles; EURO V is the future requirement for diesel cars and trucks and is currently under negotiation). Thus, the future market for RME100 will concentrate on company fleets in the transport sector.

As the blending of RME with fossil diesel increases from the current 2 per cent, competition between the distribution of pure RME100 at filling stations and the blend market (via refineries) is already occurring.

In the future, wheat, rye, barley and some sugar beets are expected to be the major sources for ethanol fuel in Germany. Firms are building new large-scale ethanol plants dedicated to fuel use, with production capacities of 590,000 cubic metres. This would increase Germany's overall ethanol production capacity to more than 900,000 cubic metres by the end of 2006.

Over the longer term, flexible-fuel vehicles (FFVs) can contribute substantially to achieving a broader market share for ethanol. In 2005, the Ford Motor Company announced that it would offer FFVs to the German market. Volvo and Saab have also announced that they will provide these vehicles, and Volkswagen, DaimlerChrysler and others are expected to follow suit (Automotive World, 2005).

In 2006, the capacity of German oil mills and transesterification plants was expected to reach some 2 million tonnes per year. Given the import potential from Eastern Europe and Asia and existing rotational restrictions in agriculture, domestic biodiesel production might well level off at this amount, which fits neatly with the 2010 EU target of 5.75 per cent when applied to total diesel consumption in Germany.

Nevertheless, it will not be possible for Germany to meet the gasoline share of the EU biofuel target with domestic biodiesel. Accordingly, the near future will probably see an increase in ethanol from biomass, especially from wheat and sugar beets, even if next-generation conversion technologies are not yet commercially available (large-scale first-generation ethanol plants using wheat as a feedstock could deliver nearly competitive biofuels if feedstock costs were below €8 (US$10) per gigajoule – that is, below €75 (US$91) per tonne of wheat). Furthermore, European sugar market subsidies will be gradually reduced, leading sugar beet farmers to seek alternative uses for their product.

Biogas, too, is increasingly being acknowledged as a potential biofuel, able to be processed to a high-methane fuel and fed into the natural gas pipeline system. Studies indicate that there is not only significant potential for biogas in Germany, but that the life-cycle emissions and costs could be lower than those for ethanol and biodiesel.

Although biodiesel has historically been the main driver for biofuels in Germany, moving beyond a 5 per cent share in all transport fuels will require the adoption of other biofuels, either from other domestic sources or from imports.

In the short term (to 2010), biodiesel and ethanol are expected to be the key biofuels in Germany, mainly as blends with fossil diesel and gasoline. Yet, their essentially large potential is restricted by limited land for energy crop cultivation (e.g. rapeseed and wheat), as well as by competing biomass use in the stationary (power and heat) sector, which currently offers higher greenhouse gas reductions and lower costs. Thus, the German Fuel Strategy envisages a total biofuel market share of 5.75 per cent – in line with the EU target – but not much more than this.

In the medium to long term, special emphasis is being given to options with the greatest impact on fossil fuel substitution – namely, increased engine efficiency, production of synthetic fuels from biomass (biomass to liquid, or BTL), hybrid power trains and hydrogen in fuel-cell vehicles. BTL is the prioritized medium-term option, while hydrogen is seen as the key long-term fuel. Since the development of both remains immature, Germany's strategy calls for continuing research and development (R&D), with a focus on gradually scaling up pilot plants, and for maintaining tax exemptions to bridge the cost gap during early stages of commercialization. Biofuel potentials are estimated to be large enough to meet the indicative EU 2020 target of 8 per cent biofuels.

Source: derived from material prepared for the original *Biofuels for Transportation* draft report, June 2006; this book marks its first publication

Appendices

Appendix 1 Per Capita Consumption of Gasoline and Diesel, 2002

Country	*Population (2002)*	*Gasoline use*	*Diesel use*
		(thousand litres per person per day)	
China	1,284,275,902	0.11	0.19
India	1,034,172,547	0.03	0.12
US	287,675,526	4.89	2.09
Indonesia	231,326,092	0.18	0.31
Brazil	179,914,212	0.25	0.59
Pakistan	153,403,524	0.03	0.15
Russia	145,266,326	0.66	0.54
Japan	127,065,841	1.29	1.52
Mexico	102,479,927	0.85	0.44
Philippines	82,995,088	0.12	0.23
Germany	82,350,671	1.22	2.28
Thailand	62,806,748	0.32	0.70
France	59,925,035	0.79	2.55
UK	59,912,431	1.30	1.35
Italy	57,926,999	1.06	1.65
Ukraine	48,058,877	0.36	0.33
Korea, South	47,969,150	0.58	1.33
South Africa	44,433,622	0.62	0.45
Spain	40,152,517	0.75	2.32
Poland	38,625,976	0.40	0.62
Argentina	38,331,121	0.26	0.73
Canada	31,902,268	3.39	2.41
Kenya	31,386,842	0.05	0.07
Saudi Arabia	24,501,530	1.66	2.61
Australia	19,546,792	2.64	1.98
Ecuador	12,921,234	0.50	0.56
Zimbabwe	11,926,563	0.09	0.15
Cuba	11,226,999	0.11	0.38
Guatemala	11,178,650	0.27	0.29
Czech Republic	10,256,295	0.69	0.93
Sweden	8,954,175	1.68	1.70
Slovakia	5,410,052	0.50	0.63
Denmark	5,374,693	1.34	2.54
Nicaragua	5,146,848	0.13	0.22
Lithuania	3,633,232	0.74	1.60
Mauritius	1,200,206	0.25	0.94
Swaziland	1,130,269	0.23	0.20
Total	4,404,764,780	0.61	0.58
World	**6,214,891,000**	**0.52**	**0.53**

Source: US Census Bureau, International Programs Center, www.census.gov/ipc/www/idbprint.html; US Department of Energy, Energy Information Administration, *International Petroleum Data*, www.eia.doe.gov/emeu/international/contents.html

Appendix 2 World Producers of Petroleum and Biofuels

Country/Region	Year 2004		Year 2002					
	PRODUCTION		CONSUMPTION		PRODUCTION	IMPORTS		
	Ethanol*	Biodiesel	Motor Gasoline	Diesel Fuel Oil	Crude Oil	Crude Oil	Motor Gasoline	Diesel Fuel Oil
Canada	4.0		680.2	484.4	2,950	-227	32.5	7.1
Mexico	0.6		550.6	282.0	3,593	-1,808	157.1	37.2
United States	230.6	1.6	8,847.8	3,775.9	9,000	9,131	498.3	267.4
North America	235.2	1.6	10,079.6	4,547.8	15,543	7,096	688.9	317.3
Argentina	2.7		62.1	176.4	827	-250	0.1	6.6
Brazil	260.2		279.1	666.8	1,761	147	2.8	110.1
Cuba	1.1		8.1	26.7	48	30	0.8	21.5
Ecuador	0.8		40.6	45.7	399	-235	9.0	11.5
Guatemala	1.1		18.9	20.2	18	-6	16.3	18.1
Nicaragua	0.5		4.1	7.2	-0	17	1.7	3.2
Venezuela			206.5	110.5	2,924	-1,622	0	0
Central & South America	266.4		1,010.6	1,495.4	6,898	-849	214.3	372.6
Austria		1.1	49.6	141.5	26	164	16.4	70.8
Belgium			48.3	233.3	13	679	28.1	129.0
Denmark		1.4	45.2	85.8	373	-211	23.1	39.5
Finland			42.6	87.3	9	217	6.5	34.3
France	14.3	6.8	299.4	960.6	81	1,632	35.7	268.5
Germany	4.6	20.3	629.6	1,179.1	155	2,114	127.1	295.8
Greece			82.0	153.8	7	375	11.9	61.2
Italy	2.6	6.3	385.1	599.7	126	1,632	13.2	19.9
Netherlands			96.5	183.2	98	932	168.1	255.7
Norway			38.6	84.2	3,334	-2,866	9.8	11.0
Portugal			47.9	110.6	4	231	0.8	11.9
Spain	5.2	0.3	190.0	587.1	24	1,132	22.3	190.9
Sweden	1.7		94.6	95.9	2	370	39.1	40.0
Switzerland			87.9	127.1	2	100	61.2	86.6
Turkey			73.0	196.2	47	478	10.4	49.2
United Kingdom		0.2	488.9	509.7	2,562	-599	50.5	63.3
Western Europe	28.4	36.4	2,825.0	5,546.9	6,910	6,573	704.0	1,781.0
Czech Republic		1.2	44.6	60.1	11	121	22.9	25.7
Lithuania		0.1	16.9	36.6	13	130	0.02	1.1
Poland	3.5		97.8	151.5	25	344	14.0	24.8
Romania			44.8	57.4	139	127	0.7	4.5
Russia	12.9		600.1	492.1	7,659	-3,831	0.02	0.1
Slovakia		0.3	17.0	21.4	4	110	7.5	5.8
Turkmenistan			17.5	20.0	195	-55	0	0
Ukraine	4.3		107.7	100.1	86	369	9.7	10.1
Uzbekistan			38.6	34.2	153	1	0	3.3
Eastern Europe & Former USSR	20.7	1.6	1,189.7	1,276.2	9,640	-3,095	144.2	178.3

Iran			296.3	431.8	3,524	-2,094	65.5	0
Iraq			79.5	128.4	2,040	-1,495	15.0	0
Kuwait			41.0	14.5	2,030	-1,138	8.0	0
Oman			18.4	18.9	899	-839	4.2	1.6
Qatar			12.7	6.7	841	-571	0	0
Saudi Arabia	5.2		256.2	402.1	8,810	-5,985	5.0	0
Syria			30.5	93.1	520	-201	0	7.9
United Arab Emirates			49.8	45.7	2,405	-1,674	0	0
Yemen			23.7	15.5	443	-342	1.1	0
Middle East	5.2		912.1	1,283.9	21,561	-13,789	140.7	65.2
Algeria			42.6	83.7	1,575	-868	0	0
Angola			4.2	15.8	896	-854	1.8	2.1
Cameroon			5.5	8.6	70	-45	0.5	0.9
Congo (Brazzaville)			1.0	1.5	249	-240	0	0
Congo (Kinshasa)			2.4	2.4	23	-23	2.4	2.4
Cote d'Ivoire (IvoryCoast)			3.8	7.0	19	44	0.9	1.9
Egypt			51.3	164.0	748	-140	5.5	46.2
Equatorial Guinea			0.2	0.8	213	-213	0.2	0.8
Gabon			1.1	3.7	251	-235	0.1	0.5
Kenya	0.2		9.0	12.9	-0	30	5.0	8.3
Libya			44.4	59.6	1,383	-984	0	0
Mauritius	0.4		1.9	7.1	0	0	1.9	7.1
Nigeria			154.8	53.4	2,123	-1,893	94.3	1.9
South Africa	7.2		174.7	126.8	211	399	6.4	10.1
Sudan			0	2.5	240	-179	0	2.5
Swaziland	0.2		1.6	1.4	0	0	1.6	1.4
Tunisia			9.9	34.6	79	-38	3.6	27.1
Zimbabwe	0.4		7.1	11.4	0	0	5.7	11.5
Africa	8.4		615.5	814.7	8,092	-5,050	198.7	243.4
Australia	2.2		324.1	243.1	744	11	25.1	24.0
Brunei			4.6	3.1	189	-179	0	0
Burma			7.9	20.1	16	8	1.2	9.0
China	62.9		876.3	1,568.0	3,530	1,242	0.001	16.1
India	30.1		176.9	791.6	813	1,610	0	2.2
Indonesia	2.9		255.2	445.2	1,340	-290	54.4	147.7
Japan	2.0		1,027.9	1,212.5	120	3,987	28.8	35.4
Korea, South	1.4		175.6	402.8	3	2,157	11.9	36.7
Malaysia			154.5	161.9	795	-246	55.9	29.3
New Zealand			56.2	47.7	41	71	18.8	5.5
Pakistan	1.7		25.5	147.9	66	143	0	83.9
Papua New Guinea			1.9	10.9	55	-55	1.9	10.9
Philippines	1.4		64.1	118.6	10	258	17.6	38.5
Sri Lanka			6.5	37.3	-1	46	1.3	22.5
Taiwan			167.7	106.5	8	787	5.8	0.6
Thailand	4.8		126.2	277.2	206	675	7.5	12.2
Vietnam			47.7	82.5	340	-322	49.6	90.7
Asia & Oceania	109.4		3,552.7	5,942.5	8,291	10,752	417.0	859.1
World Total	**673.5**	**39.6**	**20,185.2**	**20,907.4**	**76,935**	**1,639**	**2,508.0**	**3,816.8**

Note: Data shows thousands of barrels per day. Ethanol production includes fuel, industrial, and beverage production.

Source: Petroleum data from US Department of Energy, Energy Information Administration, *International Petroleum Consumption*, www.eia.doe.gov/emeu/international/contents.html; ethanol data from F.O. Licht, cited in RFA (2005b, p14); biodiesel data from Christoph Berg, senior analyst, F. O. Licht, e-mail to Peter Stair, Worldwatch Institute, 25 January 2006

Appendix 3 Biofuels as a Percentage of Gasoline and Diesel Consumption

Country/Region	Year 2004 PRODUCTION Ethanol	Year 2004 PRODUCTION Biodiesel	Year 2002 CONSUMPTION Motor Gasoline	Year 2002 CONSUMPTION Diesel Fuel Oil	Ethanol % of Gasoline & Ethanol Use	Biodiesel % of Diesel & Biodiesel Use	Biofuels % of Transport Fuel Use
Brazil	260.2		279.1	666.8	48.24%	0.00%	21.57%
Mauritius	0.4		1.9	7.1	17.39%	0.00%	4.26%
India	30.1		176.9	791.6	14.54%	0.00%	3.01%
Cuba	1.1		8.1	26.7	12.01%	0.00%	3.06%
Swaziland	0.2		1.6	1.4	11.11%	0.00%	6.25%
Nicaragua	0.5		4.1	7.2	10.95%	0.00%	4.24%
China	62.9		876.3	1,568.0	6.70%	0.00%	2.51%
Pakistan	1.7		25.5	147.9	6.25%	0.00%	0.97%
Guatemala	1.1		18.9	20.2	5.51%	0.00%	2.74%
Zimbabwe	0.4		7.1	11.4	5.35%	0.00%	2.12%
France	14.3	6.8	299.4	960.6	4.56%	0.70%	1.65%
Argentina	2.7		62.1	176.4	4.17%	0.00%	1.12%
South Africa	7.2		174.7	126.8	3.96%	0.00%	2.33%
Ukraine	4.3		107.7	100.1	3.84%	0.00%	2.03%
Thailand	4.8		126.2	277.2	3.66%	0.00%	1.18%
Poland	3.5		97.8	151.5	3.45%	0.00%	1.38%
Spain	5.2	0.3	190.0	587.1	2.66%	0.05%	0.70%
United States	230.6	1.6	8,847.8	3,775.9	2.54%	0.04%	1.81%
Kenya	0.2		9.0	12.9	2.19%	0.00%	0.91%
Philippines	1.4		64.1	118.6	2.14%	0.00%	0.76%
Russia	12.9		600.1	492.1	2.10%	0.00%	1.17%
Saudi Arabia	5.2		256.2	402.1	1.99%	0.00%	0.78%
Ecuador	0.8		40.6	45.7	1.93%	0.00%	0.92%
Sweden	1.7		94.6	95.9	1.76%	0.00%	0.88%
Indonesia	2.9		255.2	445.2	1.12%	0.00%	0.41%
Korea, South	1.4		175.6	402.8	0.79%	0.00%	0.24%
Germany	4.6	20.3	629.6	1,179.1	0.73%	1.69%	1.36%
Australia	2.2		324.1	243.1	0.67%	0.00%	0.39%
Italy	2.6	6.3	385.1	599.7	0.67%	1.04%	0.90%
Canada	4.0		680.2	484.4	0.58%	0.00%	0.34%
Japan	2.0		1,027.9	1,212.5	0.19%	0.00%	0.09%
Mexico	0.6		550.6	282.0	0.11%	0.00%	0.07%
Czech Republic		1.2	44.6	60.1	0.00%	1.96%	1.13%
Denmark		1.4	45.2	85.8	0.00%	1.61%	1.06%
Slovakia		0.3	17.0	21.4	0.00%	1.39%	0.78%
Austria		1.1	49.6	141.5	0.00%	0.77%	0.57%
Lithuania		0.1	16.9	36.6	0.00%	0.27%	0.19%
United Kingdom		0.2	488.9	509.7	0.00%	0.04%	0.02%

Notes: Data (excluding percentages) shows thousands of barrels per day. Countries are ranked in this list based on the amount of ethanol produced as a percentage of the gasoline demand plus ethanol produced. Ethanol production includes fuel, industrial and beverage production. Transport fuel use includes gasoline and diesel use, plus ethanol and biodiesel production.

Source: Petroleum data from US Department of Energy, Energy Information Administration, *International Petroleum Consumption*, www.eia.doe.gov/emeu/international/contents.html; ethanol data from F.O. Licht, cited in RFA (2005b, p14)

Appendix 4 Block diagram of ethanol, Fischer-Tropsch (F-T) fuels and gas turbine combined cycle (GTCC)

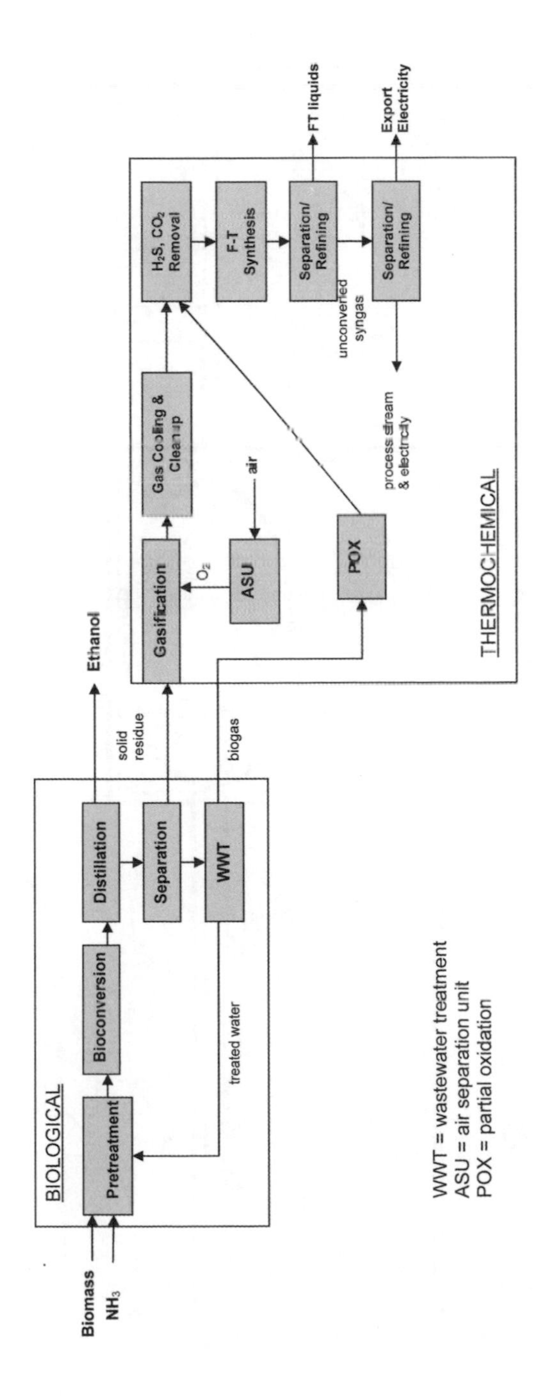

Source: Laser and Lynd (2006)

Appendix 5 Overview of Key Elements and Correlations Determining Bioenergy Potential

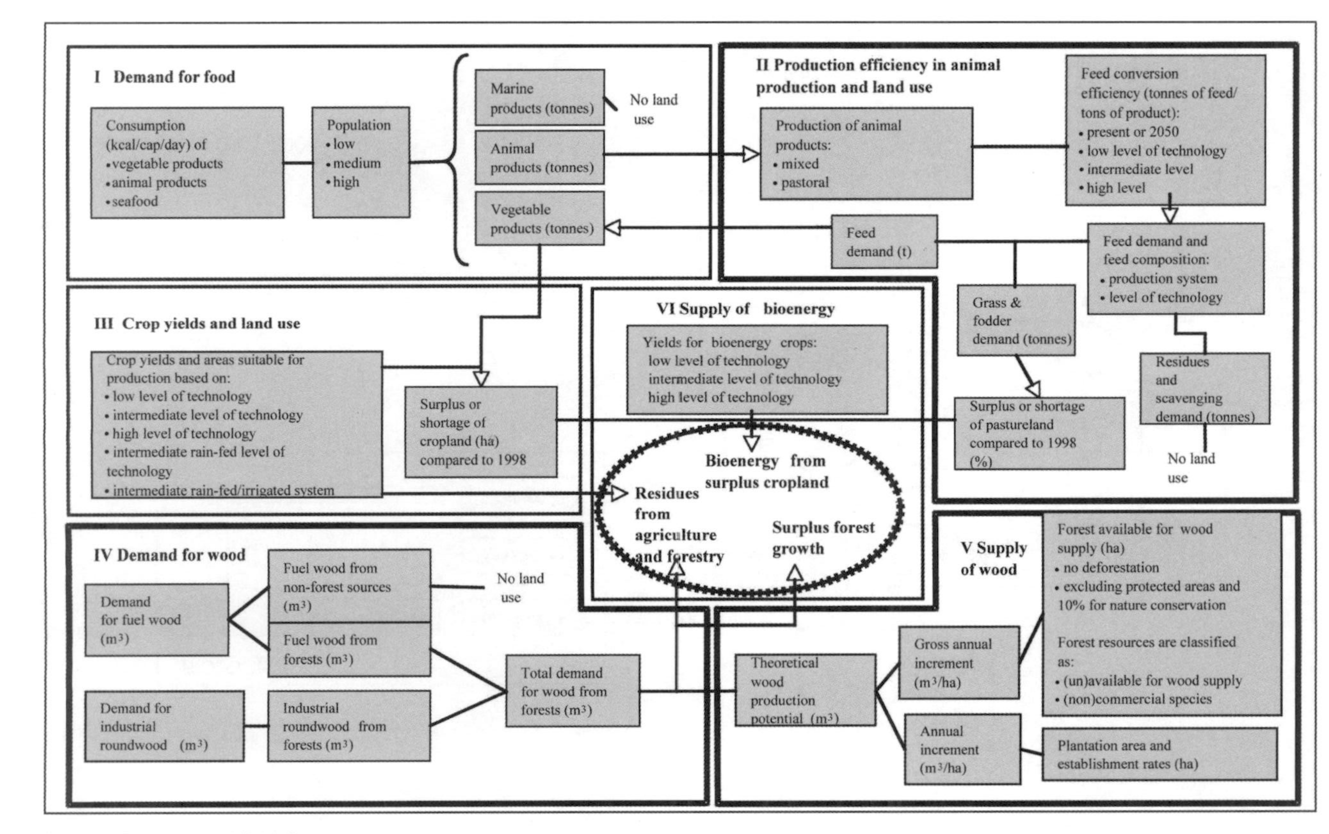

Source: Smeets et al (2004)

Appendix 6 Flow Chart of Bioenergy System Compared with Fossil Reference Energy System

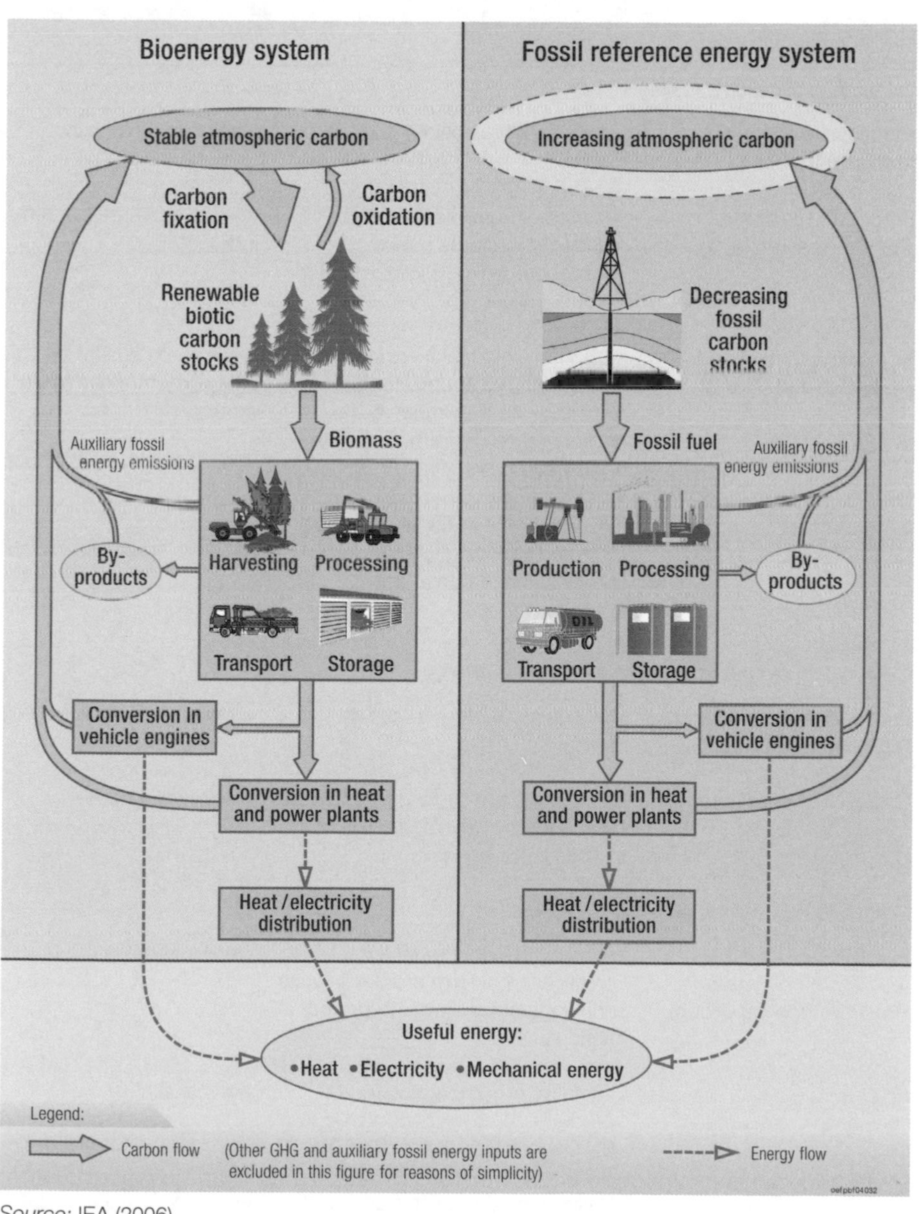

Source: IEA (2006)

Appendix 7 Selected Standards and Certification Schemes Relevant to Biofuel Production and Trade

Organization/scheme	*Description*	*Website*
General certification systems		
Clean Development Mechanism (CDM)	Approval of projects for carbon credits	www.cdm.unfccc.int
European Committee for Standardization (CEN)	Network of European national standards institutes providing voluntary technical standards at the European level	www.cenorm.be/cenorm/index.htm
Eco-Management and Audit Scheme (EMAS)	European Union (EU) voluntary instrument that acknowledges organizations that improve their environmental performance on a continuous basis	www.euroa.eu.int/comm/environment/emas/index_en.org
International Organization for Standardization (ISO)	Network of the national standards institutes providing voluntary technical standards at the international level	www.iso.org
Certification or criteria systems for biomass energy		
European Green Electricity Network (EUGENE)	Certification system for green energy	www.eugenestandard.org
Green Gold certificate	Track and trace system for biomass developed by Essent, energy utility in The Netherlands	www.skalint.com
Certification or criteria systems for agriculture		
EUREPGAP (EUREP-Euro-Retailer Produce Working Group)	A normative document for certification of farm products (fruits and vegetables) from integrated agriculture	www.eurep.org
EKO	Label for organic agricultural products produced according to EU Council Regulation No 2092/91	www.skal.nl
International Federation of Organic Agriculture Movements (IFOAM)	Basic international standard for organic agriculture and accreditation criteria for organic certification programmes	www.ifoam.org/about_ifoam/standards/ogs.html

Organization/scheme	Description	Website
Sustainable Agriculture Network (SAN)	Coalition of local non-profit conservation groups; Rainforest Alliance-certified label for bananas, coffee, cocoa, citrus and flowers/ foliage	www.rainforest-alliance. org/programs/agriculture/ certification/index.html
SQF (Safe Quality Food)	Australian certification system for farming products; criteria for good agricultural practice (GAP) in food production	www.agriholland. nl/dossiers/ kwaliteitssystemen/sqf. html
Umweltsicherungssystem (USF) (KUL)	'Environmental friendly' label for farming systems	www.tll.de//kul/kul_idx. htm
UTZ KAPEH	Certification system for fair-traded coffee; GAP guidelines for coffee	www.utzkapeh.org
Certification systems for forestry		
American Tree Farming Systems (ATFS)	Forest certification system initiated by the American Forest Foundation	www.treefarmsystem.org/ cms/pages/26_19.html
Canadian Standards Association's (CSA's) Sustainable Forest Management Standard	Forest certification system overseen by CSA, an independent, non-profit organization	www.Sfms.com/csa.htm/
Forest Stewardship Council (FSC)	International forest certification and chain-of-custody control system	www.fsc.org
Pan-European Forest Certification (PEFC)	Forest certification system initiated by 14 European countries and private national forest interest groups	www.pefc.org
Sustainable Forestry Initiative (SFI)	Forest certification system in the US and Canada, initiated by the American Forest and Paper Association, a forest trade association	goodforests.com
Certification or criteria systems for fair trade		
Agrocel Pure and Fair Indian Organic Cotton Organization	Coordinates the production of organic cotton and has developed criteria for fair-trade cotton chains	www.agrocel-cotton. com/english/en_home. html
AgroFair	Importer and distributor of organic and fair-trade tropical fresh fruit	www.agrofair.com
Fairtrade	Certification of fair-traded products	www.fairtrade.net/sites/ standards/standards.html

Organization/scheme	Description	Website
Oxfam	Chain of shops selling 'fair' products from developing countries, with criteria for selecting partners for fair trade	www.oxfam.org/eng/pdfs/strat_plan.pdf
Sustainability criteria		
Biomass Transitie Groep	Workgroup of the Dutch Ministry of Economy developing criteria for sustainable biomass trade	
Biotrade Workshop	International Workshop 2002 discussing criteria for sustainable biomass trade	
GRAIN (Global Restrictions on the Availability of biomass in the Netherlands)	Report containing criteria for sustainable biomass trade	
Greenpeace International	Environmental group with ecological criteria for sustainability	www.greenpeace.org/international/campaigns/climate-change/solutions/bioenergy
International Labour Organization (ILO)	Conventions that describe acceptable labour conditions	www.ilo.org
United Nations (UN)	Conventions and Agenda 21 provide sustainability criteria for social, economic and ecological aspects	www.un.org/esa/sustdev/csd.htm
World Wide Fund for Nature (WWF)	Environmental group with ecological criteria for sustainability	www.panda.org www.wwf.org
Indicator sets for sustainable development		
International Institute for Sustainable Development (IISD)	Indicators for sustainable development	www.iisd.org
Organisation for Economic Co-operation and Development (OECD)	Indicators for sustainable development and agro-ecological indicators	www.oecd.org/home
United Nations Development Programme (UNDP)	Indicators for sustainable livelihoods (SL)	www.undp.org
Indicator sets for assessment of project sustainability		
UN Commission for Sustainable Development (CSD)	Method for developing sustainability indicators; indicators for sustainable development; project assessment	www.un.org/esa./sustdev/csd/csd12/csd12.htm

Organization/scheme	Description	Website
Gold Standard	Tool for assessing project sustainability; best practice benchmark for Clean Development Mechanism (CDM) and Joint Implementation (JI) greenhouse gas offset projects; developed by WWF	www.panda.org/downloads/climate_change/cop8standards.pdf
World Bank	Assessment of sustainability of projects	www.worldbank.org
Guidelines for sustainable or environmentally sound management		
Canadian Council of Forest Ministers (CCFM)	Set of criteria and indicators for sustainable management of Canadian forests	www.ccfm.org
Centre for International Forestry Research (CIFOR)	Criteria for sustainable forest management; manual for developing locally adapted criteria and indicator sets	www.cifor.cigar.org/acm/pub/toolbox/html
European Union Council Regulation	Definition of organic farming and principles of organic production; certification for organic farming logo	www.europa.eu.int/eur/lex/lex/LexUriServ/LexUriserv.do?uri=CELEX:31991R2092:EN:HTML
Forum de l'Agriculture Raisonne Respectueuse de l'Environnement (FARRE)	Common codex for integrated farming; principles and indicator for GAP	www.farre.org/versionAnglaise/CommonCodex.htm
IKEA	Private company with strategy for corporate environmental and social responsibility	www.ikea.nl/ms/nl_NL/about_ikea/social/environmental/environmental.pdf
International Timber Trade Organization (ITTO)	Guidelines for the sustainable management of tropical forests; criteria for measuring sustainable tropical forest management	www.itto.or.jp/live/index.jsp
OECD	Guidelines for sustainable behaviour of multinational enterprises	www.oecd.org
Unilever	International company that has developed GAP guidelines for sustainable agriculture	www.unilever.com
International Finance Corporation (IFC)	Guidelines for environment, health and safety	www.ifcln.ifc.org/ifcest/environ.nsf/

Source: Lewandowski and Faaij (2006, pp86–87)

Glossary of Terms

Anhydrous ethanol. Ethanol that has less than 1 per cent volume water content. Most commonly used in the US for blending with gasoline (e.g. E10, E85); in Brazil, many vehicles are set up to run on hydrated ethanol.
ASTM D6751. A US standard for biodiesel that establishes fuel quality requirements, such as purity and lubricity characteristics. See CEN 14214 for European specifications.
Auto-thermal reforming (ATR). A method for extracting hydrogen from hydrocarbons. ATR breaks down hydrocarbon molecules into separate hydrogen and carbon atoms using a catalyst, steam and oxygen.
Bagasse. Sugar-cane processing residues.
Benzene. An aromatic hydrocarbon with a single six-carbon ring and no alkyl branches; a known carcinogen.
Biodiesel. A biofuel used in compression-ignition (diesel) engines containing mono-alkyl esters of long-chain fatty acids created by transesterifying plant or animal oils with a simple alcohol (typically methanol, but sometimes ethanol) and a catalyst. Biofuels for diesel engines can also be produced from lignocellulosic biomass using gasification and synthesis, pyrolysis or hydrothermal liquefaction; however, the term 'biodiesel' typically applies only to those fuels derived from renewable lipid sources.
Bioenergy. Energy produced from organic matter or biomass. Biomass may either be burned directly or converted into liquid or gaseous fuel.
Biofuels. Liquid fuels derived from organic matter or biomass.
Biomass. Organic material from plants or animals, including forest product wastes, agricultural residues and waste, energy crops, animal manures, and the organic component of municipal solid waste and industrial waste.
Biomass power and heat. Power and/or heat generation from biomass.
Biomass residues. Residue resulting from the harvesting, processing and use of biomass. Can be divided into primary residues (generated before and at harvest – for example, the tops and leaves of sugar cane), secondary residues (generated during processing – for example, sugar cane bagasse, rice husks and black liquor) and tertiary residues (generated during and after product end use – for example, demolition wood and municipal solid waste).

Biomass to liquid (BTL). Processes, such as gasification and Fischer-Tropsch synthesis, which convert biomass into liquid fuels.

Biorefinery. A refining facility where biomass is converted into fuel, chemicals, materials and other uses, all at the same plant.

Biorefining. The process by which biomass is converted into fuel, chemicals and/or biomass-based materials.

Bxx (where xx is a number – for example, B5, B10, etc.) Biodiesel blended with petroleum diesel, with biodiesel volume percentage indicated by the number.

Cellulosic biomass. Plant matter composed of linked glucose molecules that strengthen the cell walls of most plants. Next-generation biofuel conversion technologies can convert cellulosic biomass into liquids.

Cellulosic ethanol. Ethanol produced from cellulosic biomass, usually using acid-based catalysis or enzyme-based reactions to break down plant fibres into sugar, which is then fermented into ethanol.

CEN 14214. A standard for European biodiesel performance, established by the European Committee for Standardization, that sets fuel quality requirements such as purity and lubricity characteristics. See ASTM D6751 for US standards.

Cetane number. An empirical measure of a diesel fuel's self-ignition quality that indicates the readiness of the fuel to ignite spontaneously under the temperature and pressure conditions in the engine's combustion chamber (higher cetane improves the performance of diesel engines).

Clean Development Mechanism (CDM). The CDM is one of the two 'flexible' financing provisions under the Kyoto Protocol. It provides opportunities to promote biofuel development in developing countries.

Combined heat and power (CHP), or cogeneration. The use of a power station to simultaneously generate both heat and electricity. It allows for more total use of energy than conventional generation, potentially reaching an efficiency of 70 to 90 per cent, compared with approximately 50 per cent for the best conventional plants.

Combustion. A chemical reaction between a compound (fuel) and an oxidizing element (oxygen in air) that releases energy in the forms of heat and light.

Compressed natural gas (CNG). Made by compressing purified natural gas (a fossil fuel composed primarily of methane) for storage in hard containers. It is frequently used to power vehicles and is considered a cleaner alternative to more carbonaceous fuels such as diesel or gasoline.

Compression-ignition engines. Also known as diesel engines. Internal combustion engines in which atomized fuel is injected into highly compressed air. The heat and pressure of the compressed air alone causes the fuel to ignite.

Consolidated bioprocessing (CBP). A strategy for processing cellulosic biomass that involves consolidating four biologically mediated events into a single step: cellulase production, cellulose hydrolysis, hexose fermentation and pentose fermentation. This kind of processing is facilitated by micro-organisms that can simultaneously hydrolyse plant fibres and starches, and ferment the resulting sugars.

Corporate Average Fuel Economy (CAFE). A US standard that requires light vehicles to achieve a certain average mileage per gallon (3.8 litres) of gasoline. It has been a market driver for the E85 engine.
Diesel fuel. A fuel processed from petroleum that contains a mix of molecules ranging from 12 to 22 carbon atoms (C-12 to C-22). Designed to run in diesel internal-combustion engines.
Dimethyl ether (DME). Sometimes called 'methyl ether' or 'wood ether'. A gaseous ether (CH_3OCH_3) that can be manufactured as a biofuel and used as a substitute for natural gas.
Dried distillers grain (DDG). A co-product of dry-milling operations that produce ethanol, DDG is a fibrous high-protein residue that can be used as food for animals, especially cattle.
Dry mill. A type of starch-ethanol mill characterized by the method of milling grains prior to fermentation into ethanol. Dried grains are ground and all parts are introduced into the production process. Proteins and fibres are usually extracted after fermentation.
Exx (where xx is a number – for example, E10, E20, etc.) Ethanol blended with gasoline, with ethanol volume percentage indicated by the number.
Energy crops. Crops grown and harvested for use as a feedstock in the production of fuels or other energy products. In this book, energy crops are contrasted with conventional food and feed crops. Examples include perennial grasses and short-rotation forestry species such as hybrid poplar and willow.
Ethanol (CH_3CH_2OH). A vehicle fuel typically made from fermenting sugar derived from biomass (usually corn, sugar cane or wheat) that can replace ordinary gasoline in modest percentages (blends) in spark-ignition engines or can be used in pure form in specially modified vehicles. Nearly all ethanol is produced by fermenting plant sugars and starches (or hydrolysed cellulose or hemicellulose in the future); however, it can also be produced from fossil feedstocks. In this book, ethanol refers exclusively to biomass-derived ethanol (also known as 'bio-ethanol').
Ethyl ester. An alkyl ester produced by transesterfying ethanol with esters found in animal, vegetable and waste oils; used as a biodiesel fuel.
Ethyl tertiary butyl ether (ETBE – $(CH_3)_3COC_2H$). An oxygenate blend stock formed by the catalytic etherification of isobutylene with ethanol.
Exajoule (EJ). A unit of energy equal to 10^{18} joules.
Fatty acid methyl ester (FAME). Another term for biodiesel.
Feedstock. A material used as a raw material in an industrial process.
Fischer-Tropsch (F-T) process. A biomass-to-liquid (BTL) method of synthesizing hydrocarbons, specifically gasoline and diesel molecules, from 'syngas'. It passes hydrogen and carbon monoxide over a catalyst, either cobalt or iron, at high temperature and pressure. Named after German chemists Franz Fischer (1877–1948) and Hans Tropsch (1889–1935), the process is most often used to create F-T diesel, a fuel for compression-ignition engines.

Flexible-fuel vehicle (FFV). A vehicle specially designed to run on straight gasoline or any gasoline–ethanol blend up to E85 in temperate climates, and E96 in tropical climates, from a single tank.
Fossil fuel equivalent (FFE). A measure of energy potential from a given fuel or energy source relative to producing that same amount of energy with fossil fuels.
Fuel atomization. A process by which fuel atomizers, such as injectors or jets, deliver fuel in minute droplets to be mixed with air prior to combustion in an engine or turbine.
Fuel cell vehicle (FCV). A vehicle propelled by a fuel cell engine using hydrogen as a fuel (note that it is possible that on-board reformers can be used to extract the hydrogen from various fuels, such as methane, gasoline and ethanol).
Fumigation. A process by which a carburettor, fuel injector, heated vaporizer or mist generator is used to meter ethanol into a vehicle engine's air-intake manifold.
Gasification. The process of converting biomass to a mixture of carbon monoxide, carbon dioxide, hydrogen and methane by heating it in oxygen-starved conditions. The resulting 'syngas' can be used either as a fuel for heat and power production or as a feedstock for the synthesis of liquid fuels (see F-T synthesis).
Gasoline. A liquid fuel for use in internal combustion engines where the fuel–air mixture is ignited by a spark. It consists of a mixture of volatile hydrocarbons derived from the distillation and cracking of petroleum. It normally contains additives such as lead compounds or benzene to improve performance (the prevention of premature ignition) or rust inhibitors.
Gas to liquid (GTL). A route of gaseous fuel processing that results in by-products that can include liquid fuels such as naphta and diesel. The resulting BTL/GTL diesel can be used as a straight fuel or blended with ordinary diesel or biodiesel.
Gas turbine (GT). An engine that passes the products of the combustion of its fuel–air mixture over the blades of a turbine. The turbine drives an air compressor, which in turn provides the air for the combustion process. The energy of the combustion products not taken up by the compressor can be used to provide a jet of exhaust gases to drive another turbine.
Gas turbine combined cycle (GTCC). Gas-powered turbines for generating electricity that combine several components into one system to increase overall efficiency. For example, a unit may use heat from a gas turbine exhaust to produce steam to turn a second turbine for two-stage power generation. Other components, such as heat recovery steam generators or peripheral electricity generators, may also be added.
General System of Preferences (GSP). A trade policy instrument that gives developing countries preferred access to industrialized country markets, generally through lowered tariffs.
Genetically modified (GM) crops. Plants whose genetic make-up has been altered using genetic engineering technology that does not involve natural methods of reproduction. Some biomass crops, including sugar cane and corn, have been genetically modified to improve aspects of plant productivity.

Gigajoule (GJ). A unit of energy equal to 10^9 joules.
Gigawatt (GW). A unit of power-generating capacity equal to 10^9 watts.
Gigawatt hour (GWh). A unit of produced energy equal to 10^9 watt hours.
Gigawatt-thermal (GWth). A unit of heat-supply capacity equal to 10^9 watt thermal.
Greenhouse gas (GHG). Gaseous components of the atmosphere that contribute to the greenhouse effect of gases in the Earth's atmosphere (where increased amounts of solar heat are trapped in the air). Human activity contributes to the greenhouse effect by releasing GHGs such as carbon dioxide, methane and others.
Higher heating value (HHV). The amount of heat released per unit mass or unit volume of a substance when the substance is completely burned, including the heat of condensation of water vapour to liquid water.
Hydrous ethanol. Ethanol containing approximately 4 per cent water by volume.
Hydrous pyrolysis. A biomass refining process that mimics the zero-oxygen, pressurized, hot and aqueous conditions that created petroleum, although in a much shorter time period. The process removes the oxygen from the biomass and yields a solid mineral layer, gaseous fuels and a liquid 'biocrude'.
Internal-combustion engine vehicle (ICEV). A vehicle that uses either a compression-ignition or spark-ignition engine for propulsion.
Internal rate of return (IRR). A financial measure used to evaluate the return on capital investments.
Jatropha. An oilseed crop that grows well on marginal and semi-arid lands. The bushes can be harvested twice annually, are rarely browsed by livestock and remain productive for decades.
Joint Implementation (JI). A mechanism under the Kyoto Protocol designed to encourage GHG emissions reductions or carbon sequestration projects. The mechanism allows an Annex 1 party to implement such projects in other Annex 1 party territories in exchange for emissions reduction credits.
Joule (J). The SI unit of work or energy. Specifically, it is the work done or energy expended by a force of 1 Newton acting through a distance of 1 metre.
Kilowatt hour (KWh). A unit of produced energy equal to 10^3 watt hours.
Kilowatt-thermal (KWth). A unit of heat supply capacity equal to 10^3 watt thermal.
Knocking. A metallic rattling or pinging sound that results from uncontrolled combustion in an engine's cylinders. Heavy and prolonged knocking may cause power loss and damage to the engine.
Life-cycle analysis (LCA). An analysis that examines the environmental impact of a product or process from its inception to the end of its useful life.
Lignocellulosic feedstock. Biomass feedstock, such as woody materials, grasses, and agricultural and forestry residues, that contains cellulose, hemicellulose and lignin. It can be broken down in a number of ways to be used as biofuels.

Liquefied natural gas (LNG). A fossil fuel composed primarily of methane. Purified natural gas turns to liquid at –160°C and is 1/640th the volume of natural gas at standard temperature and pressure, making it easier to transport in specialized cryogenic tanks.
Liquefied petroleum gas (LPG). A fossil fuel extracted from crude oil and natural gas, comprised principally of propane (C_3H_8) and butane (C_4H_{10}). LPG turns to liquid under moderate pressure and is roughly 1/250th the volume of its gas form.
Lubricity. A measure of a substance's lubricating qualities, a property of oiliness or slipperiness. Lubricity is a concern for engine systems using liquid fuels, as many components, such as fuel pumps, depend upon fuel for lubrication.
Megawatt (MW). A unit of power equal to one million (10^6) watts.
Methanol (CH_3OH). A simple alkyl also known as methyl alcohol.
Methyl ester. Another term for biodiesel.
Methyl tertiary butyl ether (MTBE). A common oxygenate added to gasoline to help combustion and reduce emissions of carbon monoxide and troposheric ozone. MTBE is highly water soluble and has been found to contaminate groundwater. The World Health Organization has identified MTBE as a carcinogen.
Miscanthus. Also known as elephant grass. A tropical and subtropical hardy perennial grass species that originated in Asia and Africa. It is a promising source of biomass due to its high rates of growth.
Multiple-use crops. Crops that can be used for a variety of purposes, including as human food, animal feed, material inputs for products and energy (in the form of heat or electricity, or stored in liquid biofuels).
Municipal solid waste (MSW). Total waste excluding industrial waste, agricultural waste and sewage sludge; including durable goods, non-durable goods, containers and packaging, food wastes, garden wastes, and miscellaneous inorganic wastes from residential, commercial, institutional and industrial sources. Waste-to-energy combustion and landfill gas are by-products of municipal solid waste.
Neat fuel. Fuel in its pure unblended form.
Nitride glycol. An additive to alcohol fuels used to increase lubricity.
Nitrogen oxides (NO_x). The generic term for a group of highly reactive gases, all of which contain nitrogen and oxygen in varying amounts.
Nitrous oxide (N_2O). A nitrogen oxide that is a common pollutant from burning fossil fuels or organic matter. It is a powerful greenhouse gas and a known ozone-depleting substance.
Octane. A measure of a gasoline's ability to resist knocking; if a gasoline with too low an octane rating is used engine knock may result.
Particulate matter (PM). Fine particles of solids suspended in gas. Anthropogenic sources of aerosols originate from the burning of fossil fuels or from wind-blown dust from construction and agricultural areas. PM from petroleum sources, such as soot, is a human health hazard as it may carry carcinogens.

Peroxyacetyl nitrate (PAN). An organic compound formed in the atmosphere from the addition of nitrogen dioxide, NO_2, to the peroxyacyl radical formed in the oxidation of acetaldehyde. It is a component of photochemical smog and can cause irritation to the eyes and respiratory system.
Photovoltaic (PV). Photovoltaics, such as solar panels or solar cells, convert sunlight into electricity.
Polycyclic aromatic hydrocarbons (PAHs). A group of compounds, of which there are about 10,000, that result largely from the incomplete combustion of carbon-based materials, such as fossil fuels and wood. PAHs can bond with ash and become particulate matter, irritating respiratory systems when inhaled.
Polyvinyl chloride (PVC). A hard plastic commonly used in construction and pipes. While inconclusive, some studies point to PVC as a source of carcinogenic dioxins in the environment.
Potassium hydroxide (KOH). Commonly known as caustic potash, potassium hydroxide is a catalyst used with rapeseed oil in the process of transesterification to create rapeseed methyl ester (RME), a biodiesel fuel.
Proálcool. A Brazilian programme launched in 1975 that aimed to reduce the country's dependence upon petroleum by subsidizing the production of ethanol from sugar cane as a substitute for gasoline. The programme has encouraged many technological advances and contributed to the enormous increase in Brazilian ethanol production.
Prototype Carbon Fund (PCF). The first carbon-financing mechanism at the World Bank; a public–private partnership consisting of 17 companies and 6 governments.
Pyrolysis. A thermo-chemical process in which biomass is converted into liquid 'bio-oil', solid charcoal and light gases (H_2, CO, CH_4, C_2H_2, C_2H_4). Depending upon the operating conditions (temperature, heating rate, particle size and solid residence time), pyrolysis can be divided into three subclasses: conventional, fast or flash.
Rapeseed. A flowering member of the Brassicacae family and a major global source of vegetable oil. Rapeseed oil is the most common feedstock for biodiesel in Europe, especially in Germany. Canola is a common North American cultivar of rape.
Rapeseed methyl ester (RME). Biodiesel made from rapeseed oil.
Reid vapour pressure (vapour pressure). The pressure exerted by the vapours released from any material at a given controlled temperature when enclosed in a laboratory vapour-tight vessel.
Renewable Fuels Standard (RFS). A regulation requiring refiners, blenders, distributors and importers to sell increasing volumes of renewable fuels – such as ethanol and biodiesel – according to an annual schedule. The 2005 US Energy Security Act requires that the US consumes at least 4 billion gallons of these fuels in 2006, escalating to 7.5 billion gallons by 2012.
Renewable Transport Fuel Obligation (RTFO). A UK proposal requiring that 5 per cent of vehicle fuels used in the nation be derived from renewable sources by 2010.

Separate hydrolysis and fermentation (SHF). A two-step process used to convert biomass into alcohol, where cellulase enzymes break down cellulose into sugars prior to the introduction of microbes for fermentation.
Short-rotation coppice (SRC). A method of tree harvesting where the trees are harvested and the remaining tree stumps produce vigorous regrowth that is harvested after a prescribed number of years (varying by tree species and crop management priorities); three to four harvests may be possible before the trees must be replanted.
Short-rotation forestry (SRF). A forest management strategy using short-rotation coppicing (or tree harvesting and replanting) after a prescribed number of years.
Short-rotation woody crops (SRWC). Generally used to refer to tree crops grown with a short-rotation coppice approach.
Simultaneous saccharification and fermentation (SSF). A one-step process used to convert cellulosic biomass into alcohol that combines cellulase enzymes and microbes for fermentation. As enzymes break down cellulose into sugars, microbes ferment these sugars into alcohol.
Sodium hydroxide (NaOH). Commonly known as lye, this is a catalyst used to transesterify oils and an alcohol into molecules of methyl ester, or biodiesel.
Soybean methyl ester (SME). Biodiesel derived from soybean oil.
Spark-ignition engines. Internal combustion engines that use an electronic spark from a spark plug to ignite a compressed mixture of fuel and air.
Splash blending. Blending of ethanol into gasoline or biodiesel into petroleum diesel at terminals, without active mixing.
Steam-methane reforming (SMR). A process that converts methane and light hydrocarbons to carbon monoxide and hydrogen using steam and a nickel catalyst. The reforming reactions are endothermic (they absorb heat, rather than producing heat); as a result, heat must be supplied to SMR reactors, typically by a furnace surrounding a tube bundle packed with a nickel catalyst where the reforming reactions occur.
Straight vegetable oil (SVO). Known as pure plant oil (PPO) in the European Union, SVO refers to either virgin or waste vegetable oils used to fuel diesel engines. While some diesel engines can run on SVO without modification, steps must be taken to address problems in colder climates since it is generally more viscous than petro-diesel and has a higher freezing point.
Sulphur oxides (SO_x). The term for a group of compounds composed of sulphur and oxygen. SO_x, released in the burning of fuels, is a leading contributor to acid rain and can cause severe human health issues in high concentrations.
Switchgrass. A prairie grass native to North America that holds considerable promise as a feedstock for cellulosic conversion into ethanol.
Syngas (synthesis gas). A mixture of carbon monoxide, carbon dioxide, hydrogen and methane created during the gasification process of heating biomass in the presence of air, oxygen or steam. Syngas can be converted to a variety of fuels, including hydrogen, methanol, dimethyl ether and Fischer-Tropsch liquids.

Tetra-ethyl lead (TEL). An additive to gasoline introduced to reduce engine knocking and to increase efficiency and octane ratings. TEL creates lead as a highly toxic pollutant in engine exhausts and has been phased out of most vehicle fuels worldwide.

Transesterification. A reaction to transform one ester into a different ester, this process is used to transform natural oil into biodiesel by chemically combining the natural oil with an alcohol (such as methanol or ethanol).

Treated biogas (TB). Gas that has been treated to remove hydrogen sulphide (H_2S) and water (which can corrode fuel systems, engines and burners).

Tree-based oilseeds. Oilseeds grown by trees, such as jatropha, that show promise as a feedstock for biodiesel production.

Vinasse. The residue liquid from the distillation of ethanol, rich in potassium and organic matter; it is used as a fertilizer and irrigation liquid to increase sugar cane crop yields.

Volatile organic compounds (VOCs). Organic compounds comprised of carbon and hydrogen that easily vaporize into the atmosphere. VOCs can pollute soil and groundwater, and in the presence of sunlight they react with NO_x to form tropospheric ozone, a respiratory irritant.

Volumetric Ethanol Excise Tax Credit (VEETC). US legislation enacted in 2004 that establishes federal tax credits for the blending, distribution and sale of E10, E85 and biodiesel. Each gallon of renewable fuel sold through to the end of 2010 will receive a tax credit of US$0.51.

Watt (W). The SI derived unit of power (symbol W), equal to one joule per second.

Watt-hour (Wh). A unit of energy expressing energy expended by a one-watt load drawing power for one hour.

Watt thermal (Wth). A unit of heat-supply capacity used to measure the potential output from a heating plant. It represents an instantaneous heat flow and should not be confused with units of produced heat.

Well-to-wheels analysis. A life-cycle analysis of fuels that measures the efficiencies and impacts of various energy sources.

Wet mill. A type of starch-ethanol mill where grains are steeped in solutions of water and acid to break them down into separate products, such as oils, proteins and purified starch. Wet-milling operations are more complex and, generally, larger than dry-milling operations, but offer a greater variety of by-products, including high protein animal feeds, high fructose corn syrup and biomaterial feedstocks.

Notes

PREFACE

1 Christoph Berg, senior analyst, F. O. Licht, Agra Informa Ltd, Kent, UK, e-mails to Rodrigo G. Pinto, Worldwatch Institute, 20–22 March 2007. F. O. Licht has revised its worldwide fuel thanol production figures for 2005 and prior years due to the fact that the largest part of Brazilian ethanol exports did not end up in the fuel market but in beverage and industrial applications; the downward adjustment was substantial in relation to a previous preliminary estimate for 2005.
2 *F .O. Licht's World Ethanol and Biofuels Report* (2006i).
3 Brazil from Eric Martinot et al (2005) *Oil World* (2005); Martin von Lampe (2006) p15.

1 CURRENT STATUS OF THE BIOFUEL INDUSTRY AND MARKETS

1 86 per cent (based on 2006 data) from Christoph Berg, senior analyst, F. O. Licht, Kent, UK, e-mails to Rodrigo G. Pinto, Worldwatch Institute, 20–22 March 2007. F. O. Licht has revised its worldwide fuel thanol production figures for 2005 and prior years due to the fact that the largest part of Brazilian ethanol exports did not end up in the fuel market but in beverage and industrial applications; the downward adjustment was substantial in relation to a previous preliminary estimate for 2005.
2 One quarter from Christoph Berg (2004).
3 Figures 1.1 and 1.2 from Berg, op cit note 1. For biodiesel, litres per day are calculated using the approximation that 1000 litres of biodiesel equals 0.88 tonnes.
4 Oil growth from Janet Sawin and Ishani Mukherjee (2007), pp32–33.
5 Calculation based on biofuel production numbers from Berg, op cit note 1; oil production numbers from IEA (2007), page 45.
6 Bill Kovarik (1998).
7 Ibid.
8 Luiz Carlos Correa Carvalho and Usina Alto Alegre (2003).
9 Berg, op cit note 1.
10 Dehua Liu (2005).
11 Stephen Karekezi et al (2004).
12 Ibid.
13 Karin Bendz (2005c).

14 *F. O. Licht's World Ethanol and Biofuels Report* (2005a).
15 *F. O. Licht's World Ethanol and Biofuels Report* (2006j).
16 *F. O. Licht's World Ethanol and Biofuels Report* (2005f).
17 Berg, op cit note 1.
18 Ibid.
19 Ibid.
20 Berg (2004).
21 RFA (2006c).
22 Table 1.2 from Berg, op cit note 1.
23 Berg, op cit note 1.
24 Berg, op cit note 1; Berg, e-mail to Peter Stair, Worldwatch Institute, 25 January 2006; National Biodiesel Board (2005c).
25 The apparent imbalance in US production capacity versus actual output is due to several factors; in some cases, rated capacity exceeds the intended output, while in other cases developers have built extra capacity in anticipation of increased demand for biodiesel fuel.
26 Tom Bryan (2005); Berg, op cit note 1.
27 Price of crude oil from US DOE and EIA (2006c).
28 US DOE and EIA (2005a).
29 US DOE and EIA (2006c), www.eia.doe.gov/emeu/ipsr/t46.xls.
30 US DOE and EIA (2005a).
31 *F.O. Licht's World Ethanol and Biofuels Report* (2005f).
32 Bendz, op cit note 13.
33 *F. O. Licht's World Ethanol and Biofuels Report* (2005f).
34 Ibid.
35 Ibid.
36 Box 1.1 from F. Kaltner et al (2005a) *Biofuels for Transportation: Brazilian Potential and Implications for Sustainable Agriculture and Energy in the 21st Century*, Prepared for Deutsche Gesellschaft für Technische Zusammenarbeit (GTZ) GmbH, Rio de Janeiro, November.
37 *F. O. Licht's World Ethanol and Biofuels Report* (2005f) .
38 Ibid.
39 Ibid.
40 Boris Utria World Bank, e-mail to Worldwatch Institute, 21 February 2006.
41 South Africa also makes extensive use of F-T technology; however, to date it has mostly used it to produce fuel from coal.
42 Novozymes (2005); Abengoa Bioenergy (2005).

2 Liquid biofuels: A primer

1 S. W. Mathewson (1980).
2 F. Kaltner et al (2005a) *Biofuels for Transportation: Brazilian Potential and Implications for Sustainable Agriculture and Energy in the 21st Century*, Prepared for the Deutsche Gesellschaft für Technische Zusammenarbeit (GTZ) GmbH, Rio de Janeiro, November, p23.

3 National Science-Technology Roadshow Trust (2006).
4 J. N. de Vasconcelos (2004); J. Goldemberg and I. Macedo (1994).
5 Genencor International (2005); V. Singh et al (2001) NREL (2002b).
6 M. Paster, J. Pellegrino and T. Carole (2003).
7 A. McAloon et al (2000).
8 *Business Wire* (2006); NREL (2002b).
9 J. Ashworth et al (1991).
10 N. Schmitz et al (eds) (2005).
11 NREL (2002b).
12 Edgar Remmele (2004) *Position Paper: Technologie-und Forderzentrum im Kimpetenzzentrum fur Nachwachsende Rohstoffe*, p5, cited in Wuppertal Institute for Climate, Environment and Energy (ed) (2005).
13 Ibid.
14 Howard Haines, Montana Department of Natural Resources, pers comm with Worldwatch Institute, 14 December 2005.
15 Wuppertal Institute for Climate, Environment and Energy (2005). For the German filter exchange programme, see www.pflanzenoel-motor.de.
16 J. A. Kinast (2003) p2.
17 National Biodiesel Board (2002) www.biodiesel.org/pdf_files/fuelfactsheets/prod_quality.pdf.
18 National Biodiesel Board (2005a) www.biodiesel.org/pdf_files/fuelfactsheets/Production.pdf.
19 Radich (2004); AEA Technology (2003).
20 NREL (2002b), p2.
21 OECD (2006) Lew Fulton, Sustainable Transport, Division of Global Environment Facility Coordination, United Nations Environment Programme, e-mail to Peter Stair, Worldwatch Institute, July 2005.
22 Michael Hogan (2005).
23 Table 2.1 from OECD (2006).
24 Radich (2004).
25 Figures 2.1 and 2.2 based on the following sources: Lew Fulton et al (2004); Timothy Gardner (2006) DOE and EIA Energy Statistics website, www.eia.doe.gov/emeu/international/oilprice.html, accessed February 2006.
26 Cost data for dry-mill and wet-mill co-products from Fulton et al, ibid. *Business Wire* (2006); NREL (2002b).
27 Isaias de Carvalho Macedo et al (2005) p186.
28 Novozymes and BBI International (2005).

3 First-generation feedstocks

1 Table 3.1 from Lew Fulton et al (2004); F. Kaltner et al (2005).
2 Figure 3.1 from Fulton et al, ibid.
3 F. Kaltner et al (2005).
4 Isaias de Carvalho Macedo et al (2005).
5 Ibid.

6 FAO and IIASA (2000).
7 Table 3.2 from USDA, Foreign Agricultural Service (no date).
8 FAO (no date).
9 Illovo Sugar (2005).
10 Net for Cuba International (2006).
11 C. Strossman (2004) Tonnes of total sugar beets harvested are calculated based on the assumption of 18 per cent raw sugar content in the harvested sugar beet roots/plants.
12 Hosein Shapouri, senior economist, USDA, conversation with Peter Stair, 15 December 2005.
13 *F. O. Licht's World Ethanol and Biofuels Report* (2006h).
14 Jason Clay (2004) p412.
15 RFA (2005b).
16 Danielle Murray (2005a).
17 Fulton et al (2004).
18 FAO (no date).
19 Average wheat yields from FAO (2002).
20 W. Bushuk (2005); Peter Talks (2004).
21 FAO (2000).
22 *Bangkok Post* (2005); Vanguard (2005).
23 FAO (no date).
24 Table 3.3 from USDA, Foreign Agricultural Service (2006b).
25 J. Gerpen, D. Clements and G. Knothe (2004).
26 Figure 3.2 from Kaltner et al (2005), p42.
27 Average rapeseed yields from USDA, Foreign Agricultural Service (2006a); agricultural practices from Wuppertal Institute for Climate, Environment and Energy (2005); rapeseed yields from Martin Bensmann (2005).
28 FAO (no date).
29 Stat Communications (2006).
30 *F. O. Licht's World Ethanol and Biofuels Report* (2006d).
31 Table 3.4 from FAO, Statistics Division (2004).
32 Jane Earley, Thomas Earley and Matthew Straub (2005), p5.
33 FAO (no date).
34 Kaltner et al (2005).
35 Helen Buckland (2004); RSPO (2005c).
36 K. S. Sethi (2003).
37 D1 Oils (2006b).
38 R. Mandal (2005).
39 European Biomass Industry Association (2006).
40 USDA, Foreign Agricultural Service (no date), updated January 2006.
41 NREL (2002a).
42 Government of the Philippines (2005).
43 Kaltner et al (2005), pp6, 57; Multi-commodity Exchange of India, Ltd (2006).
44 W. Gehua et al (2005) p16.
45 Kaltner et al (2005); TERI (2005).
46 Subramanian et al (2005) and Azam et al (2005), cited in ibid.

47 John Sheehan et al (1998a).
48 *USA Today* (2006); GreenFuel Technologies Corporation website, www.greenfuelonline.com/index.htm.
49 Ibid.
50 Sheehan et al (1998a).
51 Kaltner et al (2005), p37.
52 Ibid.
53 Figure 3.3 from OECD (2006).
54 RFA (2005b).
55 FAO and IIASA (2000).
56 Jose Roberto Moreira (2004b).
57 Fulton et al (2004), p18.
58 *Oil World* (2005).
59 Michael C. Sutton, Nicole Klein and Gary Taylor (2005).
60 Wuppertal Institute for Climate, Environment and Energy (2005).
61 Corn from University of Illinois Extension (2005), cited in Masami Kojima and Todd Johnson (2005); soy and oil palm from Clay (2004), pp176–177, 211.
62 Table 3.5 is based on the following sources: Brazil sugar cane from Schmitz et al (2002), cited in Kojima and Johnson, ibid; corn from University of Illinois Extension (2005), cited in idem; Brazil soy and oil palm from Clay (2004), pp176–177, 211; US soy from National Biodiesel Board (2004a).
63 Robert Perlack et al (2005).
64 National Corn Growers Association (2004).
65 Box 3.1 is based on the following sources: yield increases from Angelo Bressan Filho, (2005); plant varieties, planted acreage and industry involvement from Kaltner et al, op cit note 1, p42.
66 European and German crop yield trends from R. Edwards et al (2003) and from Martin Kaltschmitt et al (2005); US corn crop yields from A. Dobermann et al (2002); biotech hybrid corn yields from National Corn Growers Association (2004); continued development from Fulton et al (2004), pp65–66; National Biodiesel Board (2004a), p2.

4 Next-generation feedstocks

1 Table 4.1 from Jonathan Scurlock (2000).
2 Table 4.2 from the following sources: ibid; Thomas P. E. Miles et al (1995); Ralph Hollenbacher (1992); Thomas Miles, Sr (1992); Thomas Milne et al (1986); Paul Gallegeher (2006).
3 M. Hoogwijk et al (2003).
4 For an overview of biomass characteristics, see A. Faaij et al (1997).
5 US DOE (2001).
6 Table 4.3 from Philippe Girard, Abigail Fallot and Fabien Dauriac (2005).
7 CTIC (no date); Robert Perlack et al (2005).
8 W. W. Wilhelm et al (2004).
9 F. Kaltner et al (2005).

10 Ibid.
11 Ibid.
12 Figure 4.1 from Perlack et al (2005), p21.
13 Ibid.
14 US DOE (2003); Wilhelm et al (2004).
15 Perlack et al (2005). Box 4.2 is based on the following sources: T. E. Devine and E. O. Hatley (1998); T. E. Devine, E. O. Hatley and D. E. Starner (1998a); T. E. Devine, E. O. Hatley and D. E. Starner (1998b); T. Devine and J. E. McMurtrey III (2004); S. Wu et al (2004).
16 Larcio Couto and David R. Betters (1995).
17 G. W. Burton (1982); G. W. Burton and B. G. Mullinix (1988) cited in Nathanael Greene et al (2004).
18 K. P. Vogel and R. A. Masters (1998).
19 D. S. Powlson, A. B. Riche and I. Shield (2005); Lew Fulton et al (2004) p136; André P. C. Faaij (2006); Greene et al (2004), p26.
20 R. van den Broek et al (2001).
21 Table 4.4 is based on the following sources: van den Broek, ibid; E. E. Biewinga and G. van der Bijl (1996); V. Dornburg, J. van Dam and A. Faaij (2004); P. Börjesson (1999); D. O. Hall et al (1993).

5 New technologies for converting biomass into liquid fuels

1 P. McKendry (2002).
2 Mark Laser and Lee Lynd (2006).
3 Ibid.
4 Figure 5.1 from Darmouth College and Natural Resources Defense Council (no date).
5 Ibid.
6 Ibid.
7 André Faaij (2003); M. Kaltschmitt, C. Rosch and L Dinkelbach (eds) (1998).
8 Mark Laser, Research Associate, Thayer School of Engineering, Dartmouth College, e-mail to Jim Easterly, Easterly Consulting, 2 March 2006.
9 C. N. Hamelinck and A. P. C. Faaij (2002).
10 Laser and Lynd (2006).
11 Ibid.
12 A. V. Bridgwater (1999).
13 S. Czernik and A.V. Bridgwater (2004).
14 DynaMotive Energy Systems Corporation website, www.dynamotive.com.
15 BTG Biomass Technology Group, BV website, www.btgworld.com.
16 E. Dinjus and A. Kruse (2004).
17 US DOE and EERE (2005c).
18 J. E. Naber and F Goudriaan (2003).
19 US DOE (2005b).
20 Arkenol website, www.arkenol.com, accessed 12 January 2006.

21 Laser and Lynd (2006).
22 P. C. Badger (2002).
23 Laser, op cit note 8.
24 Novozymes and BBI International (2005).
25 Laser and Lynd (2006).
26 Enteric bacteria from L. O. Ingram et al (1998, 1999); thermophilic bacteria from S. G. Desai, M. L. Guerinot and L. R. Lynd (1999) and from B. K. Ahring et al (1996); mesophylic bacterium from L. O. Ingram et al (1999) and from R. J. Bothast et al (1999) yeast from N. W. Z. Ho, Z. Chen and A. P. Brainard (1998) and from S. Sanchez et al (2002).
27 Timothy Gardner (2006).
28 Franz J. Kaltner et al (2005).
29 US DOE (2006).
30 EC (2005).
31 Figures 5.2 and 5.3 from Lew Fulton et al (2004).
32 US DOE and EERE (2005a); Genencor International (2004); Gardner (2006).
33 L. R. Lynd et al (2005).
34 N. C. Mosier et al (2005); M. E. Himmel et al (1997).
35 US DOE and EERE (2005b).
36 Uwe Fritsche, Öko-Institut, e-mail to Suzanne Hunt, Worldwatch Institute, 13 March 2006.
37 C. Hamelinck (2004).
38 Matthias Rudloff (2005).
39 Frank Seyfried (2005).
40 EC (2006d).
41 L. Devi, K. J. Ptasinski and F. J. J. G. Janssen (2003); M. Mozaffarian and R. W. R. Zwart (2003).
42 NREL (2005).
43 Ibid.
44 US DOE, National Energy Technology Laboratory (2005); Novozymes and BBI International (2005).
45 Hamelinck (2004).
46 Naber and Goudriaan (2003).
47 Changing World Technologies website, www.changingworldtech.com, accessed 12 January 2006.
48 Fritsche, op cit note 36.
49 Carlo N. Hamelinck and André Faaij (2006).
50 Gardner (2006).
51 US DOE and EERE (2005a).
52 Russel Heisner, BC International, conversation with Peter Stair, Worldwatch Institute, 24 August 2005; autospectator.com (2005).
53 ACB11 Eyewitness News (2007) 'Sunopta reports on cellulosic ethanol projects' Seeking Alpha, 25 January.
54 Fritsche, op cit note 36.
55 See Sino-German Workshop on Energetic Utilization of Biomass, Beijing, 8–11 October 2003, www.ipe.ac.cn/sinogerman/program.htm.

56 RenewableEnergyAccess.com (2006).
57 R. et al (2005) Wallace.
58 Fritsche, op cit note 36.
59 Darmouth College and NRDC.
60 Figure 5.4 from (no date) ibid.
61 Assumes an 8 per cent efficiency benefit for ethanol-optimized engines relative to gasoline engines, according to Dr Tom Kenney, Ford Motor Co, pers comm, 2005.
62 Laser and Lynd (2006).
63 Ibid.
64 NatureWorks, LLC website, www.natureworksllc.com, viewed 14 November 2005.
65 Darmouth College and NRDC (no date).
66 Fritsche, op cit note 36.

6 LONG-TERM BIOFUEL PRODUCTION POTENTIALS

1 Eric Martinot et al (2005).
2 André Faaij (2005a).
3 Figure 6.1 from Goran Berndes et al (2003). The following 16 studies were included in the analysis: WEC (1994); Nakicenovic et al (1998); M. Lazarus et al (1993); Edmonds et al (1996); J. Swisher and D. Wilson (1993); D. A. Lashof and D. A. Tirpak (eds) (1990); B. Sorensen (1999); T. B. Johannson et al (1993); R. H. Willams (1995); Leemans et al (1996); J. Battjes (1994); Yamamoto et al (1999); G. Fischer and L. Schrattenholyer (2000); B. Dessus, B. Devin and F. Pharabed (1992); Shell International (1995); IPCC (2000).
4 Table 6.1 from Faaij (2005a).
5 ORNL (2005).
6 Calliope Panoutsou (2005); ORNL, ibid.
7 ORNL (2005), p15.
8 Ibid, pp1–2.
9 Frank Rosillo-Calle (2001).
10 Wally Wilhelm (2005).
11 Edward Smeets, André Faaij and Iris Lewandowski (2004). Updated results obtained from the authors, March 2006.
12 D. O. Hall et al (1993).
13 Philippe Girard, Abigail Fallot and Fabien Dauriac (2005); Table 6.2 is based on International Sugar Organization (1999) and WEC (2001). The energy of 1 metric tonne of bagasse is valued at 0.264 metric tonnes of coal equivalent; 1 tonne of coal equivalent is 29.308GJ, according to UN (1997).
14 Smeets, Faaij and Lewandowski (2004).
15 Box 6.1 from Jim Easterly, Easterly Consulting, pers comm with Peter Stair, Worldwatch Institute, March 2006.
16 Rosillo-Calle (2001).
17 Faaij (2005a).
18 Smeets, Faaij and Lewandowski (2004).

19 André Faaij, Copernicus Institute, Utrecht University, pers comm with Suzanne Hunt and Peter Stair, Worldwatch Institute, 3 March 2005.
20 Smeets, Faaij and Lewandowski (2004).
21 Ibid.
22 Jose Roberto Moreira (2004a) p2.
23 Ibid.
24 Smeets, Faaij and Lewandowski (2004).
25 Rosillo-Calle (2001).
26 Faaij (2005a).
27 Ibid.
28 Smeets, Faaij and Lewandowski (2004).
29 Ibid.
30 Ibid.
31 Faaij, op cit note 19.
32 Bothwell Batidzirai et al (2006).
33 Table 6.3 from Smeets, Faaij and Lewandowski (2004).
34 Ibid.
35 Faaij, op cit note 19.
36 Smeets, Faaij and Lewandowski (2004); Faaij (2005a).
37 André Faaij et al (2005).
38 Smeets, Faaij and Lewandowski (2004).
39 Danielle Nierenberg (2005).
40 M. Hoogwijk et al (2003).
41 Smeets, Faaij and Lewandowski (2004).
42 Vaclav Smil (2002).
43 S. Peng et al (2004).
44 D. B. Lobell and G. P. Asner (2003).
45 Faaij (2005a).
46 Table 6.4 from Monique Hoogwijk (2004).
47 Ibid.
48 Ibid.
49 Faaij (2005a).
50 Ibid.
51 Ibid.
52 Ibid.
53 Ibid.
54 Ibid.
55 Ibid.
56 Ibid.
57 Ibid.
58 Ibid.
59 Ibid.
60 Ibid.
61 V. Dornburg, and A. Faaij (2005).

7 Economic and energy security

1 Suzanne C. Hunt and Janet L. Sawin, with Peter Stair (2006).
2 Figure 7.1 is based on the following sources: 1950–1993 from Worldwatch Institute database; 1993–2004 from BP (2005).
3 Milton R. Copulos (2005).
4 Jeff Rubin and Peter Buchanan (2006).
5 Figure 7.2 is based on the following sources: 1985–2004 from Platts, cited in BP (2005); 2005 from US DOE and EIA (2006d).
6 Noel Randewich (2005).
7 Masami Kojima and Todd Johnson (2005); TaTEDO (2005).
8 W. Gehua et al (2005) p98.
9 IEA (2004a).
10 *Econbrowser* (2005).
11 RWE Aktiengesellschaft (2005).
12 A. F. Alhajji (2005); M. S. Daoudi and M. S. Dajani (1984); Alan Dobson (1988).
13 *Associated Press* (2004).
14 James D. Hamilton (1983).
15 *The Economist* (2005b).
16 World Economic Forum in Russia (2005).
17 Commission of the European Communities (2006).
18 German Federal Institute for Geosciences and Natural Resources (2006).
19 Commission of the European Communities (2006), p3.
20 Jeffrey Rubin (2006).
21 Michael Renner (2002a).
22 K. Y. Amoako (2005); Boris Utria, Comments made at Biofuels Roundtable, Washington, DC, 7 July 2005.
23 Ibid.
24 TERI (2005) p46.
25 IEA (2004b); Merse Ejigu (2005).
26 André P.C. Faaij et al (2000).
27 Table 7.1 from Stephen Haley and Nydia R. Suarez (2004).
28 Ejigu (2005).
29 A. Faaij, A. Wierczorek and M. Minnesma (2002); Suani Teixeira Coelho (2005).
30 Plinio Nastari, Isaias Macedo and Alfred Szwarc (2005) p17.
31 Glen S. Hodes, Boris E. Utria and Anthony Williams (2004).
32 Ejigu (2005).
33 Hunt et al (2006).
34 Nastari, Macedo and Szwarc (2005), p21.
35 Isaias de Carvalho Macedo et al (2005) p195.
36 TERI (2005).
37 I. C. Macedo, R. Lima Verde Leal and J. Silva (2004a).
38 Randewich (2005).
39 Linda Hutchinson-Jafar (2006).
40 Vanguard (2005).
41 Table 7.2 from US General Accounting Office (GAO) (2000).

42 Masami Kojima and Todd Johnson (2005) p71.
43 Milton R. Copulos (2003).
44 J. Moore et al (1997).
45 Kojima and Johnson (2005).
46 USDA, Farm Service Agency (2005).
47 Carol Werner (2005).
48 Kojima and Johnson (2005), p28.
49 RFA (2005a).
50 US DOE and EIA (2006a).
51 L. R. Lynd et al (2003).
52 Kojima and Johnson (2005), p91.
53 Public Citizen (2001).
54 *F. O. Licht's World Ethanol and Biofuels Report* (2006e); *F. O. Licht's World Ethanol and Biofuels Report* (2006f); Barani Krishnan and Sambit Mohanty (2006).
55 OECD (2006) p31.
56 Ibid, p27.
57 A. Regmi et al (2001).
58 Coelho (2005).
59 Centre for International Economics (2005).
60 Ibid.
61 TERI (2005).
62 New Uses Council, Environmental and Energy Studies Institute and North Carolina Solar Center (2005).
63 Kojima and Johnson (2005).
64 US General Accounting Office (2000).
65 US DOE, Federal Energy Management Program (2002).
66 Council for Biotechnology Information (2005).
67 Lynd et al (2003).
68 Emily A. Heaton et al (2003).
69 Vinod Khosla (2006).
70 IEA (2004b).
71 TERI (2005), p46.
72 Danielle Murray (2005b).
73 Ibid.
74 Steven Heckeroth (1999).
75 OECD (2006), p27.
76 Ann Bordetsky et al (2005).
77 Copulos (2003).
78 Keith Bradsher (2005); Brett Clanton (2005).
79 Ed Garsten (2005).
80 United Auto Workers (2005); Sholnn Freeman and Amy Joyce (2006).
81 autoweb.com (2004).
82 Choren Industries website, www.choren.de.
83 Iogen Corporation website, www.iogen.ca.

8 Implications for Agriculture and Rural Development

1 Michael Duffy (2006).
2 Bill Kovarik (1998).
3 Ethanol share of corn crop from *F. O. Licht's World Ethanol and Biofuels Report* (2006i); US DOE and EERE (2006a); RFA (2005b).
4 This 20 per cent projection is based on the US Renewable Fuel Standard target of 7.5 billion gallons of biofuel for 2012.
5 Kevin C. Dhuyvetter, Terry L. Kastens and Michael Boland (2005); Food and Agricultural Policy Research Institute (2005).
6 In contrast, increased ethanol production results in more production and lower prices of corn by-products. Soybean meal prices are reduced by 10 per cent.
7 Lori Wilcox (2005).
8 Candida Jones (2005).
9 *Oil World* (2005).
10 Isaias de Carvalho Macedo et al (2005).
11 *F. O. Licht's World Ethanol and Biofuels Report* (2006e).
12 *ABC News Online* (2005); *F. O. Licht's World Ethanol and Biofuels Report* (2005c).
13 *Reuters* (2005f); Jones (2005).
14 Peter Apps (2005).
15 Anna Mudeva (2005).
16 *F. O. Licht's World Ethanol and Biofuels Report* (2005d); *F. O. Licht's World Ethanol and Biofuels Report* (2005j).
17 Perfecto G. Corpuz (2004); figure of 5 million from *Sun Star Magazine* (2005); *F. O. Licht's World Ethanol and Biofuels Report* (2005g); *Manila Times* (2005).
18 *F. O. Licht's World Ethanol and Biofuels Report* (2005i).
19 Centre for International Economics (2005).
20 *F. O. Licht's World Ethanol and Biofuels Report* (2005h).
21 Jones (2005); D1 Oils (2005).
22 Marie Walsh et al (2000).
23 Potential to increase farmer income from Jim Hettenhaus (2002).
24 De La Torre Ugarte et al (2003).
25 International Policy Council for Food and Agricultural Trade (2005).
26 EC (2006a).
27 Macedo et al (2005).
28 Walsh et al (2000).
29 Ibid.
30 US DOE and EERE (2006a); RFA (2005b); NREL (2000).
31 Wuppertal Institute for Climate, Environment and Energy (2005) p40.
32 Sivan Kartha, Gerald Leach and Sudhir Chella Rajan (2005) p11.
33 Wuppertal Institute for Climate, Environment and Energy (2005), p40.
34 Suani Teixeira Coelho (2005).
35 Sybille de La Hamaide (2005).
36 J. Domac, K. M. Richards and V. Segon (2005) *Old Fuel for Modern Times: Socio-Economic Drivers and Impacts of Bioenergy Use*, Paris, International Energy Agency Bioenergy Task 29, p8, www.iea-bioenergytask29.hr/pdf/Domac_Richards_Segon_2005.pdf.

37 *F. O. Licht's World Ethanol and Biofuels Report* (2005b).
38 W. Gehua et al (2005) p94.
39 Masami Kojima and Todd Johnson (2005).
40 Ibid.
41 Ibid, p129.
42 USDA, Economic Research Service (2006); Michael Duffy (2006).
43 Kartha, Leach and Rajan (2005), p11; Kojima and Johnson (2005).
44 Food and Agriculture Policy Research Institute (2005), p1; Michael Duffy (2004).
45 Figure 8.1 from Franz J. Kaltner et al (2005) p88 and from Jones (2006).
46 TERI (2005) p44.
47 Ibid, pp44–45.
48 Kaltner et al (2005).
49 Luiz Prado, Director of Brazil Programs, LaGuardia Foundation, e-mail to Suzanne Hunt, Worldwatch Institute, 28 January 2006.
50 IOL (2005).
51 Environment News Service (2005b).
52 Kaltner et al (2005).
53 *MercoPress* (2005)
54 Prado, op cit note 49.
55 De La Torre Ugarte et al (2003).
56 Kevin Watkins and Joachim von Braun (2003).
57 Bruce Gardner (2003).
58 *High Plains Journal* (2005); Wilcox (2005).
59 RFA (2005b); John Urbanchuk and Jeff Kapell (2002); David Morris (2003).
60 Nancy Novak (2002).
61 De La Torre Ugarte et al (2003).
62 Kojima and Johnson (2005), p131.
63 Macedo et al (2005), pp51–64.
64 Glen S. Hodes, Boris E. Utria and Anthony Williams (2004).
65 Energetics and NEOS Corporation (1994).
66 National Corn Growers Association (2006).
67 Jones (2005).
68 *Reuters* (2005e).
69 Prado, op cit note 49.
70 Neil Lyon (2005).
71 Kojima and Johnson (2005).
72 Rogerio Carneiro de Miranda, Winrock International, e-mail to Peter Stair, Worldwatch Institute, 16 February 2006.
73 Mali and Box 8.1 from Ibrahim Togola, Mali-Folkecenter, Bamako, Mali, visit to Worldwatch Institute, Washington, DC, 26 July 2005.
74 Ibid.
75 Ibid; Merse Ejigu (2005).
76 Tanzania Traditional Energy Development and Environment Organization (TaTEDO) (2005) *Biofuels for Transportation in Tanzania Potential and Implications for Sustainable Agriculture and Energy in the 21st Century*, Prepared for GTZ, Dar es Salaam, September.

77 Macedo et al (2005).
78 David Morris (2006).
79 Brian Halweil (2000).
80 Corn flakes from Halweil, ibid; Duffy (2004).
81 Halweil (2000).
82 Watkins and von Braun (2003).
83 Macedo et al (2005), p203.
84 Minnesota Department of Agriculture (2004); John M. Urbanchuk (2004).
85 Kurt Klein et al (2006).
86 Stephen Thompson (2004).
87 Blue Sun Biodiesel (2006).
88 Bart Minten, Lalaina Randrianarison and Johan F. M. Swinnen (2005) p1.
89 William Burnquist, Centro de Tecnologia Canavieira, Piracicaba, Brazil, conversation with Suzanne Hunt and Peter Stair, Worldwatch Institute, 24 October 2005.
90 Prado, op cit note 49.
91 Kaltner et al (2005), p36.
92 Gehua et al (2005); Wu Yong (2005).
93 Morris (2006).
94 RFA (2006b).
95 New Rules Project (2006); Dhuyvetter, Kastens and Boland (2005).
96 Joel Severinghaus (2005).
97 Joe Mead, sales associate, World Energy Alternatives, conversation with Peter Stair, Worldwatch Institute, 20 August 2005.
98 *PR Newswire* (2005).
99 *F. O. Licht's World Ethanol and Biofuels Report* (2005e).
100 Kaltner et al (2005), p36.
101 Stephanie Simon (2005).
102 IFAD (2002).
103 Kojima and Johnson (2005); biodiesel from Prado, op cit note 49.
104 Amartya Sen (2002).

9 INTERNATIONAL TRADE IN BIOFUELS

1 M. Hoogwijk et al (2003).
2 Suzanne C. Hunt and Janet L. Sawin, with Peter Stair (2006).
3 Ibid.
4 Brazil from Patrick Knight (2003); Table 9.1 from Jim Jordan and Associates (2005).
5 Christoph Berg (2004).
6 Masami Kojima and Todd Johnson (2005) p97; *Kyodo News* (2005).
7 Green Car Congress (2005c).
8 Jane Earley, Thomas Earley and Matthew Straub (2005) pp10–14.
9 Mark Ash and Erik Dohlman (2005).
10 *Rams Horn: Journal of Food Systems Analysis* (2006).
11 Agence France-Presse, (2005); John Burton (2005).

12 Green Car Congress (2006).
13 Center for Science in the Public Interest (2005).
14 EC (2006a) pp24–26.
15 Ibid, p6.
16 Figure 9.1 from Lew Fulton et al (2004) p185.
17 Ibid.
18 EC, op cit note 14, pp24–26.
19 Brent D. Yacobucci (2004).
20 Ash and Dohlman (2005).
21 HART Downstream Energy Services (2004); Suani Teixeira Coelho (2005).
22 Institute for Agriculture and Trade Policy (2005) p3.
23 HART Downstream Energy Services (2004).
24 Institute for Agriculture and Trade Policy (2005).
25 Hunt and Sawin with Stair (2006).
26 Karin Bendz (2005a).
27 Karin Bendz (2005b).
28 Table 9.2 from EC (2005c) p34.
29 Bendz (2005a).
30 EC (2005c), p34.
31 Karin Bendz (2005c); Wuppertal Institute for Climate, Environment and Energy (2005) p15.
32 MERCOSUR (2004).
33 Environmental Working Group (2006); Alexei Barrionuevo (2005).
34 Oxfam International (2002).
35 Earley, Earley and Straub (2005), p4.
36 Martin von Lampe (2006) p25.
37 US Senator Ken Salazar (2006).
38 David Borough (2005).
39 *The Economist* (2005a).
40 ICTSD (2005a).
41 EC (2005a); EC (2005b).
42 Daryll E. Ray, Daniel G. De La Torre Ugarte and Kelly J. Tiller (2003).
43 World Trade Organization website, www.wto.org.
44 David Waskow, International progamme director, Friends of the Earth-US, pers comm with Lauren Sorkin and Suzanne Hunt, Worldwatch Institute, 17 February 2006.
45 ICTSD (2005b).
46 Daniel De La Torre Ugarte (2005).
47 TERI (2003); Iris Lewanski and André Faaij (2004).
48 Robert Howse (2005).
49 Coelho (2005); Wuppertal Institute for Climate, Environment and Energy (2005), p46.
50 TERI (2005).
51 New York Board of Trade website, www.nybot.com.
52 *Reuters* (2005d).
53 EC (2006a), p9.

54 Paul Anthony A. Isla (2004).
55 Jason Clay (2005).
56 EC (2006a).
57 *Reuters* (2006).
58 Radhika Singh (undated).
59 Table 9.3 based on the following sources: Isla (2004); South Africa from D1 Oils Plc (2006a).
60 National Biodiesel Board (2005b).
61 National Biodiesel Board (2006).
62 Isla (2004).
63 Singh (undated).
64 Dennis Olson, Director, Trade and Agriculture Project, IATP, e-mail to Lauren Sorkin, Worldwatch Institute, 27 February 2006.

10 ENERGY BALANCES OF CURRENT AND FUTURE BIOFUELS

1 Table 10.2 is based on the following sources: D. Lorenz and D. Morris (1995); US DOE and EERE (2006b); M. Wang (2005); M. A. Elsayed, R. Matthews and N. D. Mortimer (2003) p20; G. Azevedo, Director, GilTech Consultants for Technology and Sustainable Development, Brazil, e-mail to Peter Stair, Worldwatch Institute, 22 November 2005; Franz Kaltner, Presentation cited in Azevedo, op cit this note; W. Gehua et al (2005); Isaias de Carvalho Macedo et al (2005) p60; J. Sheehan et al (1998c); I. R. Richards (2000); BABFO (1994) ADEME and DIREM (2002); Hosein Shapouri, James A. Duffield, and Michael S. Graboski (1995); NTB Liquid Biofuels Network (no date); Marco Aurélio dos Santos (undated); D. Andress (2002); R. Larsen et al (2004).
2 Lew Fulton et al (2004) pp51–64.
3 Hosein Shapouri, Senior economist, USDA, Washington, DC, e-mail to Peter Stair, Worldwatch Institute, 10 January 2006.
4 Robert Brown, Office of Biorenewables Program, Iowa State University (2005) 'A question of balance', e-mail to Peter Stair, Worldwatch Institute, 30 November 2005.
5 Sheehan et al (1998c).
6 Brown, op cit note 4.
7 Larsen et al (2004).
8 J. Sheehan et al (2004).
9 Richards (2000); ADEME and DIREM (2002).
10 Fulton et al (2004), p58.
11 US from Hosein Shapouri (2004); Brazil from Macedo et al (2005), p101.
12 Rockefeller Foundation and Scientific and Industrial Research and Development Centre 1998.
13 Jacob Bugge (2000).
14 ADEME and DIREM (2002); Richards (2000).
15 ADEME and DIREM (2002); Sheehan et al (1998c); Richards, (2000).
16 Ibid.

17 Wang (2005).
18 Shapouri, op cit note 3.
19 Ibid.
20 Ibid; production of fertilizer from Richards (2000).
21 Wang (2005); Macedo et al (2005).
22 Novozymes and BBI International (2005).
23 Shapouri, op cit note 3.
24 Robert Anex, Associate professor of agricultural systems and bioengineering, Iowa State University, pers comm with Peter Stair, Worldwatch Institute, 15 August 2005.
25 Brown, op cit note 4.
26 Macedo et al (2005).
27 Novozymes and BBI International (2005).
28 Chippewa Valley Ethanol Company and Frontline Bioenergy LLC (2005).
29 Macedo et al (2005).
30 Brown, op cit note 4.
31 US DOE.
32 Anex, op cit note 24.

11 Effects on Greenhouse Gas Emissions and Climate Stability

1 The Earth has warmed by 0.6°C over the past 30 years and by 0.8°C over the past 100 years, according to US National Aeronautics and Space Administration (2006).
2 Kevin A. Baumert, Timothy Herzog and Jonathan Pershing (2005); Working Group III (2001).
3 Share of global emissions in Lew Fulton et al (2004) p174; share of US emissions in Nathanael Greene et al (2004) ppiv, 8; share of European emissions in EC Directorate-General for Energy and Transport (2004) p6.
4 United Nations Framework Convention on Climate Change (2005) 'Key GHG data – GHG emission data for 1990–2003', p69, cited in EC (2006e) p4.
5 Analysis by EEA and ETC/ACC (2003).
6 Masami Kojima and Todd Johnson (2005) p17.
7 Assumptions about the relative efficiency of gasoline versus biofuel (or blended fuel) vehicles can result in up to 10 per cent variation in a study's final results, according to Lew Fulton et al (2004) p54.
8 Energy balance from M. Wang, C. Saricks and D. Santini (1999); co-products from Fulton et al (2004) p52.
9 Markus Quirin et al (2004) p12.
10 IEA, Bioenergy Task 38 (2006).
11 Quirin et al (2004).
12 See, for example, Fulton et al (2004) pp52–66; Quirin et al (2004); Eric D. Larson (2005); Alexander E. Farrell et al (2006).
13 Same vehicle efficiency from Fulton et al (2004), p64.
14 Ibid, p54.

15 André Faaij, Associate professor and coordinator, Research Energy Supply and System Studies, Copernicus Institute, Utrecht University, The Netherlands, e-mail to Suzanne Hunt, Worldwatch Institute, 3 March 2006.
16 On the other hand, anaerobic digesters for converting animal and human waste to biogas are fully commercialized and the benefits are well known. There are also serious demonstration and commercialization projects focused on hydrolysis for ethanol; demonstration and co-production (with coal/natural gas) facilities for gasification (Fischer-Tropsch fuels and dimethyl ether); and Shell and Chevron recently announced that they will soon begin F-T production from co-gasification from coal and biomass in Europe. Per André Faaij, Coordinator, Research Energy Supply and System Studies, Copernicus Institute, Utrecht University, The Netherlands, e-mail to Suzanne Hunt, Worldwatch Institute, 10 March 2006.
17 Nitrous oxide from Larson (2005) p14.
18 D. S. Powlson, A. B. Riche and I. Shield (2005).
19 Robert Edwards (2005); A. P. Armstrong et al (2002).
20 IEA (1998).
21 Jason Clay (2004) p219.
22 Franz J. Kaltner et al (2005) p50.
23 Larry Rohter (2005); Environment News Service (2005).
24 Jim Cook and Jan Beyea (undated); Larson (2005) p15.
25 Ibid.
26 Powlson, Riche and Shield (2005).
27 Ibid.
28 Larson (2005), p15.
29 Melvin Cannell (2003).
30 Powlson, Riche and Shield (2005).
31 Ibid.
32 UNEP (2000).
33 Quirin et al (2004) p19.
34 John Sheehan et al (2004) p138.
35 Powlson, Riche and Shield (2005).
36 GreenBiz.com (2005).
37 Christoph Berg (2004).
38 Furthermore, processing them into ethanol requires an additional step that is not needed for making ethanol from sugar cane and sugar beets: hydrolysing the starch into sugars at high temperatures before fermentation. This increases energy requirements.
39 Sakamoto Yakuhin Kogyo, Ltd (2005).
40 Note that biodiesel has more energy content per litre than does ethanol.
41 Journey to Forever (2005).
42 ORNL (undated a).
43 *Science Daily* (2005); ORNL (undated b).
44 Carl Trettin, Aletta Davis and John Parsons (2003); switchgrass carbon sequestration in Greene et al (2004) pv; low pesticides from S. McLaughlin et al (1999); carbon and tilling from Powlson, Riche and Shield (2005).
45 Higher yield in Martin Tampier et al (2004) p113; root system from Greene et al (2004).

46 W. W. Wilhelm and J. Cushman (2003).
47 USDA Soil Quality Institute (2003).
48 Ibid.
49 Greene et al (2004) ppv, 29.
50 USDA Soil Quality Institute (2003).
51 S. Park et al (2004).
52 Fulton et al (2004) p52.
53 Ibid, p53.
54 Kojima and Johnson (2005) p24.
55 See for example, B. J. Somera and K. K. Wu (1996).
56 M. A. Delucchi (1998) 'A model of lifecycle energy use and greenhouse-gas emissions of transportation fuels and electricity', cited in IEA (1998).
57 M. A. Delucchi (2004) cited in Farrell et al (2006).
58 Armstrong et al (2002) p10.
59 S. Kim and B. E. Dale (2002) cited in Farrell et al (2006).
60 Fulton et al (2004) p52.
61 Armstrong et al (2002) pv.
62 John Sheehan et al (1998c) p117; A. Aden et al (2002) p9.
63 Wuppertal Institute for Climate, Environment and Energy (2005) pp28–29.
64 Sheehan et al (1998c) pp48, 50.
65 Fulton et al (2004) p66.
66 Carlo N. Hamelinck, Roald A. A. Suurs and André P.C. Faaij (2005).
67 André Faaij, Associate professor and coordinator, Research Energy Supply and System Studies, Copernicus Institute, Utrecht University, The Netherlands, e-mail to Suzanne Hunt, Worldwatch Institute, 10 March 2006.
68 Sheehan et al (1998c) p221.
69 Quirin et al (2004).
70 Larson (2005) p11.
71 Farrell et al (2006).
72 Quirin et al (2004), p44.
73 Larson (2005) p2.
74 Fulton et al (2004) p52; Quirin et al (2004) pp15–18.
75 Quirin et al (2004), pp15–18.
76 Larson (2005) p2.
77 David Pimentel (1991) and David Pimentel (2001), both cited in Fulton et al (2004) p53.
78 David Pimentel (2003); D. Pimentel and T. Patzek (2005); T. Patzek (2004).
79 Farrell et al (2006) pp27–29; Quirin et al (2004) p8.
80 For example, Mark A. Delucchi (2003); M. A. Delucchi (2004); and Mark A. Delucchi (2005).
81 Mark A. Delucchi, Research scientist, ITS, University of California at Davis, e-mail to Janet Sawin, Worldwatch Institute, 10 March 2006; Delucchi (2005) p44.
82 Delucchi, op cit note 81; Delucchi (2003).
83 M. A. Elsayed, R. Matthews, and N. D. Mortimer (2003) p23; Fulton et al (2004); Kojima and Johnson (2005) pp88–89.
84 Table 11.1 is based on the following sources: Fulton et al (2004) pp53, 59; Farrell et al (2006); Wuppertal Institute (2005), p29; W. Gehua et al (2005) p84; Armstrong et

al (2002); Tom Beer et al (2001); Delucchi (2005) p100; Isaias de Carvalho Macedo, Manoel Regis Lima Verde Leal, and Joao Eduardo Azevedo Ramos da Silva (2004b); Wang, Saricks and Santini (1999); EC (1994); R. H. Levy (1993); G. Marland and A. F. Turhollow (1991); General Motors et al (2002); M. P. Gover et al (1996); Levelton Engineering Ltd (1999); Michael Wang (2001).

85 Wang, Saricks and Santini (1999) pp1–3.
86 Farrell et al (2006).
87 Larson (2005) pp2–3.
88 Delucchi, op cit note 81.
89 Kojima and Johnson (2005) pp88–89.
90 Kaltner et al (2005) p84.
91 Fulton et al (2004) pp59–60.
92 Edwards (2005) slide 25; Quirin et al (2004) p20.
93 Table 11.2 is based on the following sources: Kojima and Johnson(2005), pp89, 221; Wuppertal Institute (2005); Jean-François Larivé (2005) slide 14; Levy (1993), Armstrong et al (2002); Beer et al (2001); ETSO (1996); Delucchi (2005) (1996); Gover et al, op cit note 84; General Motors et al (2002); Levelton Engineering Ltd (1999); K. Scharmer and G. Gosse (1996); I. R. Richards (2000) Netherlands Agency for Energy and Environment (2003); John Sheehan et al (1998b).
94 Delucchi (2003).
95 Delucchi, op cit note 81.
96 Larson (2005) pp2–3.
97 Ibid, pp17–18.
98 Biofuels considered included ethanol from sugar cane, corn, wheat, sugar beets, potatoes, molasses and lignocellulose; ETBE from wheat, sugar beets, potatoes and lignocellulose; biodiesel from rapeseed, sunflowers, soybeans, canola, coconut oil, recycled vegetable oil, animal grease and used cooking grease; vegetable oil from rapeseed and sunflower; biomass to liquid (BTL), biomethanol, DME and hydrogen from lignocellulose; biogas from organic residues and cultivated biomass; and hydrogen from organic residues.
99 Quirin et al (2004), p20.
100 Fulton et al (2004) p51.
101 Wuppertal Institute (2005) p26.
102 Clay (2004) pp176–177, 211, 412; Wang et al (1999), p11; continued development in Fulton et al (2004) pp65–66; National Biodiesel Board (2004a); Wang et al (1999).
103 Clay (2004) pp176–177, 211, 412; Wang et al (1999) p11.
104 Fulton et al (2004) p136; K. P. Vogel and R. A. Masters (1998); breeding from Powlson, Riche and Shield (2005).
105 Kojima and Johnson (2005) p57.
106 Fulton et al (2004) pp60–61.
107 Plinio Mário Nastari (2005).
108 Angelo Bressan Filho (2005).
109 NOVEM and ADL (1999).
110 Quirin et al (2004) p39.
111 Tampier et al (2004) p102.
112 Wang et al (1999) p15.

113 Tampier et al (2004) p13.
114 P. Börjesson and G. Berndes (undated).
115 Quirin et al (2004).
116 Table 11.3 is based on the following sources: Beer et al (2001); Delucchi (2005), p100; General Motors et al (2002); Wang (2001); Levelton Engineering Ltd (1999).
117 Larson (2005) pp17–18.
118 Fulton et al (2004) pp61–62.
119 Delucchi, op cit note 81; Mark A. Delucchi, Research scientist, ITS, University of California at Davis, e-mail to Janet Sawin, Worldwatch Institute, 13 March 2006.
120 Fulton et al (2004) pp61–62.
121 Ibid.
122 Wang et al (1999) p3.
123 Table 11.4 from NOVEM and ADL (1999).
124 Genencor International (2005).
125 Thomas F. Riesing (2006).
126 M. Kerssen and R. H. Berends (2005).
127 Brad Lemley (2003).
128 Fulton et al (2004) p65.
129 Figure 11.1 from Lew Fulton, Sustainable Transport, Division of Global Environment Facility Coordination, United Nations Environment Programme, e-mail to Peter Stair, Worldwatch Institute, July 2005.
130 André Faaij (2005b).
131 Larson (2005) pp17–18.
132 Data and Figure 11.2 from Fulton et al (2004) pp91–94.
133 Suzanne Hunt and Janet Sawin, with Peter Stair (2006) p64.
134 Kojima and Johnson (2005) pp41, 105.
135 Jane Michael Henke, Gernot Klepper and Norbert Schmitz (2003) *Tax Exemption for Biofuels in Germany: Is Bio-Ethanol Really an Option for Climate Policy?*, Kiel Working Paper no 1184, Kiel, Kiel Institute for World Economics, September, cited in ibid, p90; abatement cost from EC (2006e) p14.
136 Larivé (2005) slide 18.
137 Kojima and Johnson (2005) p90.
138 Low end of range is for conventional production and increased derogation; high end is for conventional production and current derogation. N. D. Mortimer et al (2003) pix.
139 For every UK pound spent, biodiesel from rapeseed (depending upon the production process) could save 3.4–5.2kg of CO_2; heat from short-rotation coppice wood chips would save 18.2kg; and glass-fibre loft insulation would save 478.5kg of CO_2, according to N. D. Mortimer et al (2003) pix.
140 Ibid.
141 Fulton et al (2004) pp93–94; Carlo N. Hamelinck and André Faaij (2006).
142 Carlo N. Hamelinck (2004); Faaij, op cit note 67.
143 Hamelinck and Faaij (2006).
144 Ibid.
145 Fulton et al (2004) pp93–94; little room for cost improvements also from Hamelinck and Faaij (2006).

146 Tony Dammer (2005).
147 Sierra Club of Canada (2004); *Nature Canada* (2004).
148 For example, see M. A. Elsayed et al (2003) p23 and Kojima and Johnson (2005) pp41, 90.
149 Armstrong et al (2002) cited in Kojima and Johnson (2005) p90.
150 R. L. Graham, L. L. Wright and A. F. Turhollow (1992) cited in Robin L. Graham, Wei Liu and Burton C. English (1995).
151 Faaij (2005b).
152 Tampier et al (2004) p103.
153 IEA (1998).
154 B. Schlamadinger and G. Marland (1998) cited in IEA (1998).
155 Veronika Dornburg and André P. C. Faaij (2004).
156 Tampier et al (2004) p103.
157 Faaij, op cit note 67.
158 Larson (2005) p2.

12 ENVIRONMENTAL IMPACTS OF FEEDSTOCK PRODUCTION

1 Notably, only a relatively small number of studies look at the overall impacts of biofuels, and most consider only single variables or steps along the entire pathway. In addition, many of these studies make conclusions based on observations that are location specific, making it difficult to obtain a big picture view. Furthermore, while numerous studies focus on the environmental impacts of commonly used biofuels such as ethanol and biodiesel, research on future fuels, such as biomass to liquids, is lacking. See, for example, Markus Quirin et al (2004) Larson (2005).
2 J. S. Dukes (2003); A. Lovins et al (2005) p2.
3 SPE (2005b).
4 M. Mann and P. Spath (1997) pp43–44.
5 Ibid, p48.
6 US Geological Survey (2005).
7 Mann and Spath (1997) p57.
8 Ibid, p58.
9 S. Gerrard et al (1999).
10 SPE (2005a, b).
11 W. Baue (2004).
12 World Petroleum Council (2005).
13 Tony Dammer (2005).
14 It is estimated that US oil shale reserves total 2 trillion barrels, and Canadian tar sand reserves are 1.5 trillion barrels – far more than remaining proven and possible oil reserves from conventional sources, according to Tony Dammer (2005).
15 Robert Collier (2005).
16 Colorado Environmental Coalition et al (2006) pp18–43.
17 Franz J. Kaltner et al (2005) p50.
18 Jason Clay (2004) p165.
19 Plinio Nastari, Isaias Macedo and Alfred Szwarc (2005) pp56–57.

20 Suani Teixeira Coelho (2005) p14.
21 More than one quarter from Environment News Service (2005); sugar cane expansion from Manoel Regis Lima Verde Leal, Pesquisador Associado, e-mail to Peter Stair, Worldwatch Institute, 2 December 2005.
22 Kaltner et al (2005) p50.
23 Ibid, p49.
24 Ibid, p50.
25 *Associated Press* (2005).
26 Kaltner et al (2005) pp50–51.
27 Fred Pearce (2005).
28 Ibid.
29 Barbara Bramble, senior programe adviser for International Affairs, National Wildlife Federation, Washington, DC, e-mail to Janet Sawin, Worldwatch Institute, 23 February 2006.
30 R. Schneider et al (2000).
31 Maria Pia Palermo (2005).
32 Rob Glastra, Erik Wakker and Wolfgang Richert (2002); Pearce (2005); UK Department for Transport (2004)
33 Clay (2004) pp211–218.
34 Ibid, p219.
35 Ed Matthew (2005); Friends of the Earth-UK (2005).
36 E. P. Deurwaarder (2005) p25.
37 Bernama.com (2005).
38 David Turley, Scientist and co-author of a study commissioned by the Home Grown Cereals Authority in the UK and completed early 2005, cited in *Reuters* (2005b).
39 TaTEDO (2005) p94.
40 Michael Duffy, Professor of economics, Iowa State University, e-mails to Peter Stair, Worldwatch Institute, 2 December 2005.
41 EC Directorate-General for Energy and Transport (2004) p6.
42 Uwe Fritsche, Öko-Institut, Darmstadt, Germany, Reviewer comments, 9 March 2006.
43 See, for example, G. W. Burton (1982) and G. W. Burton and B. G. Mullinix (1998), cited in Nathanael Greene et al (2004) p26.
44 EC (2006c) p17.
45 Clay (2004) pp211, 412.
46 Box 12.1 is based on the following sources: John Vidal (2006); David Cullen (2005); *Reuters* (2003a); David Cullen (2003); Norman E. Borlaug (2000); G. Conway and G. Toennisessen (1999); Interacademies (2000); Sylvie Bonny (2003); Clay (2004) p181; Charles M. Benbrook (2004); Dennis Keeney and Steve Suppan, Institute for Agriculture and Trade Policy (IATP), comments provided in Jim Kleinschmit, IATP, e-mail to Janet Sawin, Worldwatch Institute, 21 February 2006; Paul Brown and John Vidal (2003); Michael McCarthy (2003); Pew Initiative on Food and Biotechnology (2003); Organic Consumers Association (2005); Lilian Joensen and Stella Semino (2004); Manoah Esipisu (2006) cited in Greene et al(2004) p44.
47 TERI (2005) p60.
48 Ibid.

49 Mali-Folkecenter (2004).
50 Robin L. Graham, Wei Liu and Burton C. English (1995).
51 Carlo N. Hamelinck and André Faaij (2006).
52 Kirsten Wiegmann and Uwe R. Fritsche, with Berien Elbersen (2006b) pp43–63.
53 Thomas F. Riesing (2006) p48.
54 John Sheehan et al (1998a).
55 Based on 11,000 tonnes of garbage daily and 40,000 tonnes filling 0.4ha (1 acre) of landfill space. Data from Lester R. Brown (2002).
56 Agricultural waste from Brad Lemley (2003); world oil consumption from BP (2005) p9.
57 Jim Cook and Jan Beyea (undated) p5.
58 M. J. Bullard, D. G. Christian and C. Wilkins (1996) pp61–63; O. J. F. Santos (2001) cited in D. S. Powlson, A. B. Riche and I. Shield (2005).
59 David W. Sample, Laura Paine and Amber Roth (1998).
60 Graham, Liu and English (1995); A. Schiller and V. R. Tolbert (1996).
61 TaTEDO (2005), p94.
62 Graham, Liu and English (1995).
63 Robert D. Perlack et al (1995).
64 Cook and Beyea (undated); Schiller and Tolbert, op cit note 60.
65 Perlack et al (1995).
66 TERI (2005), p60; UK Department for Transportation (2004); Graham, Liu and English (1995).
67 Schiller and Tolbert (1996).
68 IEA (2002) p8.
69 UNEP (2000).
70 Wiegmann and Fritsche, with Elbersen (2006b), p46.
71 Ibid, p57.
72 IEA (2002), p6.
73 W. W. Wilhelm and J. Cushman (2003).
74 TaTEDO (2005), p95; Wiegmann and Fritsche, with Elbersen (2006b) p46.
75 Perlack et al (1995).
76 Masami Kojima and Todd Johnson (2005) p56.
77 Television Trust for the Environment (2004).
78 TERI (2005), p57.
79 J. W. Ranney and L. K. Mann (1994) cited in T. H. Green et al (1996).
80 Ibid.
81 Kaltner et al (2005), p57.
82 D. Pimentel and J. Krummel (1987) cited in Green et al (1996).
83 Perennial grasses and trees in S. B. McLaughlin et al (2002); Perlack et al (1995), pp1–2, 4–5; more intensive crops from Greene et al (2004), pp28–29; TERI (2005), p57.
84 Clay (2004), pp418–419.
85 Green et al (1996).
86 Wiegmann and Fritsche, with Elbersen (2006b), p53.
87 Ibid, p52.
88 Ibid.

89 Pests from ibid, p57; pest and disease resistance from Powlson, Riche and Shield (2005), p194.
90 TERI (2005), p57.
91 Kaltner et al (2005), pp86–87.
92 TaTEDO (2005), p95.
93 IEA (2002) p6.
94 G. R. Benoit and M. J. Lindstrom (1987).
95 USDA, Soil Quality Institute (2003) p2.
96 Wiegmann and Fritsche, with Elbersen (2006b).
97 Wilhelm and Cushman (2003).
98 IEA (2002).
99 Graham, Liu and English (1995).
100 Sandra Postel (2006).
101 Kojima and Johnson (2005), p25.
102 Marcelo E. Dias De Oliveira, Burton E. Vaughan and Edward J. Rykiel Jr. (2005).
103 Kojima and Johnson (2005), p25.
104 UNEP (2000).
105 Michael Pollan (2002).
106 Box 12.2 is based on the following sources: natural gas and oil from Pollan, ibid; nitrogen cycle from UNEP (2000); WHO in collaboration with UNEP (1990); palm oil and Bt corn from Clay (2004), pp222, 412; Hosein Shapouri, James A. Duffield and Michael Wang (2002); comparison of fertilizer inputs from Wiegmann and Fritsche, with Elbersen (2006b), p55; nitrogen oxides from Wuppertal Institute for Climate, Environment and Energy (2005) p31; hybrids and sugar cane pesticide use and rising herbicide use from Kojima and Johnson (2005), pp80, 128; no-till practices from Kaltner et al (2005) p120–121; low pesticide needs for energy crops from IEA (2002), p4; Powlson, Riche and Shield (2005); organic manure from Francis et al, 2005, cited in TERI (2005), p56; jatropha from TERI (2005), p56; vinasse and filter cake from Coelho (2005), p14; matching with soils and yields from TaTEDO (2005), p95; careful timing and placement from University of California Agriculture and Natural Resources (2005) *Reduce Pollution with Proper Fertilizer Timing*, http://ucanr.org/delivers/impactview.cfm?impactnum=249&mainunitnum=0, accessed June 2005 strategies for lowering pesticide use from Edward Smeets, André Faaij and Iris Lewandowski (2005) *The Impact of Sustainability Criteria on the Costs and Potentials of Bioenergy Production*, Utrecht, Copernicus Institute, Utrecht University, May p61; US herbicide treatment from Clay (2004), p412; pesticide use in Brazil sugar cane from Coelho (2005) p14.
107 Powlson, Riche and Shield (2005).
108 TaTEDO (2005), p94.
109 W. Gehua et al (2005) p102.
110 Green et al (1996); Powlson, Riche and Shield (2005).
111 Powlson, Riche and Shield (2005).
112 H. A. Lyons, S. G. Anthony and P. A. Johnson (2001) cited in TaTEDO (2005), p94.
113 TERI (2005), p61.
114 Ibid, pp58–59.

115 Kojima and Johnson (2005), pp125–126.
116 Kaltner et al (2005), p85.
117 Kojima and Johnson (2005), p27.
118 Mark Peplow (2005).
119 Kojima and Johnson (2005), pp125–126.
120 Ibid, p68; I. Macedo et al (2005) p142.
121 UNEP (2000).
122 For example, Markus Quirin et al (2004) and J. Calzoni et al (2000).
123 Coelho (2005), p15.
124 Bernama.com (2005).
125 TERI (2005), p61.

13 Environmental impacts of processing, transport and use

1 Notably, only a relatively small number of studies look at the overall life-cycle impacts of biofuels, and most consider only single variables or steps along the entire pathway, making it challenging to derive a big picture view. In addition, many of these studies make conclusions based on observations that are location specific. Furthermore, while numerous studies focus on the environmental impacts of commonly used biofuels, such as ethanol and biodiesel, research on future fuels, such as biomass to liquids, is lacking. See, for example, Markus Quirin et al (2004) Eric D. Larson (2005).
2 M.A. Delucchi (1995), cited in UCS (no date).
3 UCS, ibid.
4 EC (2003d) p15.
5 Environmental Defense (1999).
6 P. C. Sherertz (1998).
7 M. Mann and P. Spath (1997) p88.
8 D. Doniger (2001).
9 US DOE and EIA (2002) p49.
10 Energetics Inc (1998) cited in Nathanael Greene et al (2004) p43.
11 E. O. van Ravenswaay (1998).
12 Robert Collier (2005).
13 Colorado Environmental Coalition et al (2006) pp18–43.
14 Mann and Spath (1997), pp53–65.
15 For a list of the biggest oil spills of the last 30 years, see Mariner Group (2000) F. A. Leighton (2000); Michael Renner (2002b).
16 Mallenbaker (2006).
17 Ilinca Bazilescu and Bret Lyhus (1997).
18 UK Onshore Pipeline Operators' Association (2006).
19 Mann and Spath (1997), p93.
20 Pennsylvania Department of Environmental Protection (2005).
21 Abt Associates Inc (2000).
22 Masami Kojima and Todd Johnson (2005).

23 Table 13.1 is based on the following sources: SO_2 exposure from Stan McMillen et al (2005) p33; Kojima and Johnson, op cit note 22, pp63–74; particulate matter deaths from US Environmental Protection Agency (EPA), cited in McMillen et al (2005), p33; lead poisoning from Paul R. Epstein and Jesse Selber (eds) (2002).
24 John Sheehan et al (1998c) p134.
25 Ibid.
26 Ibid, p135.
27 Wang Gehua et al (2005) p67.
28 Marcelo E. Dias De Oliveira, Burton E. Vaughan and Edward J. Rykiel Jr. (2005) p601, cited in Mark Peplow (2005).
29 Gehua et al (2005), p67.
30 Processing in Martin Tampier et al (2004) p32; Greene et al (2004), p43.
31 Jason Clay (2004) p162.
32 Kojima and Johnson (2005), p72.
33 William S. Saint (1982) cited in Kojima and Johnson (2005), p27.
34 Kojima and Johnson (2005), p72.
35 I. Macedo et al (2005) p167.
36 William Burnquist, Centro de Tecnologia Canavieira (CTC), Brazil, conversation with Suzanne Hunt and Peter Stair, Worldwatch Institute, 28 October 2005.
37 Sergio Pacca, University of São Paulo, reviewer comments, 13 February 2006.
38 Kojima and Johnson (2005), pp25, 72, 127.
39 Perry Beeman (2005).
40 Novozymes and BBI International (2002) cited in Kojima and Johnson (2005), p72.
41 Greene et al (2004), p43; Energetics, Inc (1998).
42 Novozymes and BBI International (2002).
43 Greene et al (2004), p43.
44 Ibid.
45 Environment News Service (2005d).
46 Beeman (2005).
47 Jim Core (2005).
48 Sheehan et al (1998c), p124.
49 Greene et al (2004), p42.
50 Kojima and Johnson (2005), p32.
51 Beeman (2005); US EPA (2005b); Greene et al (2004), pp41–42.
52 In early 2006, the US Environmental Protection Agency was considering easing air emissions standards for ethanol plants to boost the nation's ethanol supplies. See Tom Doggett (2006).
53 Novozymes and BBI International (2005).
54 Greene et al (2004), p42.
55 Kojima and Johnson, op cit note 22, p32.
56 Greene et al (2004), p42.
57 However, burning of bagasse also emits NO_x, CO and particulates.
58 Kojima and Johnson (2005), p24.
59 TaTEDO (2005) p95.
60 Lew Fulton et al (2004) p111.

61 Northeast States for Coordinated Air Use Management (2001) cited in Greene et al (2004) p52; TERI (2005) p59.
62 Randall von Wedel (1999).
63 Gehua et al (2005), pp101–102.
64 von Wedel (1999).
65 Northeast States for Coordinated Air Use Management (2001).
66 Ibid.
67 It is important to note that many toxic chemicals in gasoline are actually additives; if these were not present in gasoline, spill-related problems would be reduced.
68 Clean Air Task Force et al (2005) Annex, pp4–8.
69 Ibid, p1.
70 Greene et al (2004), pp49–51.
71 Fulton et al (2004), p113.
72 Where the RVP peaks depends upon characteristics of the base gasoline; refiners can remove more of the smaller molecules in gasoline when blending it with ethanol to keep vapour pressure low.
73 Northeast States for Coordinated Air Use Management (2001); Greene et al (2004), pp49–51; C. Hammel-Smith et al (2002) ppii, 11–12.
74 Fulton et al (2004), p113.
75 Reformulated gasoline is a gasoline that is blended to burn more cleanly and thereby reduce smog-forming and toxic pollutants.
76 Steven J. Brisby, Manager, Fuels Section, Stationary Source Division, California Air Resources Board, e-mail to Janet Sawin, Worldwatch Institute, 1 March 2006.
77 Ibid.
78 Kojima and Johnson (2005), p.5.
79 Fulton et al (2004), pp116–18.
80 Note, however, that ethanol sold at refuelling stations could contain some of these in low amounts because, by law, poisons must be added to it. They are usually gasoline components and whatever is cheapest.
81 Kojima and Johnson (2005), pp 20, 78; Gehua et al (2005), p78.
82 Kojima and Johnson (2005), p78.
83 Fulton et al (2004), p114.
84 Fulton et al (2004), pp114–115.
85 I. C. Macedo (1993) cited in Robin L. Graham, Wei Liu and Burton C. English (1995).
86 Fulton et al (2004), pp114–115.
87 Ibid.
88 AQIRP (1997) cited in Kojima and Johnson (2005), p79.
89 Uwe R. Fritsche, Katja Hünecke and Kirsten Wiegmann (2005) p32; Fulton et al (2004) pp114–115; California Air Resources Board (2004) cited in Greene et al(2004), pp49–51.
90 Fulton et al (2004), p114–115.
91 Greene et al (2004), pp49–51; Fulton et al (2004), pp114–115; Fritsche, Hünecke and Wiegmann (2005), p32.
92 Chandra Prakash (1998).
93 Kojima and Johnson (2005), pp20, 33.

94 Ibid, p112.
95 Ibid, p20.
96 C. Hammel-Smith et al (2002), pii.
97 Ibid, pp120–121.
98 Note, however, that since ETBE has toxicity levels similar to the fuel additive methyl tertiary butyl ether (MTBE). Concern in the US regarding fuel/MTBE leaks from underground storage tanks is expected to be a major barrier to acceptance of ETBE in that country (see Chapter 2 for more information).
99 CETESB (2003) cited in Kojima and Johnson (2005).
100 Kojima and Johnson (2005), p102.
101 Ibid; Suani Teixeira Coelho (2005) p10.
102 *F. O. Licht's World Ethanol and Biofuels Report* (2005k).
103 Thomas Durbin et al (2006) pp47–48.
104 Ibid; and Prakash (1998).
105 Kojima and Johnson (2005), p78.
106 K. S. Tyson, C. J. Riley and K. K. Humphreys (1993) cited in Graham, Liu and English (1995).
107 Fulton et al (2004), p116.
108 Jim Easterly, Easterly Consulting, e-mail to Suzanne Hunt, Worldwatch Institute, 4 March 2006; Marc Goodman, QSS Group, e-mail to Sergio C. Trindade, SE2T International, Ltd, 22 February 2006.
109 Easterly, ibid.
110 Goodman, op cit note 108.
111 Greene et al (2004), p51; Gehua et al (2005), p101; TERI (2005) p57.
112 Fulton et al (2004) p116.
113 US EPA (2002).
114 K. Becker and G. Francis (2005) cited in TERI (2005) p58.
115 Kojima and Johnson (2005), p81.
116 J. R. Pedersen, A. Ingemarsson, and J. O. Olsson (1999) cited in Jürgen Krahl et al (2001).
117 Krahl et al, ibid.
118 Ibid.
119 Figure 13.1 from Fulton et al (2004).
120 Liezzel M. Pascual and Raymond R. Tan (2004).
121 US EPA (no date) p49.
122 Kojima and Johnson (2005), p78.
123 US EPA (2005a).
124 Roland Hwang, Vehicle policy director, Natural Resources Defense Council, San Francisco, CA, e-mail to Janet Sawin, Worldwatch Institute, 12 March 2006.
125 Fulton et al (2004), pp111–114.
126 Robert McCormick, NREL, Golden, CO, e-mail to Peter Stair, Worldwatch Institute, 20 July 2005.
127 Robert McCormick (2005).
128 AQIRP (1997).
129 TERI (2005), p58.
130 Greene et al (2004), p49.
131 Fulton et al (2004), p118.

132 McCormick (2005).
133 Manufacturers of Emissions Controls Association (undated).
134 *DieselNet News*, May 2005, cited in *AMFI* (2005).
135 *Miljöfordon Newsletter* no 4 (2005), www.ufop.de, www.all4engineers.com, cited in *AMFI Newsletter* (2005).
136 Sean Sinico (2005).
137 Ibid.
138 Ibid; Association of the German Biofuel Industry from Karin Retzlaff of the Association of the German Biofuel Industry, cited in ibid.
139 R. L. McCormick (2003) p21; National Biodiesel Board (2004b); USDA from M. A. Hess et al (2005); World Energy Alternatives (2006).
140 Developing countries in Lew Fulton (2004a).
141 Kojima and Johnson (2005), p78.
142 Ibid, p95.
143 Greene et al (2004), p51.
144 D1 Oils Plc website, www.d1plc.com.
145 Emissions reductions from Fulton et al (2004), p111; US EPA (2002).
146 Newest vehicles in Greene et al (2004) pv.

14 INFRASTRUCTURE REQUIREMENTS

1 Figure 14.1 is derived from the following sources: EIA and DOE (2005) pp78–79; EIA and DOE (2004); ANP (2006a); ANP (2006b).
2 F. J. Kaltner et al (2005).
3 Nevertheless, it is important to note that most Brazilian sugar mills simultaneously produce sugar and ethanol, with each mill responsible for processing significant amounts of sugar cane. Thus, the largest ethanol-producing unit also uses sugar cane to produce 455,000 tonnes of sugar. Other reasons for the lower average ethanol capacity in Brazilian plants are related to significant transportation costs associated with sugar cane due to the lower alcohol yield per tonne compared with corn, and the limited duration of the operational season, which is around 180 to 210 days per year, as opposed to the ethanol from starch plants that operate year round.
4 C. E. Noon et al (1996).
5 ASSOCANA – Departamento Agrícola (2005).
6 USDA Transportation and Marketing Programs/Transportation Services Branch (2005).
7 Table 14.1 from ibid.
8 Noon et al (1996).
9 Table 14.2 from Lew Fulton et al (2004).
10 Kaltner et al (2005).
11 In reality, the cost of money in Brazil is high and requires an annual discount rate of 20 per cent in national currency to attract investors. Thus, the values quoted should be considered low and would need to be adjusted depending upon the discount criteria used by specific investors.
12 Cost estimates and Table 14.3 from Kaltner et al (2005); D. L. Van Dyne and M. G. Blasé (1998).

13 Kaltner et al (2005).
14 This value does not take into account the value of the protein meal extracted from the soybeans.
15 *F. O. Licht's World Ethanol and Biofuels Report* (2006b).
16 Fulton et al (2004).
17 Maryland General Assembly, Department of Legislative Services (2003).
18 Fulton et al (2004), p90.
19 Kaltner et al (2005).
20 Refuelling stations from *National Petroleum News* (May 2005) as quoted by National Ethanol Vehicle Coalition (2006).
21 *F. O. Licht's World Ethanol and Biofuels Report* (2006c).
22 Kaltner et al (2005).
23 Ibid.
24 Ibid.
25 Fernando Cunha (2003).
26 Transpetro (2006).
27 RFA (2006a).
28 F. O. Licht, cited in RFA (2005b) p14.
29 Table 14.4 from Downstream Alternatives Inc (2002).
30 Ibid.
31 Table 14.5 derived from the following sources: low-end data from Fulton et al (2004); high-end data from Downstream Alternatives Inc (2002).
32 Downstream Alternatives Inc (2002).
33 Table 14.6 derived from Fulton et al, op cit note 9, and Kaltner et al (2005).
34 Estimate of 85 per cent from UNICA (2005); 33 per cent from A. J. Lepach (2005).
35 Downstream Alternatives Inc (2002).
36 Donald Van Dyne and Melvin Blasé (1998).
37 K. Damen and A. Faaij (2003).
38 Ibid.
39 Figure 14.2 from USDA (2005).
40 Ibid.

15 VEHICLE AND ENGINE TECHNOLOGIES

1 Ohio Farm Bureau Federation (2005).
2 Agência Nacional do Petróleo, Gás Natural e Biocombustíveis (ANP) is the Brazilian governmental agency responsible for fuel quality specification.
3 The American Society for Testing and Materials (ASTM) elaborates upon fuel quality specifications in the US.
4 Range of 0 to 2 per cent from Apace Research Ltd (1998).
5 Brazilian law (Lei No 8.723/1993) and regulations define that ethanol content in gasoline shall be within the 20–25 per cent range. Due to the inherent characteristics of the blending process, a tolerance of ±1 per cent is allowed; therefore, the accepted range is 19–26 per cent. The actual ethanol content is established by the Inter-

Ministerial Sugar and Ethanol Council (CIMA) based on supply–demand analysis; for the last three years, it has been 25 per cent.
6 Autolex Consultoria Automotiva, São Paulo, Brazil, pers comm with Alfred Szwarc, Director, ADS Tecnologia e Desenvolvimento Sustentável, São Paulo, Brazil, 2005.
7 Hua Hin (2005).
8 National Ethanol Vehicle Coalition (undated).
9 Ford Motor Company (2007).
10 Alfred Szwarc, Director, ADS Tecnologia e Desenvolvimento Sustentável, São Paulo, Brazil, pers comm with Worldwatch Institute, 2006.
11 Fuel combustion properties include high octane, high latent heat of vaporization and high flame speed.
12 Refers to the air–fuel mixture compression level in the engine's cylinders and combustion chambers; the higher the compression ratio, the more efficient the combustion process is likely to be.
13 Ibid.
14 M. Brusstar and M. Bakenhus (2005).
15 Szwarc, op cit note 10.
16 Scania Brasil, pers comm with Alfred Szwarc, Director, ADS Tecnologia e Desenvolvimento Sustentável, São Paulo, Brazil, 2001.
17 NREL (1998).
18 G. Nagarajan et al (no date).
19 Ibid.
20 National Biodiesel Board (1998).
21 C. A. Sharp (1998).
22 US DOE (2005a).
23 A fluoroelastomer produced by E. I. Dupont de Nemours Company that is well known for its excellent resistance to heat and aggressive fuels and chemicals.
24 Sharp (1998).
25 Brazilian Reference Centre on Biomass (CENBIO), personal information of Alfred Szwarc, UNICA, 2005.
26 Elsbett website, www.elsbett.com.
27 Sweedtrack (no date).
28 Total Corporation (2006).
29 WestStart-CalStart (2005).
30 Szwarc, op cit note 10.
31 Ibid.

16 Transfer of Technology and Expertise

1 Figure 16.1 compiled by Sergio C. Trindade, President, SE2T International, Ltd, January 2006.
2 Sergio C. Trindade (1994).
3 Gavin P. Towler et al (2003).
4 Figure 16.2 compiled by Trindade, op cit note 1.
5 EC (2003a); EC (2003c).

6 Sino-German Workshop on Energetic Utilization of Biomass (2003).
7 Isaias Macedo, University of Campinas (UNICAMP), Campinas, São Paulo, Brazil, pers comm with Sergio C. Trindade, 27 December 2005; Lima Verde Leal and Manoel Régis, UNICAMP, pers comm with Sergio C. Trindade, SE2T International Ltd, 26 December 2005.
8 Sergio C. Trindade, author's estimate, 2005.
9 Sergio C. Trindade (2005) p6.

17 Biofuel policies around the world

1 Suzanne Hunt and Janet Sawin, with Peter Stair (2006) p73.
2 Share of global emissions in Lew Fulton et al (2004) p174.
3 Masami Kojima and Todd Johnson (2005) p126; RSPO (2006).
4 Table 17.1 derived from the following sources: China, France, Thailand and US from Hunt et al (2006), p75; *F. O. Licht's World Ethanol and Biofuels Report* (2006a); Brazil, Colombia and Venezuela from Kelly Hearn (2005); Canada from Ontario Ministry of Agriculture, Food and Rural Affairs (2005), Herbert Tretter (2005), *Reuters* (2006); Lars Olofsson (2005); India from Christoph Berg (2004); Japan from Danielle Murray (2005a), Green Car Congress (2005b); Philippines from *Coconut-Biodiesel for All Government Vehicles, Memorandum Circular 55* (2004) and from Green Car Congress (2005a); Environment News Service (2005a); Jeff Blend and Howard Haines (2005).
5 Sivan Kartha, Gerald Leach and Sudhir Chella Rajan (2005) p159; Francis Johnson, e-mail communication with Suzanne Hunt, Worldwatch Institute, 14 February 2006.
6 Kojima and Johnson (2005).
7 *Reuters* (2005c).
8 *F. O. Licht's World Ethanol and Biofuels Report* (2005a) p4; Stephen Karekezi (1995).
9 Peter Apps (2005); Sasol (2004).
10 Shaun Benton (2006).
11 D1 Oils (2006a).
12 *F. O. Licht's World Ethanol and Biofuels Report* (2005a).
13 Murray (2005a).
14 Commonwealth of Australia Biofuels Taskforce (2005) p63
15 *Reuters* (2003b).
16 Murray (2005a).
17 Prodyut Bhattacharya and Bharati Joshi (2003).
18 Berg (2004).
19 TERI (2005) pp31–32.
20 Bhattacharya and Joshi (2003).
21 *Reuters* (2005a).
22 *Coconut-Biodiesel for All Government Vehicles, Memorandum Circular 55* (2004); Green Car Congress (2005b).
23 Philippine Fuel Ethanol Alliance (2005).
24 Murray (2005a).

25 Commonwealth of Australia Biofuels Taskforce (2005).
26 Murray (2005a).
27 Mobil Oil Australia Pty Ltd (2005).
28 *F. O. Licht's World Ethanol and Biofuels Report* (2006a).
29 EC (2005c) p9; EC (2003b).
30 *Official Journal of the European Union* (2005).
31 Commonwealth of Australia Biofuels Taskforce (2005).
32 EC (2003c); Wuppertal Institute for Climate, Environment and Energy (2005) pp18–19.
33 USDA, Foreign Agricultural Service (2006c).
34 EC (2005c) pp9–11.
35 Ibid, p10.
36 Ibid.
37 EC (2006f).
38 Institute Français Riera (2004).
39 Germany from *Reuters* (1999); EC (2005c).
40 Jose Goldemberg et al (2003).
41 Hearn (2005).
42 Rodrigo Augusto Rodrigues (2005).
43 Angelo Bressan Filho, Director of AgriEnergy Programme, Brazil Ministry of Energy, Comments at Workshop and Business Forum on Sustainable Biomass Production for the World Market, University of Campinas, São Paulo, Brazil, November 2005.
44 Hearn (2005).
45 Ibid.
46 Agriculture policy takes shape in the so-called 'Farm Bill', which is re-authorized approximately every five years. The Farm Bill contains federal commodity support programmes, land conservation, forestry programmes, rural development, and hunger/nutrition programmes for underprivileged people, as well as a clean energy title.
47 RFA (no date) p26.
48 Under earlier laws, tax credits were geared toward blends of 5.7 per cent, 7.7 per cent and 10 per cent, per terms of obsolete requirements under the Clean Air Act.
49 RenewableEnergyAccess.com (2005a).
50 US Senate (2005) pp141–172.
51 RenewableEnergyAccess.com (2005b).
52 RenewableEnergyAccess.com (2005c).
53 Governor George Pataki (2006).
54 Government of Manitoba, Ministry of Energy, Science and Technology (2003).
55 Natural Resources Canada (2003).
56 Hunt et al (2006) p63.
57 *Miami Herald* (2005).
58 The Brazilian government required that all new vehicles should be able to run on up to 25 per cent ethanol. Government subsidies and tax breaks to develop technology and government-funded research also helped the auto industry to develop the FFVs, equipped with special motors to switch between gasoline, ethanol and natural gas.
59 Hunt et al (2006) p64.

60 Ibid.
61 EC (2005d).
62 TERI (2005) p67.
63 Hugh Dent (2005).
64 Hunt et al (2006) p73.
65 World Summit on Sustainable Development (2002).
66 Ibid.
67 Table 17.2 is based on the following sources: IEA Task 40: Sustainable International Bioenergy Trade website, www.bioenergytrade.org, Suzanne Hunt (2005); United Nations Environment Programme (UNEP), Sustainable Energy Finance Initiative website, www.sefi.unep.org; *The Gleneagles Communiqué* (2005) UNCTAD (2005) Italian Ministry for the Environment and Territory (2005) p31.
68 UNCTAD from Jana Gastellum, Energy Future Coalition, pers comm with Ladeene Freimuth, Freimuth Consulting, January 2006; IEA suggestion from André Faaij and Arnaldo Walter (2005) p18.
69 Kojima and Johnson (2005) pp7–8.
70 Boris Utria, Senior economist and co-chair of the Energy and Poverty Thematic Group, Energy Unit, Africa Region, World Bank, e-mail to Lauren Sorkin and Suzanne Hunt, Worldwatch Institute, 15 February 2006.
71 The GEF consists of the UNDP, UNEP and the World Bank as its 'implementing agencies' and regional development banks as its 'executing agencies'.
72 Global Environment Facility, Small Grants Programme (2005).
73 UNEP, op cit note 67.
74 EC (2005b) p10.
75 Natsource, Environmental Services (2006).
76 Demonstrating this 'additionality' has proven very difficult to accomplish at the methods and applications levels.
77 Lew Fulton (2004b).
78 Ben Atkinson (undated) pp2–3.
79 Jodi Brown et al (2004) p16.
80 Atkinson (undated), p3.
81 Ibid.
82 According to paragraph 20 of 'Further guidance relating to the CDM', 'a local/regional/national policy or standard cannot be considered as a clean development mechanism project activity, but project activities under a programme of activities can be registered as a single clean development mechanism project activity'.
83 Christiana Figueres (2005).
84 World Bank Carbon Finance Unit (CFU) website, www.carbonfinance.org.
85 Ibid.
86 Ibid.

18 Standards and Certification Schemes

1 Fred Pearce (2005).
2 Friends of the Earth-UK (2005) p9.

3 Rainforest Alliance (no date b).
4 Rainforest Alliance (no date a).
5 IFOAM (2006).
6 Forest Stewardship Council (FSC) website, www.fsc.org.
7 FSC (2004).
8 FSC (no date b).
9 FSC (no date a).
10 Goska Romanowicz (2006).
11 Ausilio Bauen et al (2005).
12 RSPO (2005b).
13 RSPO (2005a).
14 Ibid.
15 RSPO (no date).
16 WTO (no date).
17 A. Faaij, A. Wieczorek and M. Minnesma (2003) p65.
18 WTO (no date).
19 Faaij et al (2003).
20 Ibid.
21 Nielmar de Oliveira (2004).
22 Faaij et al (2003).
23 General Agreement on Trade and Tariffs, TBT Agreement, Article 2.4.
24 R. Howse and E. Tuerk (2006) 'The WTO impact on internal regulations: A case study of the Canada–EC asbestos dispute', in G. Bermann and P. Mavroidis (eds) *Trade and Human Health and Safety*, New York, Cambridge University Press, pp77–117.
25 General Agreement on Tariffs and Trade, Article III: 4.
26 WTO (2001b). This interpretation of 'no less favourable treatment' was applied by the panel in the EC biotech case (GMOs); see para 7.2505. For more on the EC biotech panel ruling (interim), see Friends of the Earth–UK (2006). More research on this topic is forthcoming and can be found at the website of Robert Howse, http://faculty.law.umich.edu/rhowse.
27 Ibid.
28 WTO (2001a).
29 Martin Juniger and André Faaij (2005) p23.
30 EC (2005c) p9.
31 Iris Lewandowski and André Faaij (2004).
32 André Faaij, Copernicus Institute, Utrecht University, e-mail to Suzanne Hunt, Worldwatch Institute, 3 March 2006.

References

ABC11 Eyewitness News (2007) 'Sunopta reports on cellulosic ethanol projects', Seeking Alpha, 25 January.

ABC News Online (2005) 'North coast sugar industry to benefit from biofuels decision', *ABC News Online*, 29 November

Abengoa Bioenergy (2005) *Abengoa Bioenegy R&D's First Pilot Plant Is Ready to Begin Research and Development Operations*, www.abengoabioenergy.com/feature.cfm?page=6

Abt Associates Inc (2000) *The Particulate Related Health Benefits of Reducing Power Plant Emissions*, Bethesda, MD, Abt Associates Inc, October, www.abtassociates.com/reports/particulate-related.pdf

ADEME (Agence de l'Environnment et de la Maîtrise de l'Energie) and DIREM (Direction des Ressources Energétiques et Minérales) (2002) *Energy and Greenhouse Gas Balances of Biofuels' Production Chains in France*, Paris, Ecobilan/PricewaterhouseCoopers, December, www.ademe.fr/partenaires/agrice/publications/documents_anglais/synthesis_energy_and_greenhouse_english.pdf

Aden, A. et al (2002) *Lignocellulosic Biomass to Ethanol Process Design and Economics Utilizing Co-Current Dilute Acid Prehydrolysis and Enzymatic Hydrolysis for Corn Stover*, Golden, CO, National Renewable Energy Laboratory (NREL), June

AEA Technology (2003) 'Appendix 3: Potential for production cost reduction', in *International Resource Costs of Biodiesel and Bioethanol*, Prepared for the UK Department for Transport, London

Agence France-Presse (2005) *Malaysia to Build Three Biodiesel Plants Fueled by Palm Oil*, 28 September

Ahring, B. K. et al (1996) 'Pretreatment of wheatstraw and conversion of xylose and xylan to ethanol by thermophilic anaerobic bacteria', *Bioresource Technology*, vol 58, no 2, pp107–113

Alhajji, A. F. (2005) *The Failure of the Oil Weapon: Consumer Nationalism vs Producer Symbolism*, Ada, OH, Ohio Northern University, www2.onu.edu/~aalhajji/ibec385/oil_weapon2.htm

AMFI Newsletter (Advanced Motor Fuels Information) (2005) October, http://virtual.vtt.fi/virtual/amf/pdf/amfinewsletter2005_4october.pdf

Amoako, K. Y. (2005) Former executive secretary, Economic Commission for Africa, and Commissioner, Commission for Africa, Presentation to Resources for Global Growth: Agriculture, Energy and Trade in the 21st Century conference, Washington, DC, 6 December

Andress, D. (2002) *Ethanol Energy Balances*, David Andress and Associates, Kensington, MD

ANP (Agência Nacional do Petróleo e de Biocombustíveis) (2006a) *Superintendência de Planejamento e Pesquisa, Volume de Petróleo Refinado nas Refinarias Nacionais (metros cúbicos)*, www.anp.gov.br/doc/dados_estatisticos/Processamento_de_Petroleo_m3.xls

ANP (2006b) *Superintendência de Planejamento e Pesquisa. Produção Nacional de Derivados de Petróleo (metros cúbicos)*, www.anp.gov.br/doc/dados_estatisticos/Producao_de_Derivados_m3.xls

Apace Research Ltd (1998) *Intensive Field Trial of Ethanol/Petrol Blend*, Energy Research and Development Council Project No 2511, Canberra

Apps, P. (2005) 'South Africa maize farmers see ethanol plant in 18 months', *Reuters*, 15 March

AQIRP (Auto/Oil Air Quality Improvement Research Program) (1997) *Program Final Report*, AQIRP, January

Armstrong, A. P. et al (2002) *Energy and Greenhouse Gas Balance of Biofuels for Europe – an Update*, CONCAWE Report No 2/02, Brussels, April

Ash, M. and E. Dohlman (2005) *Soybeans and Oil Crops: Trade*, Washington, DC, US Department of Agriculture (USDA), Economic Research Service, April

Ashworth, J. et al (1991) *Properties of Alcohol Transportation Fuels*, Prepared for US Department of Energy (DOE) by Meridian Corporation, Washington, DC, May 1991

ASSOCANA – Departamento Agrícola (2005) *Custo Médio Operacional da Lavoura de Cana-De-Açúcar em Reais*, Assis, Brazil, 30 September, www.assocana.com.br/custo.html

Associated Press (2005) 'Brazilian wetlands defender commits suicide in environmental protest', *Associated Press*, 15 November

Associated Press (2004) 'Oil: The flip side,' *Associated Press*, 27 October

Atkinson, B (undated) *The CDM, Kyoto Protocol and the Sugar, Ethanol and Biofuels Industry*, Ringwood, Hampshire, UK, Agrinergy, pp2–3

Automotive World (2005) 'Ford launches focus flexi-fuel in UK', www.awknowledge.com, 6 September

autospectator.com (2005) 'Biomass-to-ethanol technology could help replace half of auto fuel in US', 22 May

autoweb.com (2004) 'Toyota licenses hybrid technology to Ford', 25 March, http://autoweb.drive.com.au/cms/A_101177/newsarticle.html

BABFO (British Association for Bio Fuels and Oils) (1994) *Rationale and Economics of a British Biodiesel Industry*, London, www.biodiesel.org/resources/reportsdatabase/reports/gen/19940401_gen-294.pdf

Badger, P. C. (2002) 'Ethanol from cellulose: A general review', in J. Janick and A. Whipkey (eds) *Trends in New Crops and New Uses*, Alexandria, VA, ASHS Press, pp17–21

Bangkok Post (2005) 'Thailand looking to import ethanol to mix with petrol', *The Bangkok Post*, 5 September

Barrionuevo, A. (2005) 'A warning about trade suits over agriculture', *New York Times*, 30 November

Batidzirai, B. et al (2006) 'Biomass and bioenergy supply from Mozambique', *Energy for Sustainable Development*, March, p10

Battjes, J. J. (1994) *Global Options for Biofuels from Plantations According to IMAGE Simulations*, Groningen, The Netherlands, Rijksuniversiteit Groningen

Baue, W. (2004) *ChevronTexaco Faces Class-Action Lawsuit in Ecuador Over Environmental Damage*, SocialFunds.com, 11 May, www.socialfunds.com/news/article.cgi/article1419.html

Bauen, A. et al (2005) *Feasibility Study on Certification for a Renewable Transport Fuel Obligation*, Final report, London, Edinburgh Centre for Carbon Management, Ltd, Imperial College, June

Baumert, K. A., T. Herzog and J. Pershing (2005) *Navigating the Numbers: GHG Data and Climate Policy*, Washington, DC, World Resources Institute

Bazilescu, I. and B. Lyhus (1997) *Russia Oil Spill*, Trade and Environment Database Case Studies, Washington, DC, American University, 11 January, http://gurukul.ucc.american.edu/ted/KOMI.HTM

Becker, K. and G. Francis (2005) *Biodiesel from Jatropha Plantations on Degraded Land*, Stuttgart, University of Hohenheim

Beeman, P. (2005) 'Ethanol plans among Iowa's polluters', *Des Moines Register*, 11 September

Beer, T. et al (2001) *Comparison of Transport Fuels*, Final report submitted to the Australia Greenhouse Office on Stage 2 Study of Life-Cycle Emissions Analysis of Alternative Fuels for Heavy Vehicles, Clayton South, Victoria, Commonwealth Scientific and Industrial Research Organisation, www.greenhouse.gov.au/transport/comparison

Benbrook, C. M. (2004) *Genetically Engineered Crops and Pesticide Use in the United States: The First Nine Years*, Biotech-Info Technical Paper No 7, Sand Point, ID, Northwest Science and Environmental Policy Center, October

Bendz, K. (2005a) *EU-25 – Agriculture Situation: Pakistan, EU's Second Largest Ethanol Exporter, Loses Privileged Status*, GAIN Report, Washington, DC, USDA and Foreign Agricultural Service, 27 September

Bendz, K. (2005b) *EU-25 – Oilseeds and Products: Annual. 2005*, GAIN Report, Washington, DC, USDA and Foreign Agricultural Service, 10 June

Bendz, K. (2005c) *EU-25 – Oilseeds and Products: Biofuels Situation in the European Union 2005*, GAIN Report, Washington, DC, USDA and Foreign Agricultural Service, March

Benoit, G. R. and M. J. Lindstrom (1987) 'Interpreting tillage-residue management effects', *Journal of Soil and Water Conservation*, March–April, pp87–90

Bensmann, M. (2005) 'Bio im Tank', *Neue Energie*, vol 3

Benton, S (2006) 'Government biofuels strategy to cover multiple benefits, Says minister', *Bua News*, 10 February, http://allafrica.com/stories/200602100545.html

Berg, C. (2004) *World Fuel Ethanol Analysis and Outlook*, Ratzeburg, Germany, F. O. Licht, April

Bernama.com. (2005) *No Evidence of Threat to Wildlife Caused by Oil Palm Plantations*, Bernama.com (Malaysian National News Agency), 12 October

Berndes, G. et al (2003) 'The contribution of biomass in the future global energy supply: A review of 17 studies', *Biomass and Bioenergy*, vol 25, pp1–28

Bhattacharya, P. and J. Bharati (2003) 'Strategies and institutional mechanisms for large scale cultivation of *Jatropha curcas* under agroforestry in the context of the proposed biofuel policy of India', Abstract in *ENVIS Bulletin on Grassland Ecosystems and Agroforestry*, Indian Institute of Forest Management, December, pp58–72

Biewinga, E. E. and G. van der Bijl (1996) *Sustainability of Energy Crops in Europe: A Methodology Developed and Applied*, CLM 234-1996, Utrecht, Centre for Agriculture and Environment, February

Blend, J. and H. Haines (2005) *Economic Effects of Increased Ethanol Use in Montana, Helena, MT*, Montana Department of Environmental Quality, Air Energy and Pollution Prevention Bureau, 5 January, http://deq.mt.gov/Energy/bioenergy/ethanolUseInMT.asp

Blue Sun Biodiesel (2006) *News and Events*, www.gobluesun.com, accessed 6 March 2006

Bonny, S. (2003) 'Why are most Europeans opposed to GMOs? Factors explaining rejection in France and Europe', *Electronic Journal of Biotechnology*, 15 April

Bordetsky, A. et al (2005) *Securing America: Solving our Oil Dependence through Innovation*, New York, Natural Resources Defense Council and Institute for the Analysis of Global Security

Börjesson, P. (1999) 'Environmental effects of energy crop cultivation in Sweden', *Biomass and Bioenergy*, vol 16, pp137–154

Börjesson, P. and G. Berndes (undated) *Multi-functional Biomass Production Systems*, Lund, Sweden, Lund University

Borlaug, N. E. (2000) 'Ending world hunger: The promise of biotechnology and the threat of anti-science zealotry', *Plant Physiology*, vol 124, no 2, pp487–490

Borough, D. (2005) 'The future of ethanol in Europe hinges on state aid', *Reuters*, 6 December

Bothast, R. J. et al (1999) 'Fermentations with new recombinant organisms', *Biotechnology Progress*, vol 15, no 5, pp867–875

BP (2005) *Statistical Review of World Energy 2005*, London, BP

Bradsher, K. (2002) *High and Mighty: The Dangerous Rise of the SUV*, New York, PublicAffairs

Bridgwater, A. V. (1999) 'Principles and practice of biomass fast pyrolysis processes for liquids', *Journal of Analytical and Applied Pyrolysis*, vol 51, pp3–22

Brown, J. et al (2004) *Getting on Track: Finding a Path for Transportation in the CDM*, Final report, Winnipeg, Manitoba, Canada, International Institute for Sustainable Development, December, p16

Brown, L. R. (2002) *New York: Garbage Capital of the World*, Eco-Economy Update, Washington, DC, Earth Policy Institute, 17 April

Brown, P. and J. Vidal (2003) 'Two GM crops face ban for damaging wildlife', *The Guardian*, 17 October

Brusstar, M. and M. Bakenhus (2005) *Economical, High Efficiency Engine Technologies for Alcohol Fuels*, US Environmental Protection Agency, National Vehicle and Emissions Laboratory, XV International Symposium on Alcohol Fuels, San Diego, CA, September

Bryan, T. (2005) 'Editor's note regarding total biodiesel production capacity in the US', *Biodiesel Magazine*, October

Buckland, H. (2004) *The Oil for Ape Scandal: How Palm Oil Is Threatening the Orangutan*, London, Friends of the Earth–UK, www.foe.co.uk/resource/reports/oil_for_ape_summary.pdf

Bugge, J. (2000) *Note: Rape Seed Oil for Transport 1: Energy Balance and CO_2 Balance*, Hurup Thy, Denmark, Folkecenter for Renewable Energy, 11 September, www.folkecenter.dk/plant-oil/publications/energy_co2_balance.pdf

Bullard, M. J., D. G. Christian and C. Wilkins (1996) *Quantifying Biomass Production in Crops Grown for Energy*, ETSU B CR/0038-00-00, Harwell, Didcot, Oxon, AEA Technology Environment

Burton, G. W. (1982) 'Improved recurrent restricted phenotypic selection increases bahiagrass forage yields', *Crop Science*, vol 22, pp1058–1061

Burton, G. W. and B. G. Mullinix (1988) 'Yield distributions of spaced plants within Pensacola bahiagrass populations developed by recurrent restricted phenotypic selection', *Crop Science*, vol 38, pp333–336

Burton, J. (2005) 'Malaysia likely to legislate biofuel use', *Financial Times*, 7 October

Bushuk, W. (2005) 'Rye production and uses worldwide', *Cereal Foods World*, vol 6, no 2, p70

Business Wire (2006) 'Corn wet milling industry faces many challenges in curent global environment, says new analysis', 13 January

California Air Resources Board (2004) *Toxic Air Contaminants, Staff Report/Executive Summary*, Sacramento, 20 November

Calzoni, J. et al (2000) *Bioenergy for Europe: Which Ones Fit Best? A Comparative Analysis for the Community*, Funded by the European Commission in the Framework of the FAIR Programme, Heidelberg, Institute for Energy and Environmental Research, November

Cannell, M. (2003) 'Carbon sequestration', Presentation to The Biofuels Directive: Potential for Climate Protection?, European Climate Forum, Third Autumn Conference, Tyndall Centre, University of East Anglia, Norwich, UK, 8–10 September

Carvalho, L. C. C. and U. A. Alegre (2003) 'Brazil – state of ethanol industry', Presentation to the First World Summit on Ethanol for Transportation, Quebec, Canada, 2–4 November

Center for Science in the Public Interest (2005) *Cruel Oil: How Palm Oil Impacts Health, Rainforest and Wildlife*, Washington, DC, Center for Science in the Public Interest, May

Centre for International Economics (2005) *Impact of Ethanol Policies on Feedgrain Users in Australia*, Prepared for Meat and Livestock Australia (MLA) on behalf of the Australian Beef Industry, Canberra and Sydney, August

CETESB (São Paulo State Environment Agency) (2003) *Relatório de Qualidade do Ar no Estado de São Paulo*, CETESB, São Paulo

Chippewa Valley Ethanol Company and Frontline Bioenergy LLC (2005) *CVEC Announces Alliance with Frontline BioEnergy, LLC*, Press release, Ames, IA, and Benson, MN, 12 December

Clanton, B. (2005) 'Large SUVs lose luster, cost big 3', *The Detroit News*, 18 January

Clay, J. (2004) *World Agriculture and Environment*, Washington, DC, Island Press

Clay, J. (2005) Vice-president, Centre for Conservation Innovation, World Wide Fund for Nature, Presentation to Resources for Global Growth: Agriculture, Energy and Trade in the 21st Century conference, Washington, DC, 6 December

Clean Air Task Force et al (2005) *Prevention of Air Pollution from Ships: Reducing Shipping Emissions of Air Pollution – Feasible and Cost-Effective Options*, Submitted by Friends of the Earth International to the Marine Environment Protection Committee, International Maritime Organization, 7 April

Coconut-Biodiesel for All Government Vehicles, Memorandum Circular 55 (2004) PowerPoint presentation, Energy Centre, Manila, 30 June

Coelho, S. T. (2005) 'Biofuels – advantages and trade barriers', Presentation to the United Nations Conference on Trade and Development, Geneva, 4 February

Collier, R. (2005) 'Fueling America: Oil's dirty future – Canadian oil sands: Vast reserves second to Saudi Arabia will keep America moving, but a steep environmental cost', *San Francisco Gate*, 22 May

Colorado Environmental Coalition et al (2006) *Oil Shale and Tar Sands Resources Leasing Programmatic EIS: Scoping Comments*, Comments submitted to the US Bureau of Land Management, Argonne National Laboratory, 31 January

Commission of the European Communities (2006) *Green Paper: A European Strategy for Sustainable, Competitive and Secure Energy*, Brussels, Commission of the European Communities

Commonwealth of Australia Biofuels Taskforce (2005) *Report of the Biofuels Taskforce to the Prime Minister*, Barton, ACT, August

Conway, G. and G. Toennisessen (1999) 'Feeding the world in the twenty-first century', *Nature*, December, ppC55–C58

Cook, J. and J. Beyea (undated) *An Analysis of Environmental Impacts of Energy Crops in the USA: Methodologies, Conclusions and Recommendations*, Washington, DC, National Audubon Society, www.panix.com/~jimcook/data/ec-workshop.html, accessed 18 July 2005

Copulos, M. R. (2003) *America's Achilles Heel – The Hidden Costs of Imported Oil*, Washington, DC, National Defense Council Foundation, October

Copulos, M. R. (2005) *Economic, Security and Environmental Impacts of Alternative Fuel and Automotive Technologies: A Cost/Benefit Analysis of the Clean Cities Program*, Washington, DC, National Defense Council Foundation

Core, J. (2005) 'New method simplifies biodiesel production', *Agricultural Research*, April

Corpuz, P. G. (2004) *Philippines Oilseeds and Products: GRP Promotes Biodiesel Use 2004*, GAIN Report, Washington, DC, US Department of Agriculture (USDA) Foreign Agricultural Service, 3 May

Council for Biotechnology Information (2005) 'Answers to frequently asked questions about biotech biofuels', www.whybiotech.com/index.asp?id=5035, accessed 25 December 2005

Couto, L. and D. R. Betters (1995) *Short Rotation Eucalypt Plantations in Brazil: Social and Environmental Issues*, Oak Ridge, TN, Oak Ridge National Laboratory, February

CTIC (Conservation Technology Information Center) (no date) *Conservation Tillage and Other Tillage Types in the United States – 1990–2004*, National Crop Residue Management Survey, www.ctic.purdue.edu/Core4/CT/CT.html

Cullen, D. (2003) 'UK gene crop test results fuel demands for ban', *Reuters*, 17 October

Cullen, D. (2005) 'Biggest study of GMO finds impact on birds, bees', *Reuters*, 22 March

Cunha, F. (2003) 'A Logística Atual de Transporte das Distribuidoras e a Infra-Estrutura para Exportação de Álcool', Presentation Álcool: Potencial Gerador de Divisas e Empregos seminar, O Banco Nacional de Desenvolvimento Econômico e Social (BNDES), 25–26 August, www.bndes.gov.br/conhecimento/publicacoes/catalogo/s_alcool.asp

Czernik, S. and A. V. Bridgwater (2004) 'Overview of applications of biomass fast pyrolysis oil', *Energy and Fuels*, vol 18, pp590–598

D1 Oils Plc (2005) *D1 Oils Biodiesel Land Reclamation Project Launched in the Philippines*, Press release, London, 5 May

D1 Oils Plc (2006a) *South Africa – Biodiesel is a National Priority*, www.d1plc.com/global/africa_south_africa.php, accessed 3 March 2006

D1 Oils Plc (2006b) *The Birth of a Global Biodiesel Business*, www.d1plc.com/about/history.php, accessed 12 March 2006

Damen, K. and A. Faaij (2003) *A Life Cycle Inventory of Existing Biomass Import Chains for 'Green Electricity' Production*, Prepared for Essent Energy, Utrecht, Copernicus Institute, Department of Science, Technology and Society, Utrecht University, January

Dammer, T. (2005) 'Strategic significance of America's oil shale resource', Presentation to the 2005 Energy Information Agency Midterm Energy Outlook Conference, Office of Naval Petroleum and Oil Shale Reserves, US DOE, Washington, DC, 12 April, www.eia.doe.gov/oiaf/archive/aeo05/conf/pdf/dammer.pdf

Daoudi, M. S. and M. S. Dajani (1984) 'The 1967 oil embargo revisited', *Journal of Palestine Studies*, vol 13, no 2, pp65–90

Darmouth College and Natural Resources Defense Council (no date) *The Role of Biomass in America's Energy Future (RBAEF)*, Sponsored by US DOE, the Energy Foundation and the National Commission on Energy Policy, project website, http://engineering.dartmouth.edu/other/rbaef/index.shtml

de La Hamaide, S. (2005) 'French to boost biofuel output to meet EU target', *Reuters*, 14 June

De La Torre Ugarte, D. et al (2003) *The Economic Impacts of Bioenergy Crop Production on US Agriculture*, Agricultural Economic Report Number 815, Washington, DC, USDA, February

De La Torre Ugarte, D. (2005) *The Contribution of Bioenergy to a New Energy Paradigm*, Knoxville, TN, Agricultural Policy Analysis Center, University of Tennessee

Delucchi, M. A. (1995) *Summary of Non-monetary Externalities of Motor Vehicle Use*, Report 9 in the series *The Annualized Social Cost of Motor Vehicle Use in the US, Based on 1990–1991 Data: Summary of Theory, Methods and Data*, Draft report prepared for the Union of Concerned Scientists (UCS), Davis, California, Institute of Transportation Studies

Delucchi, M. A. (2003) *A Lifecycle Emissions Model (LEM): Lifecycle Emissions from Transportation Fuels, Motor Vehicles, Transportation Modes, Electricity Use, Heating and Cooking Fuels, and Materials. Documentation of Methods and Data*, Davis, CA, Institute of Transportation Studies (ITS), University of California Davis, December

Delucchi, M. A. (2004) *Conceptual and Methodological Issues in Lifecycle Analyses of Transportation Fuels*, Davis, CA, ITS, University of California Davis

Delucchi, M. A. (2005) *A Multi-Country Analysis of Lifecycle Emissions from Transportation Fuels and Motor Vehicles*, Davis, CA, ITS, University of California Davis, May

Dent, H. (2005) 'China: Where has all the oil gone? IEA wonders', PetroleumWorld.com, 8 December

de Oliveira, N. (2004) 'Neither FTAA nor EU: WTO is Brazil's only hope', *Brazil Magazine*, 5 November

Desai, S. G., M. L. Guerinot and L. R. Lynd (1999) 'Cloning of L-lactate dehydrogenase and elimination of lactic acid production via gene knockout in *Thermoanaerobacterium saccharolyticum*', *Applied Microbiology and Biotechnology*, vol 65, no 5, pp600–605

Dessus, B. et al (1992) *World Potential of Renewable Energies: Actually Accessible in the Nineties and Environmental Impact Analysis*, Paris, la Huille Blanche No. 1

Deurwaarder, E. P. (2005) *Overview and Analysis of National Reports on the EU Biofuel Directive: Prospects and Barriers for 2005*, Sixth Research Framework Programme of the European Union, Petten, The Netherlands: Energy Research Centre of the Netherlands, May

Deutsche Union zur Förderung von Oel- und Proteinpflanzen e.V. (UFOP) (2005) *Statusbericht Biodiesel – Biodieselproduktion und Vermarktung in Deutschland 2005*, Berlin, UFOP

de Vasconcelos, J. N. (2004) 'Continuous ethanol production using yeast immobilized on sugar cane stalks', *Brazilian Journal of Chemical Engineering*, July/September

Devi, L., K. J. Ptasinski and F. J. J. G. Janssen (2003) 'A review of the primary measures for tar elimination in biomass gasification processes', *Biomass and Bioenergy*, vol 24, pp125–140

Devine, T. E. and E. O. Hatley (1998) 'Registration of "Donegal" forage soybean', *Crop Science*, vol 38, pp1719–1720

Devine, T. E., E. O. Hatley and D. E. Starner (1998a) 'Registration of "Derry" forage soybean', *Crop Science*, vol 38, p1719

Devine, T. E., E. O. Hatley and D. E. Starner (1998b) 'Registration of "Tyrone" forage soybean', *Crop Science*, vol 38, p1720

Devine, T. E. and J. E. McMurtrey III (2004) 'Registration of "Tara" soybean', *Crop Science*, vol 44, p1020

Dhuyvetter, K. C., T. L. Kastens and M. Boland (2005) 'The US ethanol industry: Where will it be located in the future?', *Newsletter*, Davis, CA, University of California, Agricultural Issues Center, November, pi

Dias De Oliveira, M. E., B. E. Vaughan and E. J. Rykiel Jr (2005) 'Ethanol as fuel: Energy, carbon dioxide balances, and ecological footprints', *BioScience*, July, p601

Dickerson, M. (2005) 'Brazil's ethanol effort helping lead to oil self-sufficiency', *Los Angeles Times*, 17 June

Dinjus, E. and A. Kruse (2004) 'Hot compressed water – a suitable and sustainable solvent and reaction medium?', *Journal of Physics: Condensed Matter*, vol 16, ppS1161–1169

Dobermann, A. et al (2002) 'Understanding and managing corn yield potential', in *Proceedings of the Fertilizer Industry Round Table*, Charleston, SC, 28–30 October, Forest Hill, MD, Fertilizer Industry Round Table

Dobson, A. (1988) 'The Kennedy administration and economic warfare against communism', *International Affairs*, vol 64, no 2, pp599–616

Doggett, T. (2006) 'EPA seeks to ease US ethanol plant pollution rules', *Reuters*, 3 March

Domac, J., K. M. Richards and V. Segon (2005) *Old Fuel for Modern Times: Socio-Economic Drivers and Impacts of Bioenergy Use*, Paris, International Energy Agency Bioenergy Task 29, www.iea-bioenergytask29.hr/pdf/Domac_Richards_Segon_2005.pdf

Doniger, D. (2001) *Oil Companies, Making Record Profits, Seek Environmental Rollbacks*, New York, Natural Resources Defense Council, 8 May

Dornburg, V. and A. P. C. Faaij (2005) 'Cost and CO_2-emission reduction of biomass cascading: Methodological aspects and case study of SRF poplar', *Climatic Change*, vol 71, no 3, pp373–408

Dornburg, V., J. van Dam and A. Faaij (2004) 'Estimating GHG emission mitigation supply curves of large scale biomass use on a country level', *Biomass and Bioenergy*, September, pp46–65 dos Santos, M. A. (undated) *Energy Analysis of Crops Use for Producing Ethanol and CO_2 Emissions*, Rio de Janeiro, Energy Planning Program, COPPE/UFRJ, Cidade Universitária, www.ivig.coppe.ufrj.br/doc/alcofoen.pdf

Downstream Alternatives Inc (2002) *Infrastructure Requirements for an Expanded Fuel Ethanol Industry*, Oak Ridge, TN, Oak Ridge National Laboratory Ethanol Project, January

Duffy, M. (2004) 'Where is the profit?', Associate professor of economics, Iowa State University, Presentation to Southeast Research and Demonstration Farm Annual Meeting, Ames, IA, 4 March

Duffy, M. (2006) 'The changing structure of agriculture', Associate professor of economics, Iowa State University, Presentation to Independent Insurance Agents of Iowa Rural Agents Conference, Ames, IA, 26 January

Dukes, J. S. (2003) 'Burning buried sunshine: Human consumption of ancient solar energy', *Climatic Change*, vol 6, no 1–2, pp31–44

Durbin, T. et al (2006) *Final Report: Effects of Ethanol and Volatility Parameters on Exhaust Emissions*, CRC Project No E-67, Prepared for Coordinating Research Council, Inc, Alpharetta, GA, Riverside, CA, College of Engineering, Center for Environmental Research and Technology, University of California, 30 January, pp47–48

Earley, J., T. Earley and M. Straub (2005) *Specific Environmental Effects of Trade Liberalization: Oilseeds*, Washington, DC, International Policy Council for Food and Agriculture, October

EC (European Commission) (1994) *Application of Biologically Derived Products as Fuels or Additives in Combustion Engines*, Brussels, Directorate-General XII Science, Research and Development

EC (2003a) *Directive 2003/30/CE of the European Parliament and Council on 8 May 2003 Relative to the Promotion of the Utilization of Biofuels or of other renewable fuels in transportation*, L 123/42 PT, Brussels, EC, 8 May

EC (2003b) *Directive 2003/30/EC, 8 May 2003 on the Promotion of the Use of Biofuels or Other Renewable Fuels for Transport*, O.J. L123, 17/05/2003, Brussels, EC

EC (2003c) *Directive 2003/96/EC, 27 October 2003, Restructuring the Community Framework for the Taxation of Energy Products and Electricity*, O.J. L283, 31/10/2003, Brussels, EC

EC (2003d) *External Costs: Research Results on Socio-Environmental Damages Due to Electricity and Transport*, Brussels, EC

EC (2005a) *Sugar Reform Will Offer EU Producers Long-Term Competitive Future*, Press release, Brussels, EC, 22 June

EC (2005b) *EU Radically Reforms Its Sugar Sector to Give Producers Long-term Competitive Future*, Press release, Brussels, EC, 24 November

EC (2005c) *EU Biomass Action Plan*, Draft final, Brussels, EC, 7 December

EC (2005d) *Commission Urges Luxembourg, Italy, Portugal and Slovakia to Implement Biofuels Directive*, Press release, Brussels, 13 December

EC (2006a) *An EU Strategy for Biofuels*, Communication from the Commission, Brussels, EC, 7 February

EC (2006b) *An EU Strategy for Biofuels*, Draft final, Brussels, EC

EC (2006c) *Annex to the Communication from the Commission: An EU Strategy for Biofuels, Impact Assessment*, Commission Staff Working Document, Brussels, EC

EC (2006d) *Biomass Electricity – Technical Development Status*, Brussels, EC, http://europa.eu.int/comm/energy_transport/atlas/htmlu/bioetech.html

EC (2006e) *Commission Staff Working Document, Annex to the Communication from the Commission: An EU Strategy for Biofuels, Impact Assessment*, Brussels, EC

EC (2006f) *Commission Urges New Drive to Boost Production of Biofuels*, Press release, Brussels, EC, 8 February

EC Directorate-General for Energy and Transport (2004) *Promoting Biofuels in Europe: Securing a Cleaner Future for Transport*, Brussels, EC

Econbrowser (2005) 'What's up with oil prices', 4 June, www.econbrowser.com/archives/2005/06/whats_up_with_o.html

The Economist (2005a) 'Living and dying on history and artificial economic sweeteners', *The Economist*, 24 September

The Economist (2005b) 'Recycling the petrodollars', *The Economist*, 12 November

Edmonds, J. A. et al (1996) *Agriculture, Land Use, and Commercial Biomass Energy: A Preliminary Integrated Analysis of the Potential Role of Biomass Energy for Reducing Future Greenhouse Related Emissions*, Richland, WA, Pacific Northwest National Laboratory

Edwards, R. et al (2003) *Well-to-Wheel Analysis of Future Automotive Fuels and Power Trains in the EU Context, Well-to-Tank Report*, Brussels, CONCAWE, EUCAR

Edwards, R. (2005) *Updated WTW Analysis of Biomass-Based Transport Fuels*, PowerPoint presentation, Joint Research Centre, European Commission, Paris

EEA (European Environment Agency) (2005) *How Much Biomass Can Europe Use without Harming the Environment?*, EEA Briefing, Copenhagen, EEA

EEA and ETC/ACC (Topic Centre on Air Emissions and Climate Change) (2003) cited in Peter Taylor, 'Transport emission trends in Europe', Presentation to The Biofuels Directive: Potential for Climate Protection?, European Climate Forum, Third Autumn Conference, Tyndall Centre, University of East Anglia, Norwich, UK, 8–10 September

EIA (US Energy Information Administration) and US Department of Energy (DOE) (2004) *International Energy Annual 2002: World Crude Oil Refining Capacity*, 1 January, www.eia.doe.gov/pub/international/iea2002/table36.xls

EIA and DOE (2005) *EIA and DOE Petroleum Supply Annual*, Table 36: Number and capacity of operable petroleum refineries by PAD district and state as of 1 January, www.eia.doe.gov/pub/oil_gas/petroleum/data_publications/petroleum_supply_annual/psa_volume1/current/pdf/table_36.pdf

Ejigu, M. (2005) *The Biofuels Option, The Way Ahead: Some Thoughts*, Presentation at International Energy Agency Conference, Paris, 20–21 June

Elsayed, M. A., R. Matthews and N. D. Mortimer (2003) *Carbon and Energy Balances for a Range of Biofuels Options*, Report prepared for UK Department of Trade and Industry Sustainable Energy Programmes, Sheffield, UK, Sheffield Hallam University, March

Energetics Inc (1998) *Energy and Environmental Profile of the US Petroleum Refining Industry*, Report prepared for US DOE, Washington, DC, US DOE, Office of Industrial Technologies

Energetics and NEOS Corporation (1994) *Economic Impact of Ethanol Production Facilities: Four Case Studies*, Prepared for Western Regional Biomass Energy Program and Great Lakes Regional Biomass Energy Program, Golden, CO, and Chicago, June

Environment News Service (2005a) *Minnesota Aims to Be Saudi Arabia of Renewable Fuels*, 18 May

Environment News Service (2005b) *Brazil Embraces Alternative Energy*, 19 August
Environment News Service (2005c) *Brazil's New National Park Protects Vanishing Savannah*, 15 December
Environment News Service (2005d) *Illinois Ethanol Producer Must Install Air Pollution Controls*, 24 December
Environmental Defense (1999) *Oil Refining*, Main page, www.environmentaldefense.org/article.cfm?ContentID=1537, 4 May
Environmental Working Group (2006) *Top Subsidies in the US 1995–2004*, Farm Subsidy Database, www.ewg.org/farm/region.php?fips=00000, accessed 27 January 2006
Epstein, P. R. and J. Selber (eds) (2002) *Oil: A Life Cycle Analysis of its Health and Environmental Impacts*, Cambridge, MA, Center for Health and the Global Environment, Harvard Medical School, Harvard University, March
Esipisu, M. (2006) 'Gene crops no help to Africa so far – report', *Reuters*, 11 January
ETSO (1996) *Alternative Road Transport Fuels: A Preliminary Life-Cycle Study for the UK*, Oxford, ETSO
European Biomass Industry Association (2006) 'Biodiesel', *Biofuels for Transport*, http://p9719.typo3server.info/214.0.html, accessed 12 March 2006
Faaij, A. (2003) 'Bioenergy in Europe: Changing technology choices', *Energy Policy*, 5 December
Faaij, A. (2005) Report submitted to Worldwatch Institute, Copernicus Institute, Utrecht University, 17 January
Faaij, A. (2006) 'Modern Biomass Conversion Technologies', *Mitigation and Adaptation Strategies for Global Change*, vol 11, no 2, March
Faaij, A. P. C. (2006) 'Bio-energy in Europe: Changing technology choices', *Energy Policy* (Special Issue on Renewable Energy in Europe), February, pp322–342
Faaij, A. et al (1997) 'Characteristics and availability of biomass waste and residues in the Netherlands for gasification', *Biomass and Bioenergy*, vol 12, no 4, pp225–240
Faaij, A. P. C. et al (2000) *Mondiale beschikbaarheid en mogelijkheden voor import van biomassa voor energie in Nederland: Synthese van het onderzoeksproject GRAIN: Global Restrictions on Biomass Availability for Import to the Netherlands*, Utrecht, Vakgroep Natuurwetenschap and Samenleving, University of Utrecht, July
Faaij, A. et al (2005) *Biomass Potential Assessment in Central and Eastern European Countries and Opportunities for the Ukraine*, Utrecht, Copernicus Institute, Utrecht University, January
Faaij, A. and A. Walter (2005) 'Findings of the Meeting', Presentation at the Workshop and Business Forum on Sustainable Biomass Production for the World Market, Organized by IEA Task 40 Sustainable Bioenergy Trade, Tasks 30 and 31, University of Campinas, São Paulo, Brazil, November
Faaij, A., A. Wierczorek and M. Minnesma (2002) *International Debate on International Biotrade*, Summary of international workshop, Amsterdam, The Netherlands, September
Faaij, A., A. Wieczorek and M. Minnesma (2003) *International Debate on International Biotrade*, International workshop report, Office of Inter-Human Dimensions Programme – Industrial Transformation (IHDP-IT), Department of Science, Technology and Society – Utrecht University and Institute for Environmental Studies – Free University of Amsterdam, Utrecht, NOVEM, April

FAO (United Nations Food and Agriculture Organization) (2000) *Proceedings of the Validation Forum on the Global Cassava Development Strategy, Rome, 26–28 April 2000*, www.fao.org/documents/show_cdr.asp?url_file=/docrep/007/y2413e/y2413e07.htm

FAO (2002) cited in L. Gianessi, S. Sankula and N. Reigner (2003) *Plant Biotechnology: Potential Impact for Improving Pest Management in European Agriculture: Herbicide Tolerant Wheat Case Study*, Washington, DC, National Center for Food and Agricultural Policy, December

FAO (2005) 'Agriculture, trade and poverty: Can trade work for the poor?', in *The State of Food and Agriculture*, Rome, FAO

FAO (no date) *FAOSTAT electronic database*, http://faostat.fao.org, accessed 2005

FAO and IIASA (International Institute for Applied Systems Analysis) (2000) *Global Agro-Ecological Zones (Global-AEZ)*, CD-ROM, Rome and Laxenburg, Austria, www.fao.org/landandwater/agll/gaez/index.htm

FAO, Statistics Division (2004) *Major Food and Agricultural Commodities and Producers*, www.fao.org/es/ess/top/commodity.jsp?commodity=270&lang=EN&year=2004, accessed 2005

Farrell, A. E. et al (2006) 'Ethanol can contribute to energy and environmental goals', *Science*, 27 January, pp27–29

Figueres, C. (2005) *Policies and Programs under the CDM*, http://lists.iisd.ca:81/read/attachment/28306/1/Policies%20and%20programs%20under%20the%20CDM.pdf

Filho, A. B. (2005) 'The agrienergy option for the future', Director, AgroEnergy Program, Brazilian Ministry of Environment, Presentation to the Workshop and Business Forum on Sustainable Biomass Production for the World Market, São Paulo, Brazil, 30 November–2 December

Fischer, G. and L. Schrattenholyer (2000) 'Global bioenergy potentials through 2050', *Biomass and Bioenergy*, vol 20, pp151–159

Flavin, C. (2007) 'Fossil fuel use up again', in Worldwatch Institute (ed) *Vital Signs* 2007–2008, New York, W. W. Norton and Company, pp32–33

F. O. Licht's World Ethanol and Biofuels Report (2005a) 'Biofuels and the international development agenda', *F. O. Licht's World Ethanol and Biofuels Report*, 11 July

F. O. Licht's World Ethanol and Biofuels Report (2005b) 'Domestic fuel ethanol program proposed', *F. O. Licht's World Ethanol and Biofuels Report*, 25 July

F. O. Licht's World Ethanol and Biofuels Report (2005c) 'Australia's Queensland sees ethanol sector as sugar saviour', *F. O. Licht's World Ethanol and Biofuels Report*, 26 September

F. O. Licht's World Ethanol and Biofuels Report (2005d) 'Biofuels to ease economic pain', *F. O. Licht's World Ethanol and Biofuels Report*, 7 October

F. O. Licht's World Ethanol and Biofuels Report (2005e) 'EU approves French–Dutch biodiesel JV', *F. O. Licht's World Ethanol and Biofuels Report*, 7 October

F. O. Licht's World Ethanol and Biofuels Report (2005f) 'World ethanol production 2005 to be higher than expected', *F. O. Licht's World Ethanol and Biofuels Report*, 24 October

F. O. Licht's World Ethanol and Biofuels Report (2005g) 'The rest of the world', *F. O. Licht's World Ethanol and Biofuels Report*, 9 November

F. O. Licht's World Ethanol and Biofuels Report (2005h) 'Ethanol to have mixed impact on US agriculture', *F. O. Licht's World Ethanol and Biofuels Report*, 9 November

F. O. Licht's World Ethanol and Biofuels Report (2005i) 'Biofuel use spreads vegoil supplies too thin for margarine firms, say Imace', *F. O. Licht's World Energy and Biofuels Report*, 9 November

F. O. Licht's World Ethanol and Biofuels Report (2005j) 'Sugar exports may hit new low', *F. O. Licht's World Ethanol and Biofuels Report*, 1 December

F. O. Licht's World Ethanol and Biofuels Report (2005k) 'Denver winter ethanol to stay until 2008', 19 December

F. O. Licht's World Ethanol and Biofuels Report (2006a) 'Australia aims to exceed biofuel targets', *F. O. Licht's World Ethanol and Biofuels Report*, 4 January

F. O. Licht's World Ethanol and Biofuels Report (2006b) 'Malaysia to boost palm oil output by 25% by 2010', *F. O. Licht's World Ethanol and Biofuels Report*, 27 January

F. O. Licht's World Ethanol and Biofuels Report (2006c) 'Petrobras eyes ethanol pipeline with Goias', *F. O. Licht's World Ethanol and Biofuels Report*, 3 February

F. O. Licht's World Ethanol and Biofuels Report (2006d) 'German winter rapeseed plantings rise on biodiesel demand', *F. O. Licht's World Ethanol and Biofuels Report*, 3 February

F. O. Licht's World Ethanol and Biofuels Report (2006e) 'Soaring sugar prices unsettle Brazilian ethanol market', *F. O. Licht's World Ethanol and Biofuels Report*, 7 February

F. O. Licht's World Ethanol and Biofuels Report (2006f) 'Rapeseed prices continue to climb on EU biodiesel demand', *F. O. Licht's World Ethanol and Biofuels Report*, 14 February

F. O. Licht's World Ethanol and Biofuels Report (2006g) 'Soaring sugar prices unsettle ethanol market', *F. O. Licht's World Ethanol and Biofuels Report*, 20 February

F. O. Licht's World Ethanol and Biofuels Report (2006h) 'Commission cuts sugar quotas for 2006/07', *F. O. Licht's World Ethanol and Biofuels Report*, 8 March

F. O. Licht's World Ethanol and Biofuels Report (2006i) 'USDA raises corn use estimate for fuel ethanol production', *F. O. Licht's World Ethanol and Biofuels Report*, 8 March

F. O. Licht's World Ethanol and Biofuels Report (2006j) 'Rising ethanol prices may hit Brazilian flex-fuel car sales', *F. O. Licht's World Ethanol and Biofuels Report*, 9 March

Food and Agricultural Policy Research Institute (2005) *Implications of Increased Ethanol Production for US Agriculture*, Columbia, MO, University of Missouri-Columbia, August

Ford Motor Company (2007) 'Green Ford Focus fills up at first bioethanol fuel pumps', *Media.Ford.Com* press release, 29 March

Freeman, S. and A. Joyce (2006) 'Ford to cut 14 plants and up to 30,000 jobs – N. American auto division lost $1.6 billion in 2005', *Washington Post*, 24 January

Friends of the Earth–UK (2005) *The Oil for Ape Scandal: How Palm Oil is Threatening the Orang-utan*, London, FoE, September

Friends of the Earth–UK (2006) *Leaked WTO Report: US Misled the World on 'Victory'*, Press release, London, FoE, 28 February

Fritsche, U. R., K. Hünecke, and K. Wiegmann (2005) *Criteria for Assessing Environmental, Economic, and Social Aspects of Biofuels in Developing Countries*, Report commissioned by the German Federal Ministry for Economic Cooperation and Development (BMZ), Freiburg, Öko-Institut e.V., February

FSC (undated a) *About FSC*, www.fsc.org/en/about/about_fsc

FSC (undated b) *Policy and Standards*, www.fsc.org/en/about/policy_standards

FSC (2004) *FSC Principles and Criteria for Forest Stewardship*, Bonn, FSC, April

Fulton, L. (2004a) 'Driving ahead: Biofuels for transport around the world', *Renewable Energy World*, July–August, pp180–189

Fulton, L. (2004b) 'Biofuels for transport: An international perspective', Presentation at Conference of the Parties (COP)-10 for the UN Framework Convention on Climate Change, International Centre for Trade and Sustainable Development (ICTSD)/UNCTAD, Argentina, 14 December

Fulton, L., Howes, T. and Hardy, J. (2004) *Biofuels for Transport: An International Perspective*, Paris, France, International Energy Agency

Gallegeher, P. (2006) *Biomass Feedstock Composition and Property Database*, Washington, DC, USDOE, January, www1.eere.energy.gov/biomass/feedstock_databases.html

Gardner, B. (2003) *Fuel Ethanol Subsidies and Farm Price Support: Boon or Boondoggle?*, College Park, MD, University of Maryland Department of Agricultural and Resource Economics, October

Gardner, T. (2006) 'Can fungi trim the US gasoline habit?', *Reuters*, 27 February

Garsten, E. (2005) 'High gas prices scare of the sales of SUV's: Automakers depend on big vehicle sales for pumping up the bottom line', *The Detroit News*, 15 March

Gehua, W. et al (2005) *Liquid Biofuels for Transportation: Chinese Potential and Implications for Sustainable Agriculture and Energy in the 21st Century*, Report prepared for GTZ, Beijing

Genencor International (2004) *Genencor Celebrates Major Progress in the Conversion of Biomass to Ethanol*, Press release, Palo Alto, CA, 21 November

Genencor International (2005) *Transforming the Ethanol Industry*, Palo Alto, CA, June

General Motors et al (2002) *GM Well-to-Wheel Analysis of Energy Use and Greenhouse Gas Emissions of Advanced Fuel/Vehicles Systems – A European Study; ANNEX – Full Background Report*, Ottobrunn, Germany, www.lbst.de/gm-wtw

German Federal Institute for Geosciences and Natural Resources (BGR) (2006) *Total Known Potential of Conventional Petroleum*, www.bgr.bund.de/cln_030/nn_468074/EN/Home/homepage__node.html__nnn=true, accessed 28 April 2006

Gerpen, J., D. Clements and G. Knothe (2004) *Biodiesel Production Technology: August 2002 – January 2004*, Golden, CO, National Renewable Energy Laboratory, June

Gerrard, S. et al (1999) *Drill Cuttings Piles in the North Sea: Management Options During Platform Decommissioning*, Research Report 31, Norwich, UK, Centre for Environmental Risk, University of East Anglia

Girard, P., A. Fallot and F. Dauriac (2005) *Biofuels Technology: State-of-the-Art*, Prepared for the Global Environment Facility (UNEP/GEF) Technical Workshop on Liquid Biofuels, New Delhi, India, 29 August–1 September 2005

Glastra, R., E. Wakker and W. Richert (2002) *Oil Palm Plantations and Deforestation in Indonesia. What Role do Europe and Germany Play?* Zurich, WWF Germany, with WWF Indonesia and WWF Switzerland, November, www.panda.org/downloads/forests/oilpalmindonesia.pdf

The Gleneagles Communiqué Gleneagles, Scotland, 6–8 July, www.fco.gov.uk/Files/kfile/PostG8_Gleneagles_Communique.pdf

Global Environment Facility, Small Grants Programme (2005) *Climate Change*, http://sgp.undp.org/index.cfm?module=ActiveWeb&page=WebPage&s=foCC, accessed December 2005

Goldemberg, J. and I. Macedo (1994) 'Brazilian Alcohol Program: An overview', in International Energy Agency (ed) *Energy for Sustainable Development*, Paris, May

Goldemberg, J. et al (2003) 'How adequate policies can push renewables', *Energy Policy*, 22 February

Gover, M. P. et al (1996) *Alternative Road Transport Fuels – A Preliminary Life-Cycle Study for the UK*, ETSU Report R92, vols 1 and 2, Oxford, Energy Technology Support Unit

Government of India (2003) *Wasteland Atlas*, Ministry of Rural Development/Department of Land Resources, http://dolr.nic.in/fwastecatg.htm

Government of Manitoba, Ministry of Energy, Science and Technology (2003) *Energy Initiatives: Ethanol Blended Fuels*, Manitoba, Canada, www.gov.mb.ca/est/energy/initiatives/ ethanolfuels.html

Government of the Philippines (2005) *DOE Identifies Alternative Fuel Sources, Philippines*, Manila, 26 August, www.pia.gov.ph/news.asp?fi=p050826.htm&no=16

Graham, R. L., W. Liu and B. C. English (1995) 'The environmental benefits of cellulosic energy crops: A landscape scale', in *Environmental Enhancement through Agriculture: Proceedings of a Conference*, Tufts University, Boston, MA, 15–17 November 1995

Graham, R. L., L. L. Wright and A. F. Turhollow (1992) 'The potential for short-rotation woody crops to reduce US CO_2 emissions', *Climatic Change*, vol 22, pp223–238

GreenBiz.com (2005) *Researchers Explore Climate-Friendly Farming*, 27 June

Green Car Congress (2005a) *Coconut Biodiesel Blend Hits Local Philippine Market*, 13 August, www.greencarcongress.com/2005/08/coconut_biodies.html

Green Car Congress (2005b) *Malaysia to Mandate B5 Biodiesel in 2008*, 6 October, www.greencarcongress.com/2005/10/malaysia_to_man.html

Green Car Congress (2005c) *Petrobras Forms Ethanol Joint Venture in Japan*, 19 December, www.greencarcongress.com/2005/12/petrobras_forms.html

Green Car Congress (2006) *EarthFirst Importing Biodiesel to US; Soy Lobby Reacts*, 14 November, www.greencarcongress.com

Green, T. H. et al (1996) 'Environmental impacts of conversion of cropland to biomass production', *Proceedings of Bioenergy 1996 – The Seventh National Bioenergy Conference: Partnerships to Develop and Apply Biomass Technologies*, 15–20 September 1996, Nashville, TN, www.bioenergy.ornl.gov/papers/bioen96/green.html

Greene, N. et al (2004) *Growing Energy: How Biofuels Can Help End America's Oil Dependence*, Washington, DC, Natural Resources Defense Council, December

Haley, S. and N. R. Suarez (2004) *Sugar and Sweeteners Outlook*, Washington, DC, US Department of Agriculture, http://usda.mannlib.cornell.edu/reports/erssor/specialty/sss-bb/2004/sss241.pdf

Hall, D. O. et al (1993) 'Biomass for energy: Supply prospects', in B. J. Johansson et al (eds) *Renewables for Fuels and Electricity*, Washington, DC, Island Press

Halweil, B. (2000) 'Where have all the farmers gone?', *World Watch*, September/October 2000, p17

Hamelinck, C. N. (2004) *Outlook for Advanced Biofuels*, PhD thesis, Utrecht, Utrecht University, 7 June

Hamelinck, C. N. and A. P. C. Faaij (2002) 'Future prospects for production of methanol and hydrogen from biomass', *Journal of Power Sources*, vol 111, pp1–22

Hamelinck, C. N. and A. P. C. Faaij (2006) 'Outlook for advanced biofuels', *Energy Policy*, vol 34, no 17, pp3268–3283

Hamelinck, C. N., R. A. A. Suurs and A. P. C. Faaij (2005) 'International bioenergy transport costs and energy balance', *Biomass and Bioenergy*, vol 29, no 2, pp114–134

Hamilton, J. D. (1983) 'Oil and the macroeconomy since World War II', *Journal of Political Economy*, vol 91, no 2, pp228–248

Hammel-Smith, C. et al (2002) *Issues Associated with the Use of Higher Ethanol Blends (E17-E24)*, Golden, CO, NREL, October

HART Downstream Energy Services (2004) *Ethanol Market Fundamentals*, Report to the New York Board of Trade, New York, 9 March

Hearn, K. (2005) 'Buenos bios: How South American biofuels are gaining steam, and why that freaks the US out', *Grist Magazine*, 15 December

Heaton, E. A. et al (2003) *Miscanthus for Renewable Energy Generation: European Union Experience and Projections for Illinois*, Champaign, IL, University of Illinois Urban-Champaign, October

Heckeroth, S. (1999) *Toward Sustainable Agriculture*, Albion, CA, www.renewables.com/Permaculture/SustAgri.htm, accessed 25 December 2005

Henke, J., G. Klepper and J. Netzel (2003) *Tax Exemption for Biofuels in Germany: Is Bio-Ethanol Really an Option for Climate Policy? Kieler Arbeitspapiere 1136*, Kiel, Institut für Weltwirtschaft, p17

Hess, M. A. et al (2005) 'The effect of antioxidant addition on NO_x emissions from biodiesel', *Energy and Fuels*, 17 March, pp1749–1754

Hettenhaus, J. (2002) 'Talking about corn stover with Jim Hettenhaus', in *The Carbohydrate Economy*, Minneapolis, MN, Institute for Local Self-Reliance, summer

High Plains Journal (2005) 'Data shows ethanol will pass exports as #2 com use by 2008', *High Plains Journal*, 9 June

Himmel, M. E. et al (1997) *Advanced Bioethanol Production Technologies: A Perspective*, ACS Symposium Series, vol 666, p45

Hin, H. (2005) 'Ford launches ethanol-compatible car', *Mail & Guardian online*, 27 October

Ho, N. W. Z., Z. Chen and A. P. Brainard (1998) 'Genetically engineered *Saccharomyces* yeast capable of effective cofermentation of glucose and xylose', *Applied and Environmental Microbiology*, vol 64, no 5, pp1852–1859

Hodes, G. S., B. E. Utria and A. Williams (2004) *Ethanol: Re-examining a Development Opportunity for Sub-Saharan Africa*, Draft for review, Washington, DC, The World Bank Group, September

Hogan, M. (2005) 'German biodiesel sales strong as oil prices rise', *Reuters*, 19 August

Hollenbacher, R. (1992) 'Biomass combustion technologies in the United States', Paper presented the Biomass Combustion Conference, Reno, NV, 28–29 January

Hoogwijk, M. (2004) *On the Global and Regional Potential of Renewable Energy Sources*, Utrecht, Utrecht University

Hoogwijk, M. et al (2003) 'Exploration of the ranges of the global potential of biomass for energy', *Biomass and Bioenergy*, vol 25, no 2, pp119–133

Howse, R. (2005) *Post-Hearing Submission to the International Trade Commission: World Trade Law and Renewable Energy: The Case of Non-Tariff Measures*, Vienna, Renewable Energy and International Law Project, 5 May, p16

Howse, R. and E. Tuerk (2006) 'The WTO impact on internal regulations: A case study of the Canada–EC asbestos dispute', in G. Bermann and P. Mavroidis (eds) *Trade and Human Health and Safety*, New York, Cambridge University Press, pp77–117

Hunt, S. (2005) *Notes from Workshop and Business Forum on Sustainable Biomass Production for the World Market*, Organized by IEA Task 40, Sustainable Bioenergy Trade, Tasks 30 and 31, University of Campinas, São Paulo, Brazil, November, Washington, DC

Hunt, S. and J. Sawin, with P. Stair (2006) 'Cultivating renewable alternatives to oil', in Worldwatch Institute (ed) *State of the World 2006*, New York, W. W. Norton and Company

Hutchinson-Jafar, L. (2006) 'Oil-stung Caribbean looks energy alternatives', *Reuters*, 18 January

ICTSD (International Centre for Trade and Sustainable Development) (2005a) 'EU releases reform plan for sugar', *Bridges Weekly Trade Digest*, 29 June

ICTSD (2005b) 'WTO Environment Committee drafts text for Hong Kong', *Bridges Weekly Trade Digest*, 16 November

ICTSD (2006) 'Can London deliver what Hong Kong couldn't?', *Bridges Weekly Trade Digest*, January–February

IEA (International Energy Agency) (1998) *Greenhouse Gas Balances of Bioenergy Systems*, Paris, IEA Bioenergy Task 25

IEA (2002) *Sustainable Production of Woody Biomass for Energy*, Position paper prepared for IEA Bioenergy, Paris, IEA

IEA (2004a) *World Energy Outlook 2004*, Paris, IEA

IEA (2004b) *Analysis of the Impact of High Oil Prices on the Global Economy*, Paris, May

IEA (2006) *Standard Methodology for Calculation of GHG Balances*, Paris, Bioenergy Task 38

IEA (2007) *Oil Market Report*, 13 March, IEA, Paris

IEA, Bioenergy Task 38 (2006) *Standard Methodology for Calculation of GHG Balances*, Paris, IEA

IFAD (International Fund for Agricultural Development) (2002) *IFAD Strategic Framework for 2002–2006*, Rome, March

IFOAM (International Federation of Organic Agriculture Movements) (2006) *The IFOAM Organizational Structure*, www.ifoam.org/about_ifoam/inside_ifoam/organization.html, accessed 10 March 2006

Illovo Sugar (2005) *World of Sugar: International Sugar Statistics*, www.illovo.co.za/worldofsugar/internationalSugarStats.htm, accessed 12 March 2006

Ingram, L. O. et al (1998) 'Metabolic engineering of bacteria for ethanol production', *Biotechnology and Bioengineering*, vol 58, no 2–3, pp204–212

Ingram, L. O. et al (1999) 'Enteric bacterial catalysts for fuel ethanol production', *Biotechnology Progress*, vol 15, no 5, pp855–866

Institute for Agriculture and Trade Policy (2005) *CAFTA's Impact on US Ethanol Market*, Trade and Governance Program, IATP, Minneapolis, MN, June

Institute Français Riera (2004) *Panorama 2004: A Look Biofuels in Europe*, Lyon, Institute Français Riera

Interacademies (2000) *Transgenic Plants and World Agriculture*, Report prepared under the auspices of Seven Academies of Science, Brazil, China, India, Mexico, Third World, UK, US, Washington, DC

International Policy Council for Food and Agricultural Trade (2005) *A Look at the EU's Proposed Sugar Reforms*, Washington, DC, www.agritrade.org/Publications/EU%20Sugar %20Reform.pdf

International Sugar Organization (1999) *ISO Sugar Yearbook 1999*, London, ISO

IOL (2005) 'Brazil poised to become biodiesel giant', www.int.iol.co.za, South Africa, 19 November

IPCC (Intergovernmental Panel on Climate Change) (2000) *Special Report on Emission Scenarios*, Cambridge, IPCC

IPCC, Working Group III (2001) *IPCC Third Assessment Report, Climate Change 2001: Mitigation*, Cambridge, Cambridge University Press

Isla, P. A. A. (2004) 'RP, Thailand agree on regional biofuel standard', *The Manila Times*, 1 September

Italian Ministry for the Environment and Territory (2005) *Global Bioenergy Partnership White Paper*, Prepared by the Italian Ministry for the Environment and Territory with the participation of Imperial College London, Itabia and E4tech, October

Joensen, L. and S. Semino (2004) 'Argentina's torrid love affair with the soybean', *Seedling (GRAIN)*, October (excerpt from L. Joensen, S. Semino and H. Paul (2005) *Argentina: Case Study on the Impact of RoundUp Ready Soya*, Brighton, UK, Rural Reflection Group, Argentina and EcoNexus, UK, March)

Johansson, T. B. et al (1993) 'A renewables-intensive global energy scenario', Appendix to Chapter 1, in T. B. Johansson et al (eds) *Renewable Energy: Sources for Fuels and Electricity*, Washington, DC, Island Press, pp1071–1143

Jones, C. (2005) 'Europe adopts biodiesel: Can an African bean crack Europe's biodiesel blockage?', *EcoWorld*, 28 September

Jones, C. (2006) 'An African bean cracks Europe's biodiesel blockage', *Ecoworld*, 6 January, www.d1plc.com/pdf/press/ecoworld.pdf

Jordan, J. and Associates (2005) *Ethanol: Prospectus: World Markets for Fuel Ethanol*, Houston, TX, September

Journey to Forever (2005) *Vegetable Oil Yields*, www.journeytoforever.org/biodiesel_yield.html, accessed 16 May

Juniger, M. and A. Faaij (2005) *IEA Bioenergy Task 40: Country Report for the Netherlands*, Paris, International Energy Agency Task 40, Sustainable International Bioenergy Trade, July 2005, p23, www.senternovem.nl/mmfiles/IEA_Task_40_Country_Report_NL_tcm24-152474.pdf

Kaltner, F. J. et al (2005) *Liquid Biofuels for Transportation in Brazil: Potential and Implications for Sustainable Agriculture and Energy in the 21st Century*, Prepared for Deutsche Gesellschaft für Technische Zusammenarbeit (GTZ) GmbH, Rio de Janeiro, Fundação Brasileira para o Desenvolvimento Sustentável, November

Kaltschmitt, M., C. Rosch and L. Dinkelbach (eds) (1998) *Biomass Gasification in Europe*, Prepared for the European Commission, Stuttgart, Institute of Energy Economics and the Rational Use of Energy (IER), University of Stuttgart, October

Kaltschmitt, M. et al (2005) 'Analyse und Bewertung der Nutzungsmöglichkeiten von Biomasse – Arbeitspaket des IE Leipzig', in *Vorbereitung*, Leipzig, p18

Karekezi, S. (1995) 'Renewable energy technologies in sub-Saharan Africa', Presentation to Woodrow Wilson School of Public and International Affairs, Princeton University, Princeton, NJ, 19 April

Karekezi, S. et al (2004) *Traditional Biomass Energy: Improving its Use and Moving to Modern Energy Use*, Thematic background paper, Bonn, Renewables 2004, January

Kartha, S., G. Leach and S. Chella Rajan (2005) *Advancing Bioenergy for Sustainable Development: Guideline for Policymakers and Investors*, vols I, II and III, Washington, DC, World Bank Energy Sector Management Assistance Program, April

Kerssen, M. and R. H. Berends (2005) *Life Cycle Analysis of the HTU Process*, Apeldoom, The Netherlands, Dutch Research Organization for Environment, Energy and Process Innovation, www.novem.nl/default.asp?menuld=10&documentId=115819

Khosla, V. (2006) 'Biofuels: Think outside the barrel', Presentation to Center for American Progress, 10 January, www.khoslaventures.com/presentations/Biofuels_Dec2005.v3.2.ppt

Kim, S. and B. E. Dale (2002) 'Allocation procedure in ethanol production system from corn grain', *International Journal of Life Cycle Assessment*, vol 7, p237

Kinast, J. A. (2003) *Properties of Biodiesels and Biodiesel/Diesel Blends*, Prepared by Gas Technologies Institute for NREL, Golden, CO, March

Klein, K. et al (2006) *Ethanol Policies, Programs and Production in Canada*, PowerPoint Presentation, Center for Research in the Economics of Agrifood, Laval University, Ste Foy, Quebec, Canada, www.farmfoundation.org/projects/documents/klein-ethanol.ppt, accessed 6 March 2006

Knight, P. (2003) 'New flex-fuel engines transform consumer options in Brazil', *F. O. Licht's World Ethanol and Biofuels Report*, 23 July

Kojima, M. and T. Johnson (2005) *Potential for Biofuels for Transport in Developing Countries*, Washington, DC, World Bank, October

Kovarik, B. (1998) 'Henry Ford, Charles Kettering and the "fuel of the future"', *Automotive History Review*, spring, pp7–27, www.radford.edu/~wkovarik/papers/fuel.html

Krahl, J. et al (2001) *Comparison of Biodiesel with Different Diesel Fuels Regarding Exhaust Gas Emissions and Health Effects*, www.ufop.de/downloads/Biodiesel_comparison.pdf

Krishnan, B. and S. Mohanty (2006) 'Commodities – bullish palm looks to biofuel for future', *Reuters*, 23 February

Kyodo News (2005) 'Brazil's Petrobras to sell ethanol in Japan', *Kyodo News*, 20 December

Larivé, J.-F. (2005) *The Joint JCR/EUCAR/CONCAWE Well-to-Wheels Study*, PowerPoint presentation, Fourth CONCAWE Symposium, 30 November–1 December

Larson, E. D. (2005) *Liquid Biofuel Systems for the Transport Sector: A Background Paper*, Draft for discussion at the Global Environment Facility Scientific and Technical Advisory Panel Workshop on Liquid Biofuels, New Delhi, 29 August–1 September

Larsen, R. et al (2004) 'Might Canadian oil sands promote hydrogen production for transportation? Greenhouse gas emission implications of oil sands recovery and upgrading', Paper submitted to *World Resource Review*, Chicago, Center for Transportation Research, Argonne National Laboratory, December

Laser, M. and L. Lynd (2006) Report to Worldwatch Institute, Thayer School of Engineering, Dartmouth College, 13 January

Lashof, D. A. and D. A. Tirpak (eds) (1990) *Policy Options for Stabilizing Global Climate*, New York, Hemisphere Publishing Corporation

Lazarus, M. et al (1993) *Towards a Fossil Free Energy Future*, Boston, Stockholm Environmental Institute

Leemans, R. et al (1996) 'The land cover and carbon cycle consequences of large-scale utilizations of biomass as an energy source', *Global Environmental Change*, vol 6, no 4, pp335–357

Leighton, F. A. (2000) *Petroleum Oils and Wildlife*, Saskatoon, Saskatchewan, Canadian Cooperative Wildlife Health Centre, Western College of Veterinary Medicine, University of Saskatchewan, May, http://wildlife1.usask.ca/ccwhc2003/wildlife_health_topics/oil/oil1.php

Lemley, B. (2003) 'Anything into oil', *Discover*, May

Lepach, A. J. (2005) *Perspectivas Futuras Para o Etanol Combustível. Etanol Combustível: Balanço e Perspectivas*, Evento Comemorativo dos 30 Anos da Criação do Proálcool, 16–17 November 2005, Auditório da Faculdade de Ciências Médicas, University of Campinas, Campinas, Brazil, www.nipeunicamp.org.br/proalcool/Palestras/17/Andre%20j.ppt

Levelton Engineering Ltd (1999) *Engineering Assessment of Net Emissions of Greenhouse Gases from Ethanol – Gasoline Blends in Southern Ontario*, Richmond, BC, January, www.tc.gc.ca/programs/Environment/climatechange/docs/ETOH-FNL-RPTAug30-1999.htm

Levy, R. H. (1993) *Les Biocarburants*, Report to the French Government based on figures from the Commission Consultative pour la Production des Carburants de Substitution, Paris

Lewandowski, I. and A. Faaij (2004) *Steps Towards the Development of a Certification System for Sustainable Bio-Energy Trade*, FAIR Biotrade Project, Utrecht, Copernicus Institute of Sustainable Development and Innovation, July

Lewandowski, I. and A. W. C. Faaij (2006) 'Steps towards the development of a certification system for sustainable bio-energy trade', *Biomass and Bioenergy*, vol 30, pp86–87

Liu, D. (2005) 'Chinese development status of bioethanol and biodiesel', Presentation to Biofuels: An Energy Solution, 2005 World Biofuels Symposium, Tsinghua University, Beijing, 13–15 November

Lobell, D. B. and G. P. Asner (2003) 'Climate and management contributions to recent trends in US agricultural yields', *Science*, 14 February, p1032

Lorenz, D. and D. Morris (1995) *How Much Energy Does It Take to Make a Gallon of Ethanol?*, Washington, DC, Institute for Local Self-Reliance, August

Lovins, A. et al (2005) *Winning the Oil Endgame*, Snowmass, CO, Rocky Mountain Institute

Lynd, L. R. et al (2003) *Bioenergy: Background, Potential, and Policy*, Washington, DC, Center for Strategic and International Studies

Lynd, L. R. et al (2005) 'Consolidated bioprocessing of lignocellulosic biomass: An update', *Current Opinion in Biotechnology*, vol 16, pp577–583

Lyon, N. (2005) 'Taxation of on-farm biodiesel favouring fossil fuels', *The Land*, New South Wales, Australia, 13 October

Lyons, H. A., S. G. Anthony and P. A. Johnson (2001) 'Impacts of increased poplar cultivation on water resources in England and Wales', *Aspects of Applied Biology*, vol 65, pp83–90

Macedo, I. C. (1993) 'Fuel ethanol production in Brazil', in *Proceeding of the First Biomass Conference of the Americas*, Burlington, Vermont, August, Golden, CO, NREL, pp1185–1193

Macedo, I. C., R. Lima Verde Leal and J. Silva (2004a) *Perspectivas de un Programa de Biocombustibles en América Central, Proyecto Uso Sustentable de Hidrocarburos, Comisión Económica para América latina y el Caribe*, Economic Commission for Latin America, www.eclac.org.mx

Macedo, I. C., R. Lima Verde Leal and J. Silva (2004b) *Assessment of Greenhouse Gas Emissions in the Production and Use of Fuel Ethanol in Brazil*, São Paulo, April, www.unica.com.br/i_pages/files/pfd-ingles.pdf

Macedo, I. C. et al (2005) *Sugar Cane's Energy: Twelve Studies on Brazilian Sugar Cane Agribusiness and its Sustainability*, São Paulo, União da Agroindústria Canavieira de São Paulo, September

Mali-Folkecenter (2004) *The Jatropha Plant as a Tool for Combating Desertification, Poverty Alleviation, and Provision of Clean Energy Services to Local Women*, Bamako, Mali, May

Mallenbaker (2006) 'Companies in crisis – what not to do when it all goes wrong: Exxon Mobil and the *Exxon Valdez*', www.mallenbaker.net/csr/CSRfiles/crisis03.html, accessed 4 March 2006.

Mandal, R. (2005) 'The Indian Biofuels Programme – national mission on bio-diesel', Presentation to the International Conference and Expo, Biofuels 2012: Vision to Reality, New Delhi, 17–18 October

Manila Times (2005) 'Ethanol bill to become law before year-end', *The Manila Times*, 15 October

Mann, M. and P. Spath (1997) *Life Cycle Assessment of a Biomass Gasification Combined-Cycle Gasification System*, Golden, CO, National Renewable Energy Laboratory, December

Manufacturers of Emissions Controls Association (undated) *Clean Air Facts: Emission Control for Diesel Engines*, Washington, DC, www.meca.org/galleries/default-file/dieselfact%200106.pdf

Mariner Group (2000) *History*, www.marinergroup.com/oil-spill-history.htm

Marland, G. and A. F. Turhollow (1991) 'CO_2 emissions from the production and combustion of fuel ethanol from corn', *Energy*, vol 16, no 11/12, pp1307–1316

Martinot, E. et al (2005) *Renewables 2005: Global Status Report*, Bonn, Renewable Energy Policy Network for the 21st Century

Maryland General Assembly, Department of Legislative Services (2003) *Environment – Fast-Fill Compressed Natural Gas Station Program*, Senate Bill 163, Annapolis, 1 July, http://mlis.state.md.us/2003rs/fnotes/bil_0003/sb0163.doc

Mathewson, S. W. (1980) 'Yeast and fermentation', in *The Manual for the Home and Farm Production of Alcohol Fuel*, Berkeley, CA, Ten Speed Press

Matthew, E. (2005) 'Palm oil is killing off the orang-utan', Scotsman.com, 24 September

McAloon, A. et al (2000) *Determining the Cost of Producing Ethanol from Corn Starch and Lignocellulosic Feedstocks*, Golden, CO, US Department of Agriculture Agricultural Research Service and NREL-Biotechnology Center for Fuels and Chemicals, October

McCarthy, M. (2003) 'Wildlife fear leaves GM safety debate unresolved', *The Independent*, 22 July

McCormick, R. L. (2003) *NO_x Solutions for Biodiesel*, Golden, CO, NREL, and Colorado Institute for Fuels and Engine Research, Colorado School of Mines, February, p21

McCormick, R. (2005) 'Effects of biodiesel on NO_x emissions', Presentation to ARB Biodiesel Workgroup, National Renewable Energy Laboratory, Golden, CO, 8 June

McKendry, P. (2002) 'Energy production from biomass (Part 3): Gasification technologies', *Bioresource Technology*, vol 83, pp55–63

McLaughlin, S. B. et al (1999) 'Developing switchgrass as a bioenergy crop', in J. Janick (ed) *Perspectives on New Crops and New Uses*, Alexandria, VA, ASHS Press, pp282–299

McLaughlin, S. B. et al (2002) 'High-value renewable energy from prairie grasses', *Environmental Science and Technology*, vol 36, no 10, p2124

McMillen, S. et al (2005) *Biodiesel: Fuel for Thought, Fuel for Connecticut's Future*, Storrs, CT, Connecticut Center for Economic Analysis, University of Connecticut, 24 March

MercoPress (2005) 'World's first tallow-biodiesel plant in Brazil', *MercoPress*, 27 October

MERCOSUR (2004) *Fast-Tracking a 'Feasible' EU-MERCOSUR Agreement: Scenarios for Untying the Agricultural Knot*, Draft prepared for the Working Group on EU-MERCOSUR Negotiations of the MERCOSUR Chair of Sciences Po, Paris, Presented at a workshop the Faculty of Law, Barcelona, Spain, 29 March

Miami Herald (2005) 'Tons of sugar, but biofuels controversy remains', *The Miami Herald*, 1 August

Miles, T. P. E. et al (1995) *Alkali Deposits Found in Biomass Power Plants: A Preliminary Investigation of Their Extent and Nature*, Golden, CO, National Renewable Energy Laboratory, April

Miles, T. Sr (1992) 'Operating experience with ash deposition in biomass combustion systems', Presentation at Biomass Combustion Conference, Reno, NV, 28–29 January

Milne, T. et al (1986) *Thermodynamic Data for Biomass Conversion and Waste Incineration*, Golden, CO, Solar Energy Research Institute (now NREL), September

Minnesota Department of Agriculture (2006) *The Minnesota Ethanol Program*, www.mda.state.mn.us/ethanol/about.htm#ngcnote, accessed 6 March 2006

Minten, B., L. Randrianarison and J. F. M. Swinnen (2005) *Global Retail Chains and Poor Farmers: Evidence from Madagascar*, September, www.csae.ox.ac.uk/conferences/2006-EOI-RPI/papers/csae/Randrianarison.pdf

Mobil Oil Australia Pty Ltd (2005) *Mobil Submission to the Australian Government Biofuels Taskforce*, June, www.pmc.gov.au/biofuels/submissions/submission33.pdf

Moore, J. et al (1997) *Oil Imports: An Overview and Update of Economic and Security Effects*, Congressional Research Service Report for Congress 98-1, Washington, DC, 12 December

Moreira, J. R. (2004a) *Global Biomass Energy Potential*, Prepared for the Expert Workshop on Greenhouse Gas Emissions and Abrupt Climate Change: Positive Options and Robust Policy, Paris

Moreira, J. R. (2004b) *Global Biomass Energy Potential*, São Paulo, CENIBO

Morris, D (2003) 'The ethanol glass is still only half full', *Ethanol Today*, September

Morris, D. (2006) 'Ownership matters: Three steps to ensure a biofuels industry that truly benefits rural America', Institute for Local Self-Reliance, Minneapolis, MN, Presentation to the Minnesota Ag Expo 2006, Morton, MN, 25 January

Mortimer, N. D. et al (2003) *Evaluation of the Comparative Energy, Global Warming and Socio-Economic Costs and Benefits of Biodiesel*, Report prepared for the UK Department for Environment, Food and Rural Affairs, Report no 20/1, London, January, pix

Mosier, N. C. et al (2005) 'Features of promising technologies for pretreatment of lignocellulosic biomass', *Bioresource Technology*, vol 96, no 6, pp673–686

Mozaffarian, M. and R. W. R. Zwart (2003) *Feasibility of Biomass/Waste-Related SNG Production Technologies*, NOVEM Project No 249-01-03-12-0001, The Hague, Dutch Ministry of Economic Affairs

Mudeva, A. (2005) 'Interview – Malaysia IOI eyes green energy expansion in Europe', *Reuters*, 23 November

Multi-commodity Exchange of India, Ltd (2006) *Castor Oil Profile*, www.mcxindia.com/castor_oil.aspx, accessed 14 March 2006

Murray, D. (2005a) 'Ethanol's potential: Looking beyond corn', *Eco-Economy Update*, Washington, DC, Earth Policy Institute, 29 June

Murray, D. (2005b) *Rising Oil Prices Will Impact Food Supplies*, peopleandplanet.net, 13 September

Naber, J. E. and F. Goudriaan (2003) 'HTU diesel', Presented at Biofuel BV, Berlin, Germany, November

Nagarajan, G. et al (no date) *Review of Ethanol in Compression Ignition Engine*, Chennai, Institute for Energy Studies, Anna University, www.saeindia.org/saeconference/ethanolreview.htm

Nakicenovic, N. et al (1998) *Global Energy Perspectives*, Cambridge, International Institute for Applied Systems Analysis/World Energy Council

Nastari, P. M. (2005) 'Ethanol: The global overview', Presentation to the International Ethanol Conference, Brisbane, Australia, 9–10 May, www.sdi.qld.gov.au/dsdweb/v3/documents/objdirctrled/nonsecure/pdf/11584.pdf

Nastari, P., I. Macedo and A. Szwarc (2005) *Observations on the Draft Document Entitled 'Potential for Biofuels for Transport in Developing Countries'*, Report prepared for the World Bank Air Quality Thematic Group, Brazil

National Biodiesel Board (1998) *Summary of Recent Studies Citing Economic Attributes Resulting from the Successful Commercialization of Biodiesel in the US*, Reports database, Jefferson City, MO, July, www.biodiesel.org/resources/reportsdatabase/reports/gen/19980701_gen-054.pdf

National Biodiesel Board (2002) *Biodiesel Production and Quality Standards*, Fact sheet, Jefferson City, MO, updated 11 March 2002

National Biodiesel Board (2004a) *Biodiesel as a Greenhouse Gas Reduction Option*, Jefferson City, MO, www.biodiesel.org/resources/reportsdatabase/reports/gen/20040321_gen-332.pdf

National Biodiesel Board (2004b) *Study Shows NO_x Emissions Reductions in Biodiesel Blends with Additive*, Press release, Jefferson City, MO, 4 February

National Biodiesel Board (2005a) *Biodiesel Production*, Fact sheet, Jefferson City, MO, updated 29 January 2005

National Biodiesel Board (2005b) *Guidance Issued 30 November 2005*, Jefferson City, MO, 30 November, www.biodiesel.org/pdf_files/Biodiesel_Blends_Above%20_20_Final.pdf

National Biodiesel Board (2005c) *Biodiesel Production Soars*, Press release, Jefferson City, MO, 8 November

National Biodiesel Board (2006) *Fuel Fact Sheets: Standards and Warranties*, www.biodiesel.org/resources/fuelfactsheets/standards_and_warranties.shtm, accessed 2 March 2006

National Corn Growers Association (2004) *The World of Corn 2004*, http://ncga.com/WorldOfCorn/main

National Corn Growers Association (2006) *Taking Ownership of Grain Belt Agriculture*, www.ncga.com/public_policy/takingOwnership/index.htm, accessed 6 March 2006

National Ethanol Vehicle Coalition (no date) *E85 Brochure*, www.e85fuel.com

National Ethanol Vehicle Coalition (2006) *How Many E85 Compatible Vehicles are on American Roads Today?*, www.e85fuel.com/e85101/faqs/number_ffvs.php, accessed 10 March 2006

National Science-Technology Roadshow Trust (2006) *Morton Coutts – Continuous Fermentation System*, www.roadshow.org/html/resources/scientists/coutts/article.html, accessed 29 January 2006

Natsource Environmental Services (2006) 'Glossary', www.natsource.com/markets/index.asp, accessed January 2006

Natural Resources Canada (2003) *Government of Canada Launches Ethanol Expansion Program*, Press release, Ottawa, 20 October

Nature Canada (2004) 'How tempting is the Mackenzie gas project for Canada's north?', *Nature Canada*, summer

Net for Cuba International (2006) *A Brief on the Industrial Sector for Cuba*, www.netforcuba.org/InfoCuba-EN/Economy/ABrief.htm, accessed 12 March 2006

Netherlands Agency for Energy and Environment (NOVEM) (2003) *Conventional Bio-Transporation Fuels: An Update*, Report 2GAVE-03.10, Utrecht, May, www.ecn.nl/_files/bio/Report_2GAVE-03.10.pdf

New Rules Project (2006) *Ethanol Production – The Minnesota Model*, www.newrules.org/agri/ethanol.html, accessed 3 April 2006

New Uses Council, Environmental and Energy Studies Institute and North Carolina Solar Center (2005) *Development of a Workable Incentives System for Biobased Products, Biofuels and Biopower*, Project description, September

Nierenberg, D. (2005) *Happier Meals: Rethinking the Global Meat Industry*, Worldwatch Paper 171, Washington, DC, Worldwatch Institute

Noon, C. E. et al (1996) 'Transportation and site location analysis for regional integrated biomass assessment (Riba),' in *Proceedings of Bioenergy 1996 – The Seventh National Bioenergy Conference: Partnerships to Develop and Apply Biomass Technologies*, Nashville, TN, 15–20 September, http://bioenergy.ornl.gov/papers/bioen96/noon1.html

Northeast States for Coordinated Air Use Management (2001) *Health, Environmental, and Economic Impacts of Adding Ethanol to Gasoline in the Northeast States, vol 3: Water Resources and Associated Health Impacts*, Lowell, MA, New England Interstate Water Pollution Control Commission

Novak, N. (2002) 'The rise of ethanol in rural America', *Ag Decision Maker*, Iowa State University, June

NOVEM (Dutch Energy Agency) and ADL (Arthur D. Little) (1999) *Analysis and Evaluation of GAVE Chains*, vol 1–3, Utrecht, GAVE Analysis Programme, December

Novozymes (2005) *Novozymes and NREL Reduce Enzyme Cost*, Press release, Bagsvaerd, Denmark, 14 April

Novozymes and BBI International (2002) *Fuel Ethanol Production: Technological and Environmental Improvements, Its History, Its Advancements, Its Future*, Grand Forks, ND, 26 June

Novozymes and BBI International (2005) *Fuel Ethanol: A Technological Evolution*, Grand Forks, ND, June

NREL (US National Renewable Energy Laboratory) (1998) *Hannepin County's Experience with Heavy-Duty Ethanol Vehicles*, Golden, CO, January

NREL (2000) *Biomass for Sustainable Transportation*, Golden, CO, www.nrel.gov/docs/fy00osti/25876.pdf

NREL (2002a) 'Mustard seed research program', Presentation, Golden, CO, March, www.nrel.gov/docs/gen/fy02/31794.pdf

NREL (2002b) *The Biomass Economy*, Golden, CO, National Renewable Energy Laboratory 2002 Research Review, July

NREL (2005) *Thermochemical Conversion Technologies-Projects*, www.nrel.gov/biomass/proj_thermochemical_conversion.html, accessed 16 May 2005

NTB Liquid Biofuels Network (undated) *NTB Liquid Biofuels Network – The European Network for Removing Non-Technical Barriers to the Development of Liquid Biofuels*, www.biomatnet.org/secure/Ec/S1016.htm

OECD (Organisation for Economic Co-operation and Development) (2006) *Agricultural Market Impacts of Future Growth in the Production of Biofuels*, Paris, Working Party on Agricultural Policies and Markets

Official Journal of the European Union (2005) 'Directive 2003/30/EC of the European Parliament and of the Council of 6 May 2003 on the Promotion of the use of biofuels or other renewable fuels for transport', 17 May, p1

Ohio Farm Bureau Federation (2005) 'Ethanol on the right track Indy 500', News feature, Columbus, OH, 3 June

Oil World (2005) 'EU rapeoil selling out for biofuel use', *Oil World*, 19 October

Olofsson, L. (2005) 'Swedish government embraces peak oil and looks towards biofuels', *Energy Bulletin*, 17 December

Ontario Ministry of Agriculture, Food and Rural Affairs (2005) *McGuinty Government Takes Next Steps on Cleaner Air*, Press release, Toronto, Ontario, 7 October

Organic Consumers Association (2005) *How Great is the Impact of GMOs on Organic?*, Finland, MN, June

ORNL (Oak Ridge National Laboratory) (undated a) *Biofuels from Switchgrass: Greener Energy Pastures*, Brochure, Oak Ridge, TN, http://bioenergy.ornl.gov/papers/misc/switgrs.html

ORNL (undated b) *Popular Poplars: Trees for Many Purposes*, Oak Ridge, TN, http://bioenergy.ornl.gov/misc/poplars.html, accessed 10 March 2006

ORNL (2005) *Biomass as Feedstock for a Bioenergy and Bioproducts Industry: The Technical Feasibility of a Billion-Ton Annual Supply*, US Department of Energy, Office of Energy Efficiency and Renewable Energy, February

Oxfam International (2002) *The EU Sugar Scam: How Europe's Sugar Regime Is Devastating Livelihoods in the Developing World*, Oxfam Briefing Paper 27, Oxford, UK

Panoutsou, C. (2005) 'Agricultural and forestry biomass in EU-25', Center for Renewable Energy Sources, Presented to the First International Biorefinery Workshop, Washington, DC, 20–21 July

Park, S. et al (2004) 'Identifying on-farm management practices aimed at reducing greenhouse gas emissions from sugarcane primary production', in *Proceedings of the 11th Australian Agronomy Conference*, 2–6 February 2003, Geelong, Victoria, Australian Society of Agronomy.

Pascual, L. M. and R. R. Tan (2004) *Comparative Life Cycle Assessment of Coconut Biodiesel and Conventional Diesel for Philippine Automotive Transportation and Industrial Boiler Application*, Manila, College of Engineering, De La Salle University, www.lcacenter.org/InLCA2004/papers/Pascual_L_paper.pdf

Paster, M., J. Pellegrino and T. Carole (2003) *Industrial Bioproducts: Today and Tomorrow*, Prepared by Energetics, Inc, for the US Department of Energy (DOE), Office of Energy Efficiency and Renewable Energy, Office of the Biomass Program, Washington, DC, July

Pataki, Governor G. (2006) *State of the State Address*, 4 January, www.ny.gov/governor/2006_sos/sosaddress_text.html

Patzek, T. (2004) 'Thermodynamics of the corn-ethanol biofuel cycle', *Critical Reviews in Plant Sciences*, vol 23, no 519, pp519–567

Pearce, F. (2005) 'Forests paying the price for biofuels', *New Scientist*, 22 November, p19

Pedersen, J. R., A. Ingemarsson and J. O. Olsson (1999) 'Oxidation of rapeseed oil, rapeseed methyl ester (RME) and diesel fuel studied with GC/MS', *Chemosphere*, vol 38, no 11, pp2467–2474

Peng, S. et al (2004) 'Rice yields decline with higher night temperature from global warming', *Proceedings of the National Academy of Sciences*, 6 July, pp9971–9975

Pennsylvania Department of Environmental Protection (2005) *Automobile Emissions: An Overview*, www.dep.state.pa.us/dep/subject/pubs/arr/aq/fs1829.pdf, accessed June 2005

Peplow, M. (2005) 'Ethanol production harms environment, researchers claim', *Nature*, 1 July

Perlack, R. D. et al (1995) *Biomass Fuel from Woody Crops for Electric Power Generation*, Oak Ridge, TN, ORNL, 21 September, http://bioenergy.ornl.gov/reports/fuelwood/toc.html

Perlack, R. D. et al (2005) *Biomass as Feedstock for a Bioenergy and Bioproducts Industry: The Technical Feasibility of a Billion Ton Annual Supply*, Oak Ridge, TN, ORNL, April

Pew Initiative on Food and Biotechnology (2003) *Have Transgenes – Will Travel: Issues Raised by the Gene Flow from Genetically Engineered Crops*, Washington, DC, August

Philippine Fuel Ethanol Alliance (2005) *Bioethanol Benefits: Energy, Environment, Employment*, Makati City, www.bioethanol.com.ph/bioethanol/benefits.html, accessed 7 March 2006

Pia Palermo, M. (2005) 'Brazil losing fight to save the Amazon', *Reuters*, 23 May

Pimentel, D. (1991) 'Ethanol fuels: Energy security, economics and the environment', *Journal of Agricultural and Environmental Ethics*, vol 4, pp1–13

Pimentel, D. (2001) 'The limits of biomass energy', *Encyclopaedia of Physical Sciences and Technology*, September

Pimentel, D. (2003) 'Ethanol fuels: Energy balance, economics and environmental impacts are negative', *Natural Resources Research*, vol 12, no 2, pp127–134

Pimentel, D. and J. Krummel (1987) 'Biomass energy and soil erosion: Assessment and resource costs', *Biomass*, vol 14, pp15–38

Pimentel, D. and T. Patzek (2005) 'Ethanol production using corn, switchgrass and wood; biodiesel production using soybean and sunflower', *Natural Resources Research*, vol 14, no 1, p65–76

Pollan, M. (2002) 'When a crop becomes king', *New York Times*, 19 July

Postel, S. (2006) 'Safeguarding freshwater ecosystems', in Worldwatch Institute (ed) *State of the World 2006*, New York, W. W. Norton and Company, p52

Powlson, D. S., A. B. Riche and I. Shield (2005) 'Biofuels and other approaches for decreasing fossil fuel emissions from agriculture', *Annals of Applied Biology*, vol 146, pp193–201

PR Newswire (2005) 'ADM plans to expand ethanol capacity by 500 million gallons', *PR Newswire*, 4 October

Prakash, C. (1998) *Use of Higher than 10 Volume Percent Ethanol/Gasoline Blends in Gasoline Powered Vehicles*, Ottawa, Environment Canada, www.ec.gc.ca/cleanair-airpur/CAOL/transport/publications/ethgas/ethgastoc.htm

Public Citizen (2001) *Down on the Farm: NAFTA's Seven-Years War on Farmers and Ranchers in the US, Canada and Mexico*, Washington, DC, Public Citizen

Quirin, M. et al (2004) *CO_2 Mitigation through Biofuels in the Transport Sector: Status and Perspectives, Main Report*, Heidelberg, Institute for Energy and Environmental Research, August

Radich, A. (2004) *Biodiesel Performance, Costs, and Use*, Washington, DC, DOE, Energy Information Agency (EIA), updated 8 June 2004, www.eia.doe.gov/oiaf/analysispaper/biodiesel

Rainforest Alliance (no date a) *Standard for Sustainable Agriculture – Sustainable Agriculture Network*, www.rainforest-alliance.org/programs/agriculture/certified-crops/standards.html

Rainforest Alliance (no date b) *Sustainable Agriculture Network*, www.rainforest-alliance.org/programs/agriculture/san/index.html

Rams Horn: Journal of Food Systems Analysis (2006) 'Energy oil and palm', *The Rams Horn: Journal of Food Systems Analysis*, January

Randewich, N. (2005) 'Central American poor struggle as oil prices rise', *Reuters*, 14 September

Ranney, J. W. and L. K. Mann (1994) 'Environmental considerations in energy crop production', *Biomass and Bioenergy*, vol 6, pp211–228

Ray, D. E., D. G. De La Torre Ugarte and K. J. Tiller (2003) *Rethinking US Agricultural Policy: Changing Course to Secure Farmer Livelihoods Worldwide*, Nashville, Agricultural Policy Analysis Center, University of Tennessee

Regmi, A. et al (2001) *Changing Structure of Global Food Consumption and Trade*, Washington, DC, USDA Economic Research Service

RenewableEnergyAccess.com (2005a) *Biodiesel Mandate for Navy and Marine Facilities*, 22 March

RenewableEnergyAccess.com (2005b) *State Legislation Focuses on Tax Incentives for Wind*, Biofuels, 28 April

RenewableEnergyAccess.com (2005c) *New York Announces New Energy Plans*, 19 September

RenewableEnergyAccess.com (2006) *DynaMotive to Make BioOil in Ukraine and Baltic States*, 31 January

Renner, M. (2002a) *The Anatomy of Resource Wars*, Worldwatch Paper 162, Washington, DC, Worldwatch Institute

Renner, M. (2002b) 'Oil spills decline', in Worldwatch Institute (ed) *Vital Signs 2002*, New York, W. W. Norton and Company

Reuters (1999) 'Kemira to invest in new technology', *Reuters*, 6 September

Reuters (2003a) 'Most in US would shun labeled biotech foods – poll', *Reuters*, 16 July

Reuters (2003b) 'China to put corn into gas tanks to clean up', *Reuters*, 20 October

Reuters (2005a) 'Thais to invest $20 billion in energy sector in 4 years', *Reuters*, 17 May

Reuters (2005b) 'UK study sees biofuel yielding mixed green impact', *Reuters*, 17 May

Reuters (2005c) 'Zimbabwe says to reopen ethanol fuel plant', *Reuters*, 19 July

Reuters (2005d) 'Ethanol starting to trade like a world commodity', *Reuters*, 29 September

Reuters (2005e) 'Interview: Farmers are 'tomorrow's sheiks' due to biofuels', *Reuters*, 10 November

Reuters (2005f) 'Grape biofuel may lift the spirits of French vinters', *Reuters*, 16 November

Reuters (2006) 'Germany sets quotas for biofuel in fuel from 2007', *Reuters*, 3 May

RFA (no date) *Federal Regulations: VEETC*, www.ethanolrfa.org/policy/regulations/federal/veetc

RFA (Renewable Fuels Association) (2005a) *Ethanol Facts: Economy* and *Ethanol Trade Fact Sheet*, Fact sheets, Washington, DC, www.ethanolrfa.org/resource/facts

RFA (2005b) *Homegrown for the Homeland: Ethanol Industry Outlook 2005*, Washington, DC, www.ethanolrfa.org/objects/pdf/outlook/outlook_2005.pdf

RFA (2006a) *Industry Statistics*, www.ethanolrfa.org/industry/statistics/#A, accessed 10 March 2006

RFA (2006b) *Plant Locations*, www.ethanolrfa.org/industry/locations, accessed 3 April 2006

RFA (2006c) *From Niche to Nation: Ethanol Industry Outlook 2006*, Washington, DC, February.

Richards, I. R. (2000) *Energy Balances in the Growth of Oilseed Rape for Biodiesel and of Wheat for Bioethanol*, Report prepared for the British Association for Bio Fuels and Oils (BABFO), Levington Park Ipswich, Suffolk, UK, Levington Agriculture Ltd, June, www.biodiesel.co.uk/levington.htm

Riesing, T. F. (2006) 'Cultivating algae for liquid fuel production', *The Permaculture Activist*, March 2006, pp48–50

Rockefeller Foundation and Scientific and Industrial Research and Development Centre (1998) *The Potential of* Jatropha curcas *in Rural Development and Environment Protection – An Exploration*, Concept paper for a workshop sponsored by the Rockefeller Foundation and Scientific and Industrial Research and Development Centre, Harare, Zimbabwe,13–15 May 1998, www.jatropha.de/zimbabwe/rf-conf1.htm

Rodrigues, R. A. (2005) *Biodiesel in Brazil – An Overview*, PowerPoint presentation to the Third German–Brazilian Workshop on Biodiesel, Fortaleza-Ceara, Brazil, 1 July, www.ahk.org.br/inwent/palestras/rodrigo_rodrigues.ppt

Rohter, L. (2005) 'A record Amazon drought, and fear of wider ills', *New York Times*, 11 December

Romanowicz, G. (2006) *UK and Europe Push for 'Right Kind' of Biofuels*, edie.net, 3 March

Rosillo-Calle, F. (2001) 'Biomass (other than wood)', in World Energy Council (ed) *Survey of Energy Resources 2001*, London, WEC

RSPO (Roundtable on Sustainable Palm Oil) (2005a) *RSPO Principles and Criteria for Sustainable Palm Oil Production*, Public release version, Kuala Lumpur, 17 October

RSPO (2005b) *RSPO Adopts the Principles and Criteria for Sustainable Palm Oil Production*, Press release, Singapore, 23 November

RSPO (2005c) *The Malaysian Palm Oil Association (MPOA)*, Singapore, www.sustainable-palmoil.org/profile.htm

RSPO (2006) *Sustainable Palm Oil Production Principles Set Out 15 December 2005*, www.sustainable-palmoil.org, accessed 7 March 2006

RSPO (no date) *Frequently Asked Questions*, www.sustainable-palmoil.org/FAQs.htm

Rubin, J. (2006) *The Time of Sands*, CIBC World Markets Inc, Monthly Indicators, 9 January

Rubin, J. and P. Buchanan (2006) *The Global Crude Supply Outlook: Tighter Markets Ahead*, CIBC World Markets Inc, Monthly Indicators, 9 January

Rudloff, M. (2005) 'Biomass-to-liquid Fuels (BTL) – made by CHOREN. Process, environmental impact and latest developments', Presentation to the Automobile and Environment Belgrade EAEC Congress, May

RWE Aktiengesellschaft (2005) *World Energy Report 2005: Determinants of Energy Prices*, Essen, p36

Saint, W. S. (1982) 'Farming for energy: Social options under Brazil's national alcohol programme', *World Development*, vol 10, no 3, pp223–238

Sakamoto Yakuhin Kogyo, Ltd (2005) *Glyerol*, www.sy-kogyo.co.jp/English/sei/1_gly.html, accessed 22 May 2005

Salazar, K. (2006) *Report on the 2006 Colorado Renewable Energy Summit*, Denver, 16 February, http://salazar.senate.gov/images/pdf/SummitReport060216.pdf

Sample, D. W., L. Paine and A. Roth (1998) 'Harvested switchgrass fields provide habitat for declining grassland birds', Paper presented at BioEnergy 1998: Expanding Bioenergy Partnerships, Madison, WI, 4–8 October 1998

Sanchez, S. et al (2002) 'The fermentation of mixtures of D-glucose and D-xylose by *Candida shehatae*, *Pichia stipitis* or *Pachysolen tannophilus* to produce ethanol', *Journal of Chemical Technology and Biotechnology*, vol 77, no 6, pp641–648

Santos, O. J. F. (2001) 'Environmental aspects of *Miscanthus* production', in M. B. Jones and M. Walsh (eds) *Miscanthus for Energy and Fibre*, London, James and James, pp46–47

Sasol (2004) *Explore Sasol: Sasol in Fuels*, Johannesburg, 12 August, www.sasol.com/sasol_internet/frontend/navigation.jsp?navid=600003&rootid=2, accessed 7 March 2006

Scharmer, K. and G. Gosse (1996) *Energy Balance, Ecological Impact and Economics of Vegetable Oil Methylester Production in Europe as a Substitute for Fossil Diesel*, Report commissioned by GET Germany and INRA France, Brussels, EU ALTENER Programme

Schiller, A. and V. R. Tolbert (1996) *Hardwood Energy Crops and Wildlife Diversity: Investigating Potential Benefits for Breeding Birds and Small Mammals*, Oak Ridge, TN, Oak Ridge Associated Universities and Biofuels Feedstock Development Program, Oak Ridge National Laboratory (ORNL), http://bioenergy.ornl.gov/papers/bioen96/schiler1.html

Schlamadinger, B. and G. Marland (1998) 'Some results from the Graz/Oak Ridge Carbon Accounting Model (GORCAM)', in *Greenhouse Gas Balances of Bioenergy Systems*, Paris, International Energy Agency

Schmitz, N. et al (eds) (2005) *Innovationen bei der Bioethanolerzeugung. Schriftenreihe, Nachwachscende Rohstoffe, Band 26*, Bonn, German Federal Ministry of Food, Agriculture and Consumer Protection and Germany Agency of Renewable Resources

Schneider, R. et al (2000) *Sustainable Amazon: Limitations and Opportunities for Rural Development*, Washington, DC, World Bank and Imazon, Brazil

Science Daily (2005) 'Hybrid grass may prove to be valuable fuel source', *Science Daily*, 9 September

Scurlock, J. (2000) *Bioenergy Feedstock Characteristics*, Oak Ridge, TN, ORNL

Sen, A. (2002) 'Why half the planet is hungry', *Observer*, 16 June

Sethi, K. S. (2005) *Biofuels 2012: Vision to Reality*, Planning Commission, www.teriin.org/events/docs/5biofuel.htm, accessed 18 October 2005

Severinghaus, J. (2005) *Why We Import Brazilian Ethanol*, Trade report, West Des Moines, IA, Iowa Farm Bureau Federation, 14 July

Seyfried, F. (2005) *RENEW – Renewable Fuels for Advanced Powertrains*, First Year Report for EU Sixth Framework Integrated Project, www.renew-fuel.com

Shapouri, H. (2004) *The 2001 Net Energy Balance of Corn-Ethanol*, Washington, DC, USDA

Shapouri, H., Duffield, J. A. and Graboski, M. S. (1995) *Estimating the Net Energy Balance of Corn Ethanol*, US Department of Agriculture, Economic Research Service, Office of Energy, Agricultural Economic Report No. 721, July

Shapouri, H., J. A. Duffield and M. Wang (2002) *The Energy Balance of Corn Ethanol: An Update*, Agricultural Economic Report No 813, Washington, DC, USDA Office of Energy Policies and New Uses

Sharp, C. A. (1998) *Exhaust Emissions and Performance of Diesel Engines with Biodiesel Fuels*, San Antonio, TX, Southwest Research Institute, July

Sheehan, J. et al (1998a) *A Look Back the US Department of Energy's Aquatic Species Program: Biodiesel from Algae*, Golden, CO, NREL

Sheehan, J. et al (1998b) *An Overview of Biodiesel and Petroleum Diesel Life Cycles*, Golden, CO, NREL, May

Sheehan, J. et al (1998c) *Life Cycle Inventory of Biodiesel and Petroleum Diesel for Use in an Urban Bus*, Golden, CO, National Renewable Energy Laboratory, May

Sheehan, J. et al (2004) 'Energy and environmental aspects of using corn stover for fuel ethanol', *Journal of Industrial Ecology*, vol 7, no 3–4, pp117–146

Shell International (1995) *The Evolution of the World's Energy System 1860–2060*, London, Shell Center

Sherertz, P. C. (1998) *Petroleum Products*, Richmond, VA, Virginia Department of Health, 30 June www.vdh.state.va.us/HHControl/petrofac.PDF

Sierra Club of Canada (2004) *Mackenzie Pipeline to Fuel America's Gas Tank*, Press release, Ottawa, 14 April

Simon, S. (2005) 'To replace oil, US experts see amber waves of plastic', *Los Angeles Times*, 28 June

Singh, R. (undated) *Literature Review on Biodiesel*, University of the South Pacific, written for South Pacific Applied Geoscience Commission, Suva, Fiji

Singh, V. et al (2001) *Modified Dry Grind Ethanol Process*, Champaign, IL, University of Illinois Urbana-Champaign Agricultural Engineering Department, 18 July

Sinico, S. (2005) *Fill it Up with Natural*, dw-world.de, 22 September, hwww.dw-world.de/dw/article/0,2144,1717299,00.html

Sino-German Workshop on Energetic Utilization of Biomass (2003) Beijing, 8–11 October, www.ipe.ac.cn/sinogerman/program.htm

Smeets, E., A. Faaij and I. Lewandowski (2004) *A Quickscan of Global Bioenergy Potentials to 2050*, Utrecht, Copernicus Institute, Utrecht University, March, Updated results obtained from the authors, March 2006

Smeets, E., A. Faaij and I. Lewandowski (2005) *The Impact of Sustainability Criteria on the Costs and Potentials of Bioenergy Production*, Utrecht, Copernicus Institute, Utrecht University, May

Smil, V. (2002) 'Eating meat: Evolution, patterns, and consequences', *The Population and Development Review*, vol 28, no 4, pp588–639

Somera, B. J. and K. K. Wu (1996) *Energy Inventory of Hawaiian Sugar Plantations – 1995*, Energy Report 35, Alea, HI, Hawaii Agriculture Research Center, August

Sorensen, B. (1999) *Long-Term Scenarios for Global Energy Demand and Supply: Four Global Greenhouse Mitigation Scenarios*, Roskilde, Roskilde University, Insitute 2, Energy and Environment Group

SPE (Society of Petroleum Engineers) (2005a) *How Is the Industry Working to Protect the Environment?*, www.spe.org/spe/jsp/basic/0,,1104_1008218_1109940,00.html, accessed June 2005

SPE (2005b) *Seismic Technology*, www.spe.org/spe/jsp/basic/0,,1104_1714_1004089,00.html, accessed June 2005

Stat Communications (2006) *Biodiesel Lifts German Rapeseed Area*, statpub.com, 21 February www.statpub.com/open/183421.phtm

Strossman, C. (2004) *European Union Sugar Annual 2004*, GAIN Report, Washington, DC, USDA and FAS, 8 April

Sun Star Magazine (2005) 'Biodiesel plant seen to help coco industry', *Sun Star Magazine*, 9 June

Sutton, M. C., N. Klein and G. Taylor (2005) 'A comparative analysis of soybean production between the United States, Brazil and Argentina', *Journal of the American Society of Farm Managers and Appraises*, www.asfmra.org/documents/Sutton33_41.pdf

Sweedtrack (no date) 'Reducing our dependence on oil', www.sweedtrack.com/eflwa22a.htm

Swisher, J. and D. Wilson (1993) 'Renewable energy potentials', *Energy*, vol 18, no 5, pp437–459

Talks, P. (2004) *EU-25 Grain and Feed Semi-Annual 2004*, GAIN Report, Washington, DC, USDA, Foreign Agricultural Service, 8 December

Tampier, M. et al (2004) *Identifying Environmentally Preferable Uses for Biomass Resources, Stage 2 Report: Life-Cycle GHG Emissions Reduction Benefits of Selected Feedstock-to-Product Threads*, Prepared for Natural Resources Canada and National Research Council Canada, North Vancouver, BC, Envirochem Services Inc, 19 July

TaTEDO (Tanzania Traditional Energy Development and Environment Organization) (2005) *Biofuels for Transportation in Tanzania: Potential and Implications for Sustainable Agriculture and Energy in the 21st Century*, Prepared for the Deutsche Gesellschaft für Technische Zusammenarbeit (GTZ) GmbH, Dar es Salaam, September

Television Trust for the Environment (2004) *Power Pods – India*, Hands On-Ideas to Go, Series 4, London, February, www.tve.org/ho/doc.cfm?aid=1433

TERI (The Energy and Resources Institute) (2003) *Seminar on Trade Liberalization in Environmental Goods and Services*, New Delhi, 16 May

TERI (2005) *Liquid Biofuels for Transportation: India Country Study on Potential and Implications for Sustainable Agriculture and Energy*, Report prepared for the Deutsche Gesellschaft für Technische Zusammenarbeit (GTZ) GmbH, New Delhi, October, www.gtz.de/de/dokumente/en-biofuels-for-transportation-in-india-2005.pdf

Thompson, S. (2004) *Biodiesel with Attitude: Colorado's Blue Sun Biodiesel Grows Rapeseed for Biodiesel Production*, Rural Cooperatives (USDA), November/December

Total Corporation (2006) *Alternative Energies, Promising New Technologies*, www.total.com/en/group/corporate_social_responsibility/future_energy/energy_supply/alternative_energies_6676.htm, accessed 10 March 2006

Towler, G. P. et al (2003) *Development of a Sustainable Liquid Fuels Infrastructure Based on Biomass*, Des Plaines, Illinois, UOP Innovation Group

Transpetro (2005) Condições Gerais de Serviço – Álcool, Rio de Janeiro, November, www.transpetro.com.br/portugues/negocios/dutosTerminais/files/CGSA.pdf

Tretter, H. (2005) *Development of Biodiesel in Austria*, PowerPoint presentation, Austrian Energy Agency, Vienna, 28 February

Trettin, C., A. Davis and J. Parsons (2003) *Sustainability of High-Intensity Forest Management with Respect to Water and Soil Quality and Site Nutrient Reserves*, Prepared by US Forest Service, Southern Research Station and North Carolina State University, Oak Ridge, TN, US Department of Energy (DOE), Office of Biomass, Oak Ridge National Laboratory, August

Trindade, S. C. (1994) 'Transfer of clean(er) technologies to developing countries', in R. U. Ayres and U. E. Simonis (eds) *Industrial Metabolism: Restructuring for Sustainable Development*, New York, United Nations University Press, pp319–336

Trindade, S. C. (2005) *Beyond Petroleum and Towards a Biomass-Based Sustainable Energy Future: Opportunities for Financing Sustainable Development and Carbon Trade*, Renewable Energy Partnerships for Poverty Eradication and Sustainable Development, Partners for Africa, May

Tyson, K. S., C. J. Riley and K. K. Humphreys (1993) *Fuel Cycle Evaluations of Biomass-Ethanol and Reformulated Gasoline*, Golden, CO, NREL

UCS (no date) 'Subsidizing big oil', www.ucsusa.org/clean_vehicles/fuel_economy/subsidizing-big-oil.html

UK Department for Transportation (2004) *Towards a UK Strategy for Biofuels – Public Consultation*, London, 26 April, www.dft.gov.uk/stellent/groups/dft_roads/documents/page/dft_roads_028393.hcsp

UK Onshore Pipeline Operators' Association (2006) *Worldwide Pipeline Incidents–April 2002*, www.ukopa.co.uk/publications/pdf/020035.pdf, accessed 5 March 2006

UN (United Nations) (1997) *Energy Statistics Yearbook*, New York, UN

UNCTAD (United Nations Conference on Trade and Development) (2005) *UNCTAD Launches the Biofuels Initiative*, Press release, Geneva, 21 June

UNEP (United Nations Environment Programme) (2000) *Global Environment Outlook, GEO-2000*, Nairobi, UNEP

UNICA (Uniao da Agroindustria Canavieira do Estado de São Paulo) (2005) *Brasil Agroindustria da Cana de Acucar*, São Paulo

United Auto Workers (2005) 'The American auto industry in crisis: The threat to middle-class jobs, wages, health care and pensions', *United Auto Workers News and Views*, 8 December

United Nations Framework Convention on Climate Change (2005) *Key GHG Data – GHG Emission Data for 1990–2003*, Bonn, p69, cited in EC (2006e) *Commission Staff Working Document, Annex to the Communication from the Commission: An EU Strategy for Biofuels, Impact Assessment*, Brussels, EC, p4

University of California Agriculture and Natural Resources (2005) *Reduce Pollution With Proper Fertilizer Timing*, http://ucanr.org/delivers/impactview.cfm?impactnum=249&mainunitnum=0, accessed June 2005

University of Illinois Extension (2005) cited in M. Kojima and T. Johnson (eds) *Potential for Biofuels for Transport in Developing Countries*, Washington, DC, World Bank

Urbanchuk, J. M. (2004) *The Contribution of the Ethanol Industry to the American Economy in 2004*, Wayne, PA, 12 March, www.ncga.com/ethanol/pdfs/EthanolEconomicContributionREV.pdf

Urbanchuk, J. and J. Kapell (2002) *Ethanol and the Local Community*, Mt Laurel, NJ, and Boston, AUS Consultants and SJH and Company, 20 June

USA Today (2006) 'Algae – like a breath of mint for smokestacks', *USA Today*, 10 January

USDA (United States Department of Agriculture) (2005) *Grain Transportation Report*, Washington, DC, USDA, 15 December

USDA, Economic Research Service (2006) *Rural Population and Migration: Rural Population Change and Net Migration*, www.ers.usda.gov/briefing/population/popchange, accessed 6 March 2006

USDA, Farm Service Agency (2005) *FY 2005 Production of Bioenergy Program Participants and Payments Earned by Commodity*, Kansas City, MO, 12 December, www.fsa.usda.gov/daco/bioenergy/2005/FY2005ProductPayments.pdf

USDA, Foreign Agricultural Service (2006a) *Oilseeds: World Markets and Trade*, Circular Series FOP106, Washington, DC, January

USDA, Foreign Agricultural Service (2006b) *Counselor and Attaché Reports Official Statistics, Estimates for March 2004/2005*, Washington, DC, Cotton, Oilseeds, Tobacco and Seeds Division, March, www.fas.usda.gov/oilseeds/circular/2006/06-03/table1t3.pdf

USDA, Foreign Agricultural Service (2006c) *EU-25 Agricultural Situation – European Commission Publishes Biofuel Strategy 2006*, Global Agricultural Information Network (GAIN) report, 10 March

USDA, Foreign Agricultural Service (no date) *Production, Supply and Distribution Database*, www.fas.usda.gov/psd/complete_files/HTP-0612000.csv

USDA, Soil Quality Institute (2003) *Crop Residue Removal for Biomass Energy Production: Effects on Soils and Recommendations*, Washington, DC, 4 March, pp2–6

USDA, Transportation and Marketing Programs/Transportation Services Branch (2005) *Grain Transportation Report*, Washington, DC, USDA, 22 December

US DOE (US Department of Energy) (2001) *Black Liquor Gasification Expected to Yield Energy, Environmental, and Environmental Benefits* ,Washington, DC, fall, www1.eere.energy.gov/industry/bestpractices/fall2001_black_liquor.html

US DOE (2003) *Roadmap for Agriculture Biomass Feedstock Supply in the United States*, Washington, DC, November

US DOE (2005a) *Biodiesel Blends Fact Sheet*, Washington, DC, April

US DOE (2005b) *Dilute Acid Hydrolysis*, www.eere.doe.gov/biomass/dilute_acid.html, accessed 16 May 2005

US DOE (2006) *DOE Announces $160 Million for Biorefinery Construction*, Press release, Washington, DC, 22 February

US DOE and EERE (Office of Energy Efficiency and Renewable Energy) (2003) *Roadmap for Agricultural Biomass Feedstock Supply in the United States*, Washington, DC, November

US DOE and EERE (2005a) 'Novozymes and NREL cut cost of converting biomass to ethanol', *EERE Network News*, 20 April

US DOE and EERE (2005b) *Pretreatment Technology Evaluation*, www1.eere.energy.gov/biomass/technology_evaluation.html, accessed 16 May

US DOE and EERE (2005c) *Pyrolysis and Other Thermal Processing*, Washington, DC, www1.eere.energy.gov/biomass/pyrolysis.html

US DOE and EERE (2006a) *Biofuels and the Economy*, http://permanent.access.gpo.gov/websites/www.ott.doe.gov/biofuels/economics.html

US DOE and EERE (2006b) *Net Energy Balance for Bioethanol Production and Use*, www1.eere.energy.gov/biomass/net_energy_balance.html, updated 8 February 2006

US DOE and EIA (Energy Information Administration) (2002) *Annual Energy Review 2001*, Washington, DC

US DOE and EIA (2004) *International Energy Annual 2002*, Washington, DC

US DOE and EIA (2005a) *Annual Energy Outlook 2005*, Washington, DC, February

US DOE and EIA (2005b) *Petroleum Supply Annual 2004*, vol 1, Washington, DC

US DOE and EIA (2006a) *Annual Energy Outlook 2006*, Washington, DC, February

US DOE and EIA (2006b) *International Petroleum (Oil) Prices and Crude Oil Import Costs*, www.eia.doe.gov/emeu/international/oilprice.html, accessed February 2006

US DOE and EIA (2006c) *Petroleum Information Monthly*, Washington, DC, February

US DOE and EIA (2006d) *Short Term Energy Outlook*, Washington, DC, 7 February

US DOE and EIA (2006e) 'West Texas Intermediate crude oil price', *Short-Term Energy Outlook*, Washington, DC, www.eia.doe.gov/emeu/steo/pub/gifs/Slide2.gif.

US DOE, Federal Energy Management Program (2002) *Biomass and Alternative Methane Fuel Super ESPCs Helping Federal Facilities Turn Waste into Energy*, FEMP Focus, November

US DOE, National Energy Technology Laboratory (2005) *Current Industry Perspective: Gasification, World Survey Results 2004*, Pittsburgh, PA, www.netl.doe.gov/publications/brochures/pdfs/Gasification_Brochure.pdf

US EPA (US Environmental Protection Agency) (2002) *A Comprehensive Analysis of Biodiesel Impacts on Exhaust Emissions*, Draft Technical Report, Washington, DC, October

US EPA (2005a) *EPA Upholds Reformulated Gas Requirement in California, New York, and Connecticut*, Press release, Washington, DC, 2 June

US EPA (2005b) *Ethanol Plant Clean Air Act Enforcement Initiative*, Washington, DC, updated 1 September 2005, www.epa.gov/compliance/resources/cases/civil/caa/ethanol/

US EPA (no date) *Health and Environmental Impacts of* NO_x, www.epa.gov/air/urbanair/nox/hlth.html

US General Accounting Office (2000) 'Petroleum and ethanol fuels: Tax incentives and related GAO work', Letter to Senator Tom Harkin, 25 September, www.gao.gov/new.items/rc00301r.pdf

US Geological Survey (2005) *Subsidence and Fault Activation Related to Fluid Energy Production, Gulf Coast Basin Project*, Washington, DC, Coastal and Marine Geology Program, 21 March, http://coastal.er.usgs.gov/gc-subsidence/overview.html

US National Aeronautics and Space Administration (2006) '2005 warmest year in over a century', News feature, Washington, DC, 24 January www.nasa.gov/vision/earth/environment/2005_warmest.html

US Senate (2005) *US Renewable Fuels Standard*, Energy Policy Act of 2005 http://energy.senate.gov/public/_files/H6_EAS.pdf

van den Broek, R. et al (2001) 'Potentials for electricity production from wood in Ireland', *Energy*, vol 26, no 11, pp991–1013

Van Dyne, D. L. and M. G. Blasé (1998) 'Cheaper biodiesel through a reduction in transaction costs', in *Proceedings of Bioenergy 1998: Expanding Bioenergy Partnerships*, www.biodiesel.org/resources/reportsdatabase/reports/gen/19981001_gen-109.pdf

van Ravenswaay, E. O. (1998) *Pollution Prevention Case Study on Petroleum*, East Lansing, MI, Michigan State University, 15 October, www.msu.edu/course/eep/255/PetroleumP2CaseStudy.htm

Vanguard (2005) 'Nigeria's search for an alternative to petrol', *Africa News*, 6 September

Vidal, J. (2006) 'America's masterplan is to force GM food on the world', *The Guardian*, 13 February

Vogel, K. P. and R. A. Masters (1998) 'Developing switchgrass into a biomass fuel crop for the Midwestern USA', Paper presented to BioEnergy 1998: Expanding Bioenergy Partnerships, Madison, WI, 4–8 October 1998

von Lampe, M. (2006) *Agricultural Market Impacts of Future Growth in the Production of Biofuels*, Paris, OECD

von Wedel, R. (1999) *Technical Handbook for Marine Biodiesel in Recreational Boats*, second edition, Point Richmond, CA, CytoCulture International, www.cytoculture.com/Biodiesel%20Handbook.htm.

Wallace, R. et al (2005) *Feasibility Study for Co-Locating and Integrating Ethanol Production Plants from Corn Starch and Lignocellulosic Feedstocks*, Washington, DC, US Department of Agriculture and DOE, January

Walsh, M. et al (2000) *Biomass Feedstock Availability in the United States: 1999 State Level Analysis*, Oak Ridge, TN, Oak Ridge National Laboratory, updated January 2000

Wang, M. (2001) *Greet Model Version 1.5a*, Revision June 2001 (calculations made by IEA for reference case using the downloadable model, in consultation with author), http://greet.anl.gov

Wang, M. (2005) 'The debate on energy and greenhouse gas emissions impacts of fuel ethanol', Presentation to Energy Systems Division Seminar, Argonne National Laboratory, Chicago, IL, 3 August 2005, www.transportation.anl.gov/pdfs/TA/347.pdf

Wang, M., C. Saricks and D. Santini (1999) *Effects of Fuel Ethanol on Fuel-Cycle Energy and Greenhouse Gas Emissions*, Argonne, IL, Center for Transportation Research, Energy Systems Division, Argonne National Laboratory, January

Watkins, K. and J. von Braun (2003) *Time to Stop Dumping on the World's Poor*, 2002–2003 IFPRI Annual Report, Washington, DC, International Food Policy Research Institute

WEC (World Energy Council) (2001) 'Biomass (other than wood)', in *Survey of Energy Resources 2001*, London, WEC

WEC (1994) *New Renewable Energy Resources: A Guide to the Future*, London, Kogan Page Ltd

Werner, C. (2005) *Subsidies: Historic, Current and the Skewing of Market Signals*, Environmental and Energy Study Institute, Presentation to Conference on Policy, Institutions and the True Costs of Carbon, Albuquerque, NM, 29 July, www.eesi.org/publications/Presentations/NCC%20Energy%20Subsidies%207.29.05.pdf, accessed 7 March 2006

WestStart-CalStart (2005) *California Biogas Industry Assessment White Paper*, Pasadena, CA, April

WHO (World Health Organization) in collaboration with UNEP (United Nations Environment Programme) (1990) *Public Health Impact of Pesticides Used in Agriculture*, Geneva, WHO

Wiegmann, K., U. R. Fritsche and B. Elbersen (2006a) *Environmentally Compatible Biomass Potentials in Europe*, Study sponsored by the EEA, Darmstadt/Wageningen

Wiegmann, K. and U. R. Fritsche, with B. Elbersen (2006b) *Environmentally Compatible Biomass Potential from Agriculture*, Darmstadt, Öko-Institut, February

Wiegmann, K. et al (2006) *Environmentally-Compatible Bioenergy Potentials in the EU 25*, Final report prepared for the European Environment Agency, Darmstadt/Wageningen/Kopenhagen

Wilcox, L. (2005) 'Energy outlook: Impacts on biofuels and implications for the farm sector', Presentation to 2005 Crop Management Conference, Food and Agricultural Policy Research Institute, Missouri, 14 December, www.fapri.missouri.edu/outreach/presentations/2005/cropmang_dec05.pdf

Wilhelm, W. (2005) 'Sustainability of crop residues as feedstock for bio-based products', US Department of Agriculture, Agricultural Research Service, Presentation to the First International Biorefinery Workshop, Washington, DC, 20–21 July

Wilhelm, W. W. and J. Cushman (2003) 'Implications of using corn stalks as a biofuel source: A joint ARS and DOE project', *Eos. Trans. Agu*, vol 84, no 46, ars.usda.gov/research/publications/publications.htm?SEQ_NO_115=166979

Wilhelm, W. W. et al (2004) 'Crop and soil productivity response to corn residue removal: A literature review', *Agronomy Journal*, vol 96, no 1, p117

Willams, R. H. (1995) *Variants of a Low CO_2-Emitting Energy Supply System (LESS) for the World*, Prepared for the IPCC Second Assessment Report Working Group IIa, Energy Supply Mitigation Options, Pacific Northwest Laboratories

World Economic Forum in Russia (2005) *In-depth: Economic Diversification in Russia*, World Economic Forum in Russia, Geneva, World Economic Forum, 18 November, www.weforum.org/site/homepublic.nsf/Content/Report+Russia+2005+-+theme+2

World Energy Alternatives (2006) *Biodiesel Emissions*, www.worldenergy.net/product/emissions.asp, accessed 29 January 2006

World Petroleum Council (2005) *Exploration and Production in the Marine Environment*, www.world-petroleum.org/education/exprod/, accessed June 2005

World Summit on Sustainable Development (2002) *Executive Summary of the Brazilian Energy Initiative*, Presented to the World Summit on Sustainable Development, Johannesburg, South Africa, 26 August–4 September, 2002, www.worldenergy.org/wec-geis/focus/wssd/goldemberg.pdf

Worldwatch Institute (2005) *Vital Signs 2005*, New York, W. W. Norton and Company

WTO (World Trade Organization) (2001a) *Canada – Certain Measures Affecting the Automotive Industry*, Geneva, WTO, 11 February

WTO (2001b) *European Communities – Measures Affecting Asbestos and Asbestos-Containing Products*, Geneva, WTO, 12 March

WTO (no date) *Domestic Support in Agriculture: The Boxes*, www.wto.org

Wu, S. et al (2004) 'Soil conservation benefits of large biomass soybean (LBS) for increasing crop residue cover', *Journal of Sustainable Agriculture*, vol 24, no 1, pp107–128

Wuppertal Institute for Climate, Environment and Energy (2005) *Synopsis of German and European Experience and State of the Art of Biofuels for Transport*, Prepared for Deutsche Gesellschaft für Technische Zusammernarbeit (GTZ) GmbH, Wuppertal, July

Yacobucci, B. D. (2004) *Ethanol Imports and the Caribbean Basin Initiative*, Washington, DC, Energy Policy Resources, Science, and Industry Division, Congressional Research Service, September, http://ncseonline.org/NLE/CRSreports/04Sep/RS21930.pdf

Yamamoto, H. et al (1999) 'Evaluation of bioenergy resources with a global land use and energy model formulated with SD technique', *Applied Energy*, vol 63, pp101–113

Yong, W. (2005) 'Liaoning adopts ethanol gasoline', *China Daily*, 1 November

Index